T0146190

Powering American Farms

Powering American Farms

The Overlooked Origins of Rural Electrification

Richard F. Hirsh

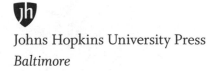

Johns Hopkins University Press

Baltimore

© 2022 Johns Hopkins University Press
All rights reserved. Published 2022
Printed in the United States of America on acid-free paper
9 8 7 6 5 4 3 2 1

Johns Hopkins University Press
2715 North Charles Street
Baltimore, Maryland 21218-4363
www.press.jhu.edu

Library of Congress Cataloging-in-Publication Data
Names: Hirsh, Richard F., author.
Title: Powering American farms : the overlooked origins of rural
    electrification / Richard F. Hirsh.
Description: Baltimore : Johns Hopkins University Press, [2022] |
    Includes bibliographical references and index.
Identifiers: LCCN 2021029175 | ISBN 9781421443621 (hardcover) |
    ISBN 9781421443638 (ebook)
Subjects: LCSH: Rural electrification—United States—History—20th
    century. | Electric utilities—United States—History—20th century. |
    Farmers—United States—History—20th century.
Classification: LCC HD9688.U4 H57 2022 | DDC 333.793/20973—dc23
LC record available at https://lccn.loc.gov/2021029175

A catalog record for this book is available from the British Library.

*Special discounts are available for bulk purchases of this book. For more
information, please contact Special Sales at specialsales@jh.edu.*

Contents

PART III: GROWTH OF RURAL ELECTRIFICATION EFFORTS
IN THE 1930S

Preface

I never thought I would write a book about the electrification of farms in the early twentieth century. As someone who studies the evolution of modern technological systems and contemporary energy policy, I normally draw on interviews (among other sources) that I conduct on historical actors who are often younger than I am. To study long-passed people and events usually does not appeal to me.

Yet, as an unanticipated outgrowth of a book chapter I wrote to celebrate the 150th anniversary of the Morrill Land-Grant Act of 1862, I rediscovered the enlightening narrative of rural electrification, which I first encountered in my high school and college readings several decades ago. From venerated historians, I learned how the federal government strung power lines to rural America during the depths of the Great Depression in the 1930s, giving isolated and often destitute farmers a sense of the modern life that their urban cousins had enjoyed for many years. Of course, ruralites knew about electricity and often wanted access to it, but private electric power companies saw little profit in extending lines outside of cities. As presented in these accounts, federal rural electrification efforts demonstrated (at least in this case) that government can play an important role in addressing social inequities that insensitive private businesses chose to neglect.

Archival research at my university, Virginia Tech, made me question this standard narrative, as I found evidence of significant rural electrification work performed by agricultural engineers, who won support from those supposedly covetous power companies. Further research convinced me that the appealing traditional account contained serious flaws. Private companies did not neglect the farm market. Rather, they pursued it in a conservative fashion that nevertheless saw huge strides made in the decade before the New Deal began.

When I realized that the conventional story of rural electrification no longer proved tenable, I felt compelled to present an alternative narrative. This book constitutes an attempt to do so. Though offering an account less sensational than the standard version, it describes the still-moving process by which rural Americans began to obtain an exceptionally versatile form of energy.

My research and writing benefited from the assistance of old and new friends. Alan I. Marcus at Mississippi State University gave me the initial impetus to pursue this project, having asked me to write a book chapter on electrical engineering at land-grant institutions. Work on that piece metamorphosed into a study of agricultural engineers at the schools and their work with industry partners who pursued rural electrification. When I gained the courage to present the first results of my research, I received valuable advice from A. Roger Ekirch, Daniel B. Thorp, Kevin L. Borg, Terry S. Reynolds, Thomas J. Misa, Ronald R. Kline, Ruth Schwartz Cowan, Jonathan G. Koomey, and David E. Nye. Roger, Jonathan, and David also showed their kindness by reading drafts of my manuscript and by suggesting useful changes and additions. Happily, research on this new topic also brought me into contact with another group of scholars, some of whom offered not just advice but also primary resources. These generous colleagues included W. Cully Hession, Brent Cebul, Mark Luccarelli, John L. Neufeld, and Abby Spinak.

I am especially grateful to the many professionals who helped me in archives and libraries. They included Marc Brodsky, Aaron D. Purcell, and Kayla Sweet at Virginia Tech; Jennifer Baker and Todd Kosmerick at North Carolina State University; Jennie Russell at Michigan State University; David White and Melanie Bazil at Kettering University; David Null and Cat Phan at the University of Wisconsin–Madison; Cheryl Gunselman, Mark O'English, and Gayle O'Hara at Washington State University; Laura Guedes, Amy Thompson, and Robert Perret at the University of Idaho; Becky S. Jordan and Olivia Garrison at Iowa State University; Sara Gunasekara and Dawn Collings at the University of California, Davis; Chris Burns and Jeffrey D. Marshall at the University of Vermont; John Verner and Aaron Trehub at Auburn University; Clint Pumphrey, Bradford Cole, and Robert Parson at Utah State University; Susan Hoffman and Erin George at the University of Minnesota; Adrian Fischer at the Bakken Museum; Caroline J. White at the University of Massachusetts–Amherst; Lynne Belluscio at the LeRoy (New York) Historical Society; Tyrone Corn at Idaho Power Company; and Chris Belena at the Franklin D. Roosevelt Presidential Library.

Family members provided substantive support in many ways. My charming wife, Margene, worked with me in the dusty archives (which really aren't so dusty anymore!) and helped me locate wonderful resources upon which much of this text draws. Sister-in-law Debra Bingham listened patiently to summaries of my work and gave me useful advice about recent popular histories that continue to relate the conventional story of rural electrification. My sons, Stephen and David, heard me talk too much about my topic at family gatherings but nevertheless offered encouragement to finish this task. David also provided invaluable technical assistance in preparing many of the delightful images used here. To these loved ones and my extended family, I dedicate this book.

Introduction

E ven after decades of retelling, the story of rural electrification in the United States remains dramatic and affecting. As textbooks and popular histories inform us, farmers obtained electric service only because a compassionate federal government established the Tennessee Valley Authority (TVA) and the Rural Electrification Administration (REA) during the Great Depression of the 1930s. The agencies' success in raising standards of living for millions of Americans contrasted with the failure of the greedy metropolitan utility companies, which showed little interest in the apparently unprofitable nonurban market. Traditional accounts often describe the nation's population as split in two, separated by access to a magical form of energy: just past the city limits, a bleak, preindustrial class of citizens endured, literally in near-darkness at night, and envied their urban cousins, who enjoyed electrically operated lights, refrigerators, radios, and labor-saving appliances.

A visit in 1979 to a rustic general store in Florida reminded me of the significance of rural electrification. Upon entering the ramshackle structure, I encountered the elderly proprietress sitting in front of a framed photograph of President Franklin D. Roosevelt. When I asked her why she kept this picture on the wall rather than that of then-president Jimmy Carter, she reminded me—with gravitas in her voice—that Roosevelt's REA brought electricity to her community. This response, occurring decades after electricity had become a common and largely invisible element of Americans' reality, reinforced the truth of the standard narrative. In stringing electric lines to previously neglected citizens, the generous federal government brought modernity and a measure of parity between rural and city folk.

Almost by accident and several decades after my encounter with the Florida store owner, I began finding evidence that challenged the conventional account. I discovered, for example, that from the end of 1923 through 1933, during a period of supposed utility disdain for rural residents, the proportion of California farms that obtained electricity jumped from 23.5 percent to 60.0 percent. In the same time span, Washington State agriculturalists saw electrification soar from 18.3 percent to 52.3 percent, while Massachusetts farmers went from 7.3 percent electrified to 56.3 percent (figure I.1). Nationally, the number of farms obtaining so-called central station (utility) power quadrupled in this period, from 2.8 percent in 1923 to 11.4 percent (out of about 6.3 million total) electrified at the end of 1933 (figure I.2).[1] These gains occurred, moreover, while farmers suffered an economic recession beginning in 1920, only to be followed by more severe distress during the Great Depression, hardly a time when one would expect rapid electrification rates.

I also learned that power companies collaborated with researchers at land-grant colleges to conduct studies on augmenting farm production using electricity. As

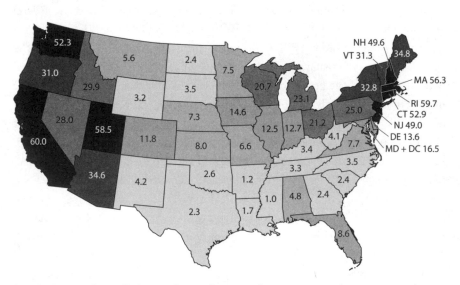

Fig. I.1. Farms electrified in each state by central stations, 1933 (in percentages). *Source*: Drawn from data in George W. Kable and R. B. Gray, *Report on C.W.A. National Survey of Rural Electrification* (Washington, DC: USDA, 1934), 53, which used information provided by the National Electric Light Association and the Edison Electric Institute.

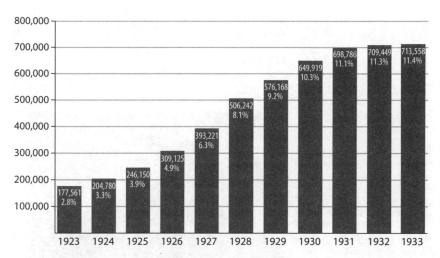

Fig. I.2. Numbers of US farms (and percentages of national total) receiving central station electricity, 1923–33. *Source*: Drawn from information in National Electric Light Association, *Progress in Rural and Farm Electrification for the 10 Year Period 1921–1931* (New York: NELA, 1932), 5; and from Edison Electric Institute data in "Add 533,000 New Customers," *Electrical World* 106, no. 1 (4 January 1936): 62.

a result, ruralites in the mid-1920s began wiring their farmsteads and purchasing electrical equipment at a staggering rate compared with earlier periods, enabling many to keep their children from fleeing to the electrified and more enticing cities. And portending the efforts of government-funded programs several years later, the utility industry and its partners produced colorful artwork (such as plate 1) that depicted a prosperous farm crisscrossed by electric "high lines" (a colloquial term for overhead distribution wires coming from power plants). But the image did not appear on the walls of a New Deal–era Federal Art Project exhibit. Rather, it embellished a 1924 report from an industry-financed group of college professors who performed rural electrification studies. Likewise, a national group of utility-funded researchers employed photographs (such as figure I.3) illustrating the benefits of farm electrification. With the caption "The Electric wires—a sign of agricultural progress," the picture appeared in a 1931 engineering bulletin and indicated that electricity on the farm had great technical and cultural meaning to those who gained access to it.

In other words, I concluded that the established story of how farmers obtained electricity is wrong—or, to say the least, severely exaggerated. Because it perpetuates the notion that the US government acted to enhance farmers' lives while private enterprise did little, the account needs serious revision.

C. R. E. A. Bulletin

E. A. WHITE, Editor   LEE C. PRICKETT, Assistant Editor

Published by Committee on the Relation of Electricity to Agriculture

1120 GARLAND BUILDING   -:-   CHICAGO, ILLINOIS

| Volume VII | November, 1931 | Number 1 |

PREFACE —

The material presented in this bulletin represents a summary of the best available information at the command of the Committee on the Relation of Electricity to Agriculture. It is a revision of the editorial material published in *C. R. E. A. Bulletin Volume IV, Number 1, January, 1928.* Its pages reveal satisfactory progress in the art and science of the use of electricity on the farm. A large amount of information is presented, yet this promises to be only a substantial beginning.

Compiled by
LEE C. PRICKETT

The Electric wires—a sign of agricultural progress

Fig. I.3. Cover of the *CREA Bulletin* featuring a photograph captioned "The Electric wires—a sign of agricultural progress." *Source*: NELA, *Electricity on the Farm and in Rural Communities, CREA Bulletin* 7, no. 1 (1931): 1. Used with permission of the Edison Electric Institute.

This book reinterprets the history of rural electrification. It does so by high-lighting the environment in which utility companies and farmers acted in the 1920s and early 1930s—in the years before the federal government engaged sub-stantially in the nation's economy.[2] In this effort, I do not seek to minimize the work of the TVA or REA, which accomplished much by taking advantage of the extensive resources made available during a national crisis. But one needs to re-alize that, until the Great Depression, Americans (including prominent political leaders) put their faith in private businesses to achieve social and economic "pro-gress." As profit-seeking enterprises, electric utility companies did not pursue money-losing business (among rural customers or others)—nor did many people think they should.

Refuting the standard narrative and confronting orthodox New Deal scholar-ship, this study documents extensive and previously dismissed farm electrification work performed in the years before the federal government's efforts. To appreci-ate this accomplishment, I focus on three sets of actors: (1) utility managers, who steeped themselves in the era's laissez-faire business culture but who, in some cases, sought to expand electrification to rural customers; (2) farmers, who (con-trary to stereotypes) adopted new hardware in innovative ways that expressed a sense of self-reliance and technical savvy; and (3) professors of agricultural engi-neering at land-grant colleges, who promoted rural applications of power even as farmers failed to share in the prosperity savored by others during the "roaring twenties." My emphasis on the last group of stakeholders appears especially orig-inal: gaining prestige in the 1910s within a new subfield of engineering, these prag-matic and infrequently recognized agricultural professionals demonstrated ways to make electricity use profitable to farmers and utilities alike, and they helped set the stage for rural electrification work later performed by private companies and federal agencies.

Consisting of three parts, this book starts by providing a context—in histori-ography, economics, and culture—for an understanding of rural electrification in the early twentieth century. It first describes the traditional scholarship that dis-parages the utility industry for its failure to electrify farms. Chapter 1 offers exam-ples from the literature—some of which the REA produced itself—demonstrating how the government bestowed on farmers a technology and leading-edge lifestyle that had been withheld from them by villainous utility moguls. This part includes, in chapter 2, an explanation of power companies' general reluctance to seek the business of rural customers. Having begun producing electricity in the 1880s for densely populated urban areas, utilities simply could not earn a good return on

their investments by extending lines to scattered farms. Going beyond this customary economic rationale for the industry's apprehensive behavior, chapter 3
presents evidence of attitudinal reasons for managers' scorn of farmers. Many
company officials held condescending views of the agriculturalists as independent,
dim-witted folk who resisted efforts for improvement offered by business and educational experts.

Despite the utility men's patronizing attitude, ruralites became more aware of the
promise of electricity. Quickly after the introduction of electric lighting, chapter 4
explains, they realized that the new energy form could improve farm life. They
learned of potential uses of electricity—some overstated and others realistic—
that stimulated a latent demand and interest in rural electrification. Chapter 5
explores how that interest motivated some farmers to obtain electricity themselves, rather than wait for power firms to provide it, often using isolated power
plants such as those manufactured by the Delco-Light Company. As even utility
executives observed at the time, these generating units (which we now call "distributed generation" systems) served a useful function by providing an educated
and ready customer base in anticipation of the time when companies could erect
lines to them.[3] A description of the farmers' use of isolated sets, which generated
power on about 4 percent of farms in 1931, adds more dissonance to the mainstream story by suggesting that many rural folk should not have been portrayed
as so uniformly unsophisticated.[4] Rather, farmers constituted a non-monolithic
and discerning population that refused to commit to a single notion of technological advancement. By employing historical tools of contingency and counterfactual thinking, we can imagine another logical course of rural electrification:
it could have evolved successfully without the involvement of either private power
companies or government agencies. That alternative approach, abandoned in the
United States, still has policy relevance to today's almost one billion people
worldwide who live without electricity.

The book's second part explores the utility industry's pursuit of rural electrification—an activity that standard histories overlook or diminish. Chapter 6 describes
the efforts of individuals who saw value in providing electricity to farming districts. For example, Wisconsin utility executive Grover Neff argued in the early
1920s that his colleagues should stop regarding service to ruralites as "a troublesome and unprofitable business."[5] Instead, he worked with like-minded colleagues
to create groups within the utility industry's predominant trade organization, the
National Electric Lighting Association (NELA), to examine means to boost rural
electrification. The next chapter explains why some leaders, who generally ridi-

culed the idea of serving farmers, nevertheless tolerated such endeavors as part of a campaign to retain support from government officials and the public. Especially during the 1920s, the industry fought municipal takeovers by private companies through public relations battles—conflicts in which utility managers saw rural electrification playing an increasingly prominent role. Perhaps most significant, industry executives established the Committee on the Relation of Electricity to Agriculture (CREA) in 1923, an organization (described in chapters 8 and 9) that partnered with land-grant college agricultural engineers, farm associations, government agencies, and utility companies in more than half the states. The CREA and its affiliates performed fruitful research on farm uses of electricity that made service profitable to both power companies and customers. Such a sense of mutual self-interest, the stakeholders felt, would propel further increases in the number of electrified farms and without the need for significant government intervention.

But growth in rural electrification required more than just technical advances. As described in chapters 10 and 11, farmers' increasing power use benefited from the contributions of numerous stakeholders who created a legal, social, and educational infrastructure. The states' regulatory commissions in the 1920s, for example, worked with rural advocates to simplify and reduce the cost of connecting farmsteads to utility power lines. And as some farmers showed they would become good customers, companies established a growing number of rural service departments, staffed by agricultural engineers and other experts to help farmers raise their electricity consumption and increase productivity. To augment knowledge about the benefits of farm electrification, land-grant colleges established short courses for interested parties while utilities collaborated with manufacturers and magazine publishers to demonstrate the value of electricity on the farm.

The book's final part examines the impact of government efforts to spur rural electrification after President Roosevelt and Congress created the Tennessee Valley Authority (in 1933) and the Rural Electrification Administration (in 1935). Chapters 12 and 13 take a different approach from that of standard histories by highlighting the federal agencies' exploitation of resources that utilities did not possess. The REA, for example, loaned tens of millions of dollars for rural electrification at below-market rates, while the Electric Home and Farm Administration, an offshoot of the TVA, provided attractive funding for home wiring and appliance purchases. At the same time, the organizations established a competitive force that stimulated private firms to offer rural service at a previously unimagined pace. Employing techniques that (rightly, in most cases) elicited scorn from government administrators, politicians, and the press, utilities nevertheless wired up more farm customers in the years prior to 1950 than did the REA, when about 86 percent of rural citizens enjoyed central station power.[6]

Overall, this book represents a corrective for academic and general readers whose views have been distorted by poignant accounts describing the successes of the TVA and the REA. Soon after the organizations' creations, supporters began producing narratives of downtrodden farmers who belatedly entered a progressive electrical age with the assistance of a benevolent federal government. Though not dismissing people's sentiments, like those of the rural Florida business matron who genuinely felt that the REA brought modernity to her life, this book presents a more nuanced and less ideologically imbued history that better explains the process of rural electrification in America. It provides insights into the minds and behaviors of the big-city corporate executives and the business community at large while also illuminating the ingenuity and creativity of farmers who sought to electrify their farms—with (and sometimes without) the assistance of utility managers and their associates in land-grant colleges.

Methodologies

This book employs interdisciplinary methodologies that, I hope, give my rural electrification account more meaning. The first draws on an understanding of sociotechnical systems as explicated by historian Thomas Hughes, who demonstrated that electric power networks consist not only of generation plants, transmission lines, and associated technologies; they also reflect considerations described as cultural, economic, financial, political, legal, educational, and regulatory.[7] He showed that the systems created by inventor- and entrepreneur-managers established "momentum" toward achieving the goal of widespread urban electrification by obtaining support from various stakeholders and the public at large. Industry leaders reduced uncertainty and eliminated challenges to their authority by overseeing a congeries of social and technical variables such that they ultimately dictated the direction of their system. As a major contribution, Hughes's approach emphasizes the importance of corporate and institutional cultures and reduces the explanatory power of engineering concerns alone.

Modifying Hughes's approach slightly, I suggest that the framework for establishing rural electrification can be considered a *sub*system within the larger electric utility system, sharing some features of the overall system but also exhibiting noteworthy differences. Doing so enables a focus on the same analytical categories (in the social, technical, and business domains) but with different actors. The managers who championed rural electrification, for example, sometimes consisted of "big names" and founding fathers of the larger system, such as Commonwealth Edison's president, Samuel Insull. But their interest in the rural electrification subsystem remained peripheral, unlike that of others who appear marginal if one looks only at events occurring in the larger system. These latter participants included Grover Neff, a Wisconsin power company official who integrated modestly sized utilities serving small (and often rural) communities. And though both sets of stakeholders interacted with major manufacturing firms, the subsystem managers encouraged a distinctive type of technology, such as low-cost utility poles, wires, and transformers to meet the needs of rural customers; that development thrust contrasted with the demand created by large-system actors for increasingly powerful turbine-generators and high-voltage transmission technologies.[8] Furthermore, the subsystem stakeholders—similar to the manager-entrepreneurs who created a national network of power companies—made alliances with university engineers, but with those in different specialties. While the big-system players established relationships with electrical engineers, the rural electrification advocates cultivated bonds with agricultural engineers, who worked with a utility-sponsored institution (the CREA) and other narrowly focused institutions such as the US Department of Agriculture. The promoters also engaged with magazine publishers and other parties who catered to the farming audience and who rarely interacted with urban utility company magnates.

Simply put, the subsystem actors developed a social and technical infrastructure that constituted part of the larger system, but they sought the at-first unpopular goal of electrifying nonurban customers. As several historians have noted, the overall utility system by the 1930s had become populated by stakeholders within educational, financial, and regulatory institutions and had successfully resisted radical innovations. Viewed as closed (in Hughes's terms), that relatively stable system had effectively become controlled by elite business leaders who reduced uncertainties in almost all elements of its environment.[9]

Such closure had not yet arrived in the rural electrification subsystem, and utility managers such as Grover Neff never achieved total authority. Unlike the larger system, which had obtained support from the financial community for funding an expensive central station and transmission network, the subsystem failed to create a financial model that satisfactorily allowed utilities to serve all farm customers profitably. Expressed differently, the rural subsystem remained contested, in flux, and open longer than the overall system, such that people like Neff would not achieve the closure they expected. Rather, the federal government introduced new actors, such as the TVA, REA, and farmer-owned rural electric cooperatives while also initiating radical innovations in the financial realm. These novelties upended the utility industry's dominance over the rural electrification subsystem. Differing from their counterparts in the larger system, power company managers ceded authority of the subsystem to nonutility entities, with federally financed organizations overtaking private companies as the principal supplier of electricity to rural customers after 1950. As late as 2019, such actors—largely the cooperatives that originally drew on REA resources—prevailed as major players, providing power to more than 20 million businesses, schools, homes, and farms while delivering 12 percent of the nation's electrical energy. Geographically, they supplied power to 56 percent of the nation's land area.[10]

This book also relies on the literature of social movements. It shows that the push toward rural electrification was driven by individuals and groups that had been largely marginalized by utility system elites but who acquired means to gain leverage within the management hierarchy. As part of an undercurrent campaign, subsystem actors appreciated electricity's value to magnify farm productivity and elevate everyday life, as it had done in cities, and they touted the market's potential. Urban utility managers in the early 1920s largely dismissed rural customers, however, as too expensive to serve. Besides, they still had a large-enough job to complete in electrifying the more lucrative cities. But even they recognized the growing social and political pressure to electrify farms. Proponents of "public power"—government-owned energy systems—also argued that ruralites would more likely see high lines strung to them if nonutility providers secured greater control over electricity generation and distribution networks. Consequently, some leaders in the private power industry realized in the 1910s and 1920s that they could mitigate criticism by establishing rural electrification study groups within their trade organization. With this formal acknowledgment, rural backers gained recognition and resources to pursue their undercurrent activities. They formed allegiances with other stakeholders, such as manufacturing firms, farm lobbyists, government agencies, and agricultural engineers at land-grant colleges. Perhaps only interested in allaying criticism, industry leaders nevertheless supplied modest support for rural electrification work.

By the mid-1920s, the rural advocates had acquired a sense of legitimacy and influence within the utility community that enabled them to constitute a true movement. Academics sometimes describe such a campaign as the result of work performed by players in political, corporate, and cultural arenas who establish a knowledge base and belief system that transforms "discontent into collective action."[11] Other theorists consider the importance of people who capitalize on political opportunity and who mobilize existing social groups. Within this explanatory framework, actors like Grover Neff advanced the rural electrification subsystem by successfully exploiting mechanisms such as "social organization, strategizing, reasoning, analyses, and rationality."[12]

More generally, this book highlights the importance of narratives—and not just among historians. Stories that carry substantive interpretations of events often influence the attitudes and behaviors of those who accept them.[13] The narrative that gained credence after the American Civil War of the "Lost Cause," in which heroes of the South fought valiantly against huge odds, helped sustain the losers of the conflict. It also gave license to people (at least in their own minds) to continue political and social practices that privileged white men and disenfranchised Black citizens into the twenty-first century. A similar narrative of Canadians defeating the Americans in the War of 1812 bolsters an image of strength and independence against a stronger political and economic rival.[14] Narratives after World War I of German leaders being "stabbed in the back" by Jews provided the intellectual basis for the rise of the Nazis. More recently, different narratives of the causes of Middle East tensions—taught and popularized in several countries— justify nations' political and military postures in the Arab-Israeli conflict.[15]

In this book's case, the revision of the standard narrative of rural electrification disabuses the notion that only the federal government could have resolved problems stemming from the Great Depression. To be explicit, I do not seek to make a political point about big government's ability (or inability) to pursue socially valuable endeavors. Indeed, I acknowledge that the Roosevelt administration likely undertook initiatives that ameliorated the Depression and restored confidence in American capitalist institutions. But the assertion that big government's creation of the REA constituted the *sole* means of electrifying rural America may not be true. One can imagine other approaches, also involving government support perhaps, that could have yielded a greatly increased number of electrified farms.

To write a revisionist narrative, though, may make me appear to be an apologist for the utility industry, which assuredly I am not. Rather, I seek to focus more on contemporaneous circumstances, noting particularly the exuberant expectations held by engineers and managers about their ability to make a better world. Moreover, most businesspeople working within the political and social environment of the 1920s dismissed government intervention in the market, believing that dynamic corporations (and sometimes their allies in related institutions, such as land-grant colleges) could better meet society's needs. In an era of laissez-faire government and public disavowal of socialistic approaches employed by some European and

Canadian governments, elite engineers and businesspeople sought to electrify farms using methods that appeared prudent at the time.

Of course, we know that within two decades of the REA's creation, high-line rural electrification rates jumped from about 11 percent to more than 90 percent. That commendable accomplishment suggests the failure of private industry to provide what we now consider a life-enriching form of energy. But we must be careful not to compose narratives based on our understanding of events that occurred later, with history written in reverse.[16] Rather, we should consider the context of the times when actors made choices and avoid projecting backward the radically altered views, emerging in the New Deal 1930s, concerning the proper roles of the private and public sectors. Otherwise, we produce bad history that offers inadequate accounts of people's attempts to manage social, technical, and economic challenges.

PART I / Historical Context of Rural Electrification

Historians and advocates of New Deal programs have done a good job in creating a narrative that disparages the efforts of power company leaders who dismissed the rural market. The account conforms with the notion that the federal government, operating in an era of profound turmoil, improved the lives of a neglected segment of the American population. At the same time, it demonstrates the pitfalls of leaving progress in the hands of private businesses. Such a view, repeated in popular histories as well as in academic scholarship, resonates well among people who expect government to play a prominent role in society.

During the first several decades of the twentieth century, however, government intervention in everyday life remained limited, with corporations largely holding responsibility for bringing new goods and services to people within a free enterprise framework. Power company managers naturally wanted to earn profits, and for widely accepted reasons, those profits did not appear likely to come from rural customers. Big-city utility leaders also considered farmers as backward and resistant to innovation, diminishing the probability that nonurban citizens would ever become worthy of service.

Putting a lie to such views, however, many farmers became interested in the possible comfort- and productivity-improving uses of electricity. They read about the ways electric power could not only illuminate their homes but also increase efficiency in fields and barns. Just as important, some farmers demonstrated their technical aptitude by harnessing natural resources and purchasing isolated power sets to generate electricity. These farmers sought modernity and did not wait for a reluctant industry to provide it for them.

Chapter One

The Standard Narrative and Its Defects

> If it had not been for Gifford Pinchot, George W. Norris, and Franklin D.
> Roosevelt, we certainly would not have had the rural electrification
> development we have today.
> —MORRIS L. COOKE, FIRST REA ADMINISTRATOR, 1948

A s a college student in the early 1970s, I fondly remember reading Arthur
Schlesinger's book *The Politics of Upheaval.* Running more than seven hundred pages, the tome commanded respect; it contained uplifting prose detailing President Franklin Roosevelt's efforts to fight the Great Depression and restore the American economy and spirit.[1] Reinforcing the message promoted in my high school texts, the book described one significant organization, the Rural Electrification Administration, which brought light and power to farmers who had previously been neglected by the avaricious utility companies. The compelling account suggested that liberal government sometimes could effect positive change better than private enterprise.

The standard narrative of rural electrification emerged in the work of REA officials and supporters who constructed an attractive account of the organization's origins. Academics appear to have accepted much of this narrative and perhaps did poor history themselves by uncritically repeating the claims and statements of REA actors without carefully investigating the work of utilities and their allies.

Morris Cooke and the Rural Electrification Narrative

As REA's first administrator, Morris Llewellyn Cooke did much to establish the conventional history of the organization's genesis, which included colorful descriptions of its electric utility adversaries. After receiving a degree in mechanical engineering from Lehigh University in 1895, Cooke came under the influence of

Frederick W. Taylor, the originator of "scientific management," a supposedly logical approach for obtaining high productivity from industrial laborers.[2] Appointed by Philadelphia's reformist mayor in 1911 to serve as director of the Department of Public Works, Cooke applied efficiency-raising principles to eliminate wasteful practices and to lower electric rates. Though the Progressive Era in American politics effectively ended during World War I, Cooke continued to pursue technological programs that prioritized social betterment. For example, as the adviser to the governor of Pennsylvania, Gifford Pinchot, Cooke became involved in designing a "Giant Power" proposal that sought to provide cheap electricity to cities and farms alike.[3] The Pennsylvania legislature rejected the proposal in early 1926, as utility industry representatives, consultants, and academics disabused the plan, with at least one calling it "communistic."[4]

Despite Giant Power's failure, Cooke established himself as a vigorous advocate for rural electrification and public power.[5] He partnered with Senator George Norris, a Progressive Republican, who proposed to use the Wilson Dam at Muscle Shoals, Alabama (built initially to supply power for making explosives and fertilizer during World War I), for public and rural use during the 1920s rather than let for-profit companies exploit it.[6] Cooke gained another forceful ally in Franklin Roosevelt, who as New York governor established the state's Power Authority to produce hydroelectric energy along the St. Lawrence River, some of which would aid agricultural interests. Roosevelt appointed Cooke as a member of the Authority's board and also consulted with him after becoming president in March 1933. About two months later, Roosevelt signed legislation, largely crafted by Senator Norris, that established the Tennessee Valley Authority, using the Muscle Shoals dam as its first power source.[7]

Seeking to establish a larger program to serve rural citizens, Cooke won appointment in 1933 as head of a Public Works Administration committee striving to enhance life in the Mississippi Valley, partly by extending electrification to farmsteads.[8] Drawing on this work, Cooke authored in 1934 a "National Plan for the Advancement of Rural Electrification," which called for the founding of a government agency that would yield social and economic benefits by extending power lines to farms.[9] Apparently impressed, President Roosevelt created the Rural Electrification Administration by executive order on 11 May 1935, with Cooke serving as its first head.[10] A little more than a year later, Congress established the REA as a statutory entity.[11] To achieve its goal of offering electricity to unserved rural areas, the organization loaned funds largely to cooperative electric corporations (known as co-ops), owned by their members and unrelated to the existing utility firms that constructed power distribution networks.

Likely as a way to build support for his new organization, Cooke began establishing a narrative that differentiated the REA from private power companies. In an essay titled "The New Viewpoint" in the second issue of the REA's monthly *Rural Electrification News* magazine, for example, Cooke stated simply that the organization strove to "electrify as many American farms and farm homes as possible and to do this in the shortest possible time." The REA sought to accomplish this task because for-profit firms had left America "a 'backward country' by failing to furnish its rural population with the comforts and necessities made possible by electricity."[12]

In subsequent issues of the *Rural Electrification News*, the administrator amplified this message. Responding publicly to a 1935 letter from Senator Norris asking about means to augment farm electrification, Cooke observed that the federal government constituted the only entity that could energize half of all farms within a decade, a huge increase from the existing situation, in which "only a pitiable percentage" of ruralites enjoyed service.[13] Cooke and the periodical's editors continued to press the notion that the new agency would do what utility firms could not (or would not). The magazine reprinted favorable editorials and news accounts from various publications, including one from the *Washington Post* in late 1935 commenting that "[p]rivate companies have been all too short-sighted in restricting their distribution lines to the most profitable areas."[14] In his own presentations and writings, Cooke maintained the theme of utility company neglect, noting in 1936 that private firms generally refrained from serving ruralites; in the rare cases in which companies sold to farmers, they charged "notoriously high" rates. Moreover, since utilities expected farmers to pay steep construction costs in advance, unlike urban manufacturers and residential customers who had lines built for them by utilities with no up-front expense, most farmers "could never hope to enjoy the advantages of electricity."[15]

Cooke left his post as REA administrator in February 1937, but he continued promoting the agency, distinguishing its work from that of the power companies.[16] In 1948, well after the REA had proven its mettle and when the rural electrification rate had reached almost 69 percent,[17] he authored a retrospective article in the academic *American Political Science Review*. Describing efforts to extend high lines to farms in the years before 1936, Cooke belittled the work of the utility industry. He admitted that, as early as 1910, there was "increasing talk—hardly to be described as representing any deep-seated conviction or interest—within the commercial industry as to the possible future of electricity in agriculture." He also acknowledged the industry's construction of "[s]hort experimental lines" that "were installed amid much hullabaloo," such as those built in Red Wing, Minnesota (see chapter 8).[18] But more prominently, he highlighted cases in which advocates for the farmer operated against almost impossible odds, combatting the entrenched interests of private companies. Continuing with his criticism, Cooke argued that the power industry established the Committee on the Relation of Electricity to Agriculture (CREA) in 1923 as a ploy to "keep rural electrification in commercial hands." Complimentary portrayals followed of Governor Pinchot's Giant Power plan, efforts to spur rural electricity consumption in Canada, the New York Power Authority, and ultimately the Tennessee Valley Authority. Cooke offered a case study of a rural co-op distributing electricity to a Mississippi community beginning in June 1934. The article next described the creation of the REA by executive order, which included a brief account of failed efforts to gain cooperation of commercial utility companies. He concluded the historical survey by praising the heroes of rural electrification, namely Pinchot, Norris, and Roosevelt.[19]

Rural Electrification History as Told by Other REA Advocates

Other REA leaders extended this favorable and self-serving origin story. Harry A. Slattery, the third REA administrator (from 1939 to 1944), for example, wrote of the utility industry's disappointing efforts to electrify farms.[20] Though remarkably charitable about the goals and useful research performed by the national Committee on the Relation of Electricity to Agriculture, he nevertheless pointed out that it failed to meet its own objectives. Shortly after the CREA's founding, he noted, the head of the NELA Rural Lines Committee, Grover Neff, set the goal of connecting one million farms to central station power within ten years (coming off a base of 177,000 farms, or 2.6 percent of all farms in 1923). The REA leader observed that only about 744,000 farms had been electrified by the beginning of 1935, just 10.9 percent of all farms. Presenting data in an unflattering manner, he incorrectly claimed this number represented a "gain of 8.3 percent in eleven years, with over 6,000,000 farms left unserved!" (Slattery's number of unenergized farms proved accurate, but, in fact, the gain was not 8.3 percent, but 320 percent—equivalent to an average annually compounded rate of 14 percent between 1923 and 1935. He probably meant an increase of 8.3 percentage *points*.)[21] In his annual report to the Department of Agriculture in 1940, Slattery summarized the reasons for the lack of rural electrification before 1935: "The suppliers of power were not interested in rural electrification. They were concentrating on urban markets, because rural areas did not promise the lucrative returns of urban developments."[22]

At the same time, Slattery suggested that the utilities and the REA took fundamentally different approaches to electrify farms. Power companies insisted that they would serve farmers only when rural income and demand for electricity reached high-enough levels to provide reasonable profits. Ruralites needed to purchase new equipment and elevate consumption before utilities could reduce rates, they argued. But the government contended that lower rates must occur first, to encourage farmers to employ electricity initially for a few purposes; they would later increase usage as the new energy form yielded improved farm productivity and income.[23] It remained for President Roosevelt and the REA to break the impasse over which approach would prevail, according to the agency head.[24]

Slattery's successor, Claude Wickard (serving from 1945 to 1953) continued to praise his organization's accomplishments in a 1951 *New York Times Magazine* article, "Power Revolution on the Farm." Aside from highlighting the positive changes in farmers' lives after they acquired electricity, he noted that the REA program exemplified the nation's finest principles in action. The federal government, he observed, did not make outright grants to co-ops to erect distribution lines to their members. Rather, it offered loans—the huge majority of which were paid off on time or early.[25] The REA acted largely as a banker and not as a heavy-handed bureaucratic institution. But unlike traditional financial enterprises, the agency did not demand collateral from its borrowers, accepting instead "the integrity of the American farmer as loan security."[26] Moreover, the collective behavior of the membership-owned co-ops became an "object lesson" in making "democracy a living and vital force," such that almost 90 percent of farms enjoyed electric power by the REA's sixteenth birthday.[27]

Besides employing books and articles, the REA promoted itself with novel media such as movies that contrasted the standards of living before and after rural electrification. At the beginning of *Power and the Land*, a 1940 agency film, viewers learned that "[o]ur cities glow with light, but most of our farms, even now, still rely on the kerosene lantern, [and] the iron cook stove." Of course, that situation began changing after creation of REA co-ops, in which "there are no private investors" and no requirements to earn profits. Best of all, farmers "get power at cost" without dealing with big utility companies.[28] Depression-era efforts to make work for artists also benefited the REA, such as Lester Beall's creation of inspirational posters for the organization starting in 1937. The graphic designs, which attained status as true art (and were put on display in March 2012 at New York's Museum of Modern Art), offered a simple message, according to an exhibit preparator: the REA brought "a bright and shiny future to the [rural] youth of America" (plate 2).[29] Meanwhile, photographers such as Peter Sekaer illustrated with his pictures how REA light conquered nighttime darkness.[30] Other New Deal programs financed the work of David Stone Martin, who produced inspiring murals for post office walls depicting the triumph of rural electrification made possible by the Tennessee Valley Authority and the REA.[31]

Using another art form, the Works Progress Administration's Federal Theatre Project enhanced the reputation of the government's rural electrification efforts at the expense of utility companies. It did so by producing Arthur Arent's play, *Power*, described in a promotional flyer as an "exciting dramatization of the development and use of electric energy in the USA" and advertised with a Federal Theatre Project color poster (plate 3).[32] This popular production, first staged in many American cities in 1937, took aim at the private power industry, which since 1928 had come under investigation by the Federal Trade Commission for financial and propagandistic abuses. Touted by a *New York Times* reviewer as "one of the most exuberant shows in town," the dramatization directly attacked utility firms for high power rates and their aversion to serve most farmers until the creation of the Tennessee Valley Authority and (without naming it specifically) the Rural Electrification Administration.[33] Setting a scene in Dayton, Tennessee (a city that received one of the earliest REA loans in July 1935),[34] the play showed excited farmers who had been refused service by the local electric company but who later realized that they could borrow money from the government and set up their own distribution network.[35] Another scene depicted desperate, profit-hungry power company managers scrambling to meet the new government competition by lowering rates and seeking farm customers while also fighting the existence of the TVA in the courts.[36] With heroes such as Senator George Norris and villains such as Wendell Willkie, president of a company that contested the TVA's legality, the production clearly tarnished the image of private power firms.[37]

Promotional literature for co-op members also reinforced the origin story in which the REA fought against greedy utility companies. A 1936 administration-produced book began by explaining many of the wondrous benefits of electricity, illustrating, for example, that a single kilowatt-hour of electricity could provide twenty hours of lighting (with a 50-watt bulb) or six hours of work from a washing machine. And yet "nine-tenths of the farms of America are still deprived of some of the conveniences that are enjoyed in the villages and the towns."[38] But governments in Europe and elsewhere (such as New Zealand) had become actively involved in rural electrification and had allegedly energized a greater percentage of farms than in the United States. Happily, the text continued, the American government recently intervened to overcome impediments to electrification, such as by lowering power line construction costs and providing capital at attractive interest rates.[39] The author acknowledged that farm electrification had increased substantially in the pre-REA years to 1935, but, in general, "[u]tility men were content to think that the farm market for electricity was an unprofitable one, and that they therefore had no responsibility to the farmer." Furthermore, the "utilities, in dealing with rural service, were not inclined to risk any large sums of money in the hope of developing the possibilities of the agricultural market."[40]

Similar themes wove through other publications, such as REA's 1939 booklet, *The Electrified Farm of Tomorrow*, which explained that "[f]or decades[,] electricity was denied rural people for the simple reason that they lived in the country." But the arrival of the government agency altered the situation, such that "[f]or the first time[,] farmers no longer had to beg for electricity from uninterested private utilities." Until the REA's creation, "the refusal of the private utilities to supply electricity to the bulk of American farmers acted as a log jam in the stream of agricultural progress."[41] Criticism of power companies continued in the REA's 1950 pamphlet, *A Guide for Members of Rural Electric Co-Ops*, which explained how "[t]he Government of the world's richest country believed that rural people should have the same electrical help, for production and for convenience, which city people had had for many years." Further, the publication observed that commercial utilities avoided the farm business because they could never "make as much profit from rural lines as from urban lines." Some firms, of course, did build lines to farms, but only lucrative ones close to towns. "In other words," the brochure continued, they "skimmed the cream" in a manner that left "many farms without any hope of ever getting electricity."[42]

Histories written by REA allies also strove to discredit central station power companies. In describing the origin of the National Rural Electric Cooperative Association (NRECA), created in 1942 as a trade organization of co-ops, director Clyde Ellis recalled growing up in rural Arkansas and experiencing firsthand the greed and "sheer arrogance of utility executives in their dealings with rural people."[43] The officials acted in a condescending manner toward farmers, his 1966 book averred, telling applicants for power that they had too little money to make it worth the utilities' efforts to erect distribution lines to them.[44] As evidence of their callousness, Ellis reported that the companies quoted prices of

$2,000 to $5,000 per mile of line, "far more than the actual cost."[45] "In desperation," he continued, farmers in some areas began establishing nonprofit cooperatives, with thirty-one existing in nine states by 1923.[46] That message of utility disregard for farmers still resonates on the NRECA website today: it observes (correctly) that only one out of ten farm homes enjoyed central station service in the mid-1930s. Consequently, the "farmer milked his cows by hand in the dim light of a kerosene lantern" and his "wife was a slave to the wood range and washboard"—the result of the many years that "power companies ignored the rural areas of the nation."[47]

The beautifully illustrated, oversize volume, *The Next Greatest Thing*, may offer the clearest and most dramatic version of this anti-utility narrative. Commissioned in 1985 by the NRECA to celebrate the fiftieth anniversary of the REA's creation, Richard Pence's tome described the application of science to improve the state of humanity. Most impressively, it highlighted the invention of electrical lights and equipment that constituted "the powerful forces transforming America from an agrarian to an industrial and urban nation."[48] But all was not well, the book continued, because "part of the society of Humankind, the rural people of America, was not to know electricity. They were told that, for them, it was not a commercial proposition. There was no profit in it. And because there was no profit, there were no lights for rural people. Sadly, what electricity did for them was to illuminate difference."[49]

The author went on to tell a story of two Americas: "Because there was no electric connection, because it was unattainable under the established economic order of the time, a great gulf developed. Two nations, two classes, two centuries: One of light, one of darkness. One 'backward,' one 'enlightened.'"[50] The following twenty-one pages continued this theme, using lyrical prose (with refrains of "because there was no electricity") that was written by Robert Caro (and credited appropriately), along with striking pictures of farmers doing painstaking work under primitive pre-electric conditions.[51] Pence then told the stirring story of the emerging "seeds of the rural electric dream" early in the twentieth century, farmers' despair during the Depression, the origins of the New Deal, and eventual salvation with the REA's creation in 1935. The first year saw loan applications and inquiries pile up as ruralites sought to obtain service for the millions of farms subsisting without electricity.[52] Rejecting the utilities' offer to participate in the process, the REA employed "the best and the brightest" professionals in engineering, accounting, the law, and other realms;[53] the book told inspiring stories of the administration's innovative spirit and the ability to empower previously powerless farmers. Concluding the first half of the book, the author noted that the "REA had truly brought power to the people."[54]

Academic and Popular Treatments of the REA

Academic studies of rural electrification perpetuated the laudatory themes expressed by REA advocates. Historian Arthur M. Schlesinger Jr. in his 1960 *The Politics of Upheaval*, for example, portrayed utilities as unwilling to help electrify the farmland in the 1920s and early 1930s. "So long as profit determined power policy," he wrote, "it appeared increasingly evident that the farmers would remain at the end of the queue."[55] The discussion implicitly suggested that company officials—unlike the New Dealers—felt no moral or social obligation to electrify farms, which would have enabled rural citizens to enjoy a standard of living comparable to that of urbanites. "It seemed more and more obvious," the historian continued, "that rural electrification, if it was to come in anyone's lifetime, would have to be brought about by government."[56] After describing the hundreds of millions of dollars spent by the federal government to create cooperative organizations and build power lines to farms, Schlesinger summarized the REA's contributions: "Where farm life had been so recently drab, dark, and backbreaking, it now received in a miraculous decade a new access of energy, cleanliness, and light. No single event, save perhaps for the invention of the automobile, so effectively diminished the aching resentment of the farmers and so swiftly closed the gap between country and city. No single public agency ever so enriched and brightened the quality of rural living."[57] Not unexpectedly, perhaps, the endnotes for the book's section on the REA demonstrated Schlesinger's reliance on sources that fostered the traditional narrative. They included publications written by Morris Cooke, other REA administrators and employees, and similarly sympathetic authors.[58]

The congratulatory view of the REA and deprecating picture of the electric utility industry continued in William Leuchtenburg's authoritative *Franklin D. Roosevelt and the New Deal: 1932–1940*.[59] Published in 1963, this book (and the entire series in which it was published) won wide readership and critical acclaim. The reviewer in the *Journal of American History* praised the monograph as "the most thoroughly researched and comprehensive one-volume history of the New Deal," and "one of the permanent landmarks of the Roosevelt literature."[60] Referring to the revolutionary nature of the New Deal, Leuchtenburg's book served as a standard interpretation of the era in the decades following its publication.[61] Though devoting only one paragraph to the REA's creation in a chapter highlighting FDR's "Second New Deal," the author nevertheless related a moving story:

> Perhaps no single act of the Roosevelt years changed more directly the way people lived than the President's creation of the Rural Electrification Administration in May, 1935. Nine out of ten American farms had no electricity. The lack of electric power divided the United States into two nations: the city dwellers and the country folk. "Every city 'white way' ends abruptly at the city limits," wrote one public-power advocate. "Beyond lies darkness." Farmers, without the benefits of electrically powered machinery, toiled in a nineteenth-century world; farm wives, who enviously eyed pictures in the *Saturday Evening Post* of city women with washing machines, refrigerators, and vacuum cleaners, performed their backbreaking chores like peasant women in a preindustrial age. Under Morris Llewellyn Cooke, a veteran of the public power fight in Pennsylvania, the REA revolutionized rural life. When private power companies refused to build power lines, even when offered low-cost government loans, Cooke sponsored the creation of nonprofit co-operatives. In the next few years, farmers voted, by the light of kerosene lamps, to borrow hundreds of thousands, even millions, from the government to string power lines into the countryside. Finally, the great moment would come: farmers, their wives and children, would gather at night on a hillside in the Great Smokies, in a field in the Upper Michigan peninsula, on a slope of the Continental Divide, and, when the switch was pulled on a giant generator, see their homes, their barns, their schools, their churches, burst forth in dazzling light. Many of them would be seeing electric light for the first time in their lives. By 1941, four out of ten American farms had electricity; by 1950, nine out of ten.[62]

Though gushing with praise for FDR and the REA, the legendary historian may not have done good academic research when preparing this paragraph. His listed sources consisted of REA-friendly authors (such as Morris Cooke), a flattering 1938 article in *Time* magazine, and the complimentary 1952 book by newspaperman Marquis Childs, *The Farmer Takes a Hand*. He also cited a penciled memorandum from Judson King, a public power supporter and REA consultant.[63]

Scholars produced few book-length monographs on rural electrification for several decades. The first one (though perhaps tinged with more ideology than most academic treatises) appeared in 1963 as *The Rural Electrification Administration: An Evaluation*, produced by the American Enterprise Institute for Public Policy, a conservative "think tank" that generally opposes government overreach. The seventy-two-page volume observed that the REA emerged during the "economic emergency in the depths of the depression," though, by the 1960s, its "original need has been all but exhausted."[64] A three-page description of rural electrification efforts before 1935 simply suggested that, despite poor profit potential, "private enterprise was probing, testing, and pushing rural electrification without any governmental encouragement (or interference)," but without great success.[65]

Philip Funigiello, a historian at the College of William and Mary, wrote more impartially about rural electrification before the REA's existence in his 1973 book, *Toward a National Power Policy: The New Deal and the Electric Utility Industry.* This well-reviewed treatise concentrated mostly on the conflict between private utilities and advocates of public power, with President Franklin Roosevelt seeking to eliminate or regulate the financial entities, known as holding companies, that held so much sway in American politics and business. Two chapters out of nine dealt with rural electrification—one focusing on the REA's origins and another on the organization's internal politics during its formative years. Given his emphasis on federal policy, the author understandably did not expend too much effort on utilities' rural electrification work before the Depression. In fact, one reviewer commented on Funigiello's omission of discussions of utility economics, though the author repeated the traditional (and in this case, accurate) rationale about high costs incurred to serve sparsely settled farmers.[66]

Funigiello offered brief accounts of rural electrification work done by power companies between 1911 and the creation of the CREA in 1923. He further mentioned some of CREA's test projects, though he drew on assessments made by Cooke in observing that they "were conducted under ideal conditions, closely supervised by experts, and used large quantities of freely loaned equipment." (That description seems to characterize the Red Wing, Minnesota, experiment starting in 1923; see chapter 8.) CREA's model farm experiments, utility critics argued, did not duplicate real-life situations, and therefore the efforts seemed inadequate, with the nation's low farm electrification rate before the REA's creation serving as proof.[67] The author balanced this analysis with a more understanding approach toward utility executives, who still saw untapped markets in urban and small-town settings, where investments would yield better returns than in farming districts. However, he also directly cited the works of Leuchtenburg and Schlesinger to give the impression that, overall, the industry had disappointed rural residents.[68]

In a policy-oriented study performed by the Environmental Policy Institute in 1979, Jack Doyle offered a brief history of rural electrification that also diminished the work of any organization besides the REA. It accurately noted that electrification began in the cities; on the other hand, "[r]ural America . . . had neither the advantage of an easily served population nor willing financiers interested in backing the development of electric power." Consequently, "[i]t took more than 50 years from the opening of Edison's Pearl Street Station in New York City [in 1882] before the nation began to apply electricity to its farms and rural areas."[69] Prior to the REA's establishment, the author explained, "the idea of providing electric service to rural Americans was thought to be highly impractical by all but a few forward-thinking individuals."[70] Clearly, these statements showed little appreciation for the work of utility companies that brought central station power to about 11 percent of the farms. They also exhibited no knowledge of the firms that sold isolated power plants to another 4 percent by the early 1930s.

The first academic study devoted almost entirely to the REA was D. Clayton Brown's *Electricity for Rural America: The Fight for the REA*, published in 1980 by Greenwood Press.[71] Authored by a history professor at Texas Christian University, the book largely constituted a political account of REA's creators, who sought to make the agency thrive despite opposition from utility companies. Perhaps because of his use of conflict as a theme, Brown downplayed utilities' contributions to rural electrification. Though noting (and even giving grudging credit to) the industry's Committee on the Relation of Electricity to Agriculture and state affiliates, he summarized their work as "only a half-hearted attempt" and concluded that "[b]y 1930 CREA was regarded as a failure."[72]

Use of the passive voice in the last assessment, of course, hid the actors who considered industry efforts as unsuccessful. However, a quotation in the same paragraph (suggesting that the CREA had "been little more than a 'window-front'" and that the organization "never had any real driving force") came from Harry Slattery's book *Rural America Lights Up*—hardly an unbiased source given that the author served as the third REA administrator. Another of Brown's references consisted of a memorandum sent by Mercer Johnston, an REA employee, to Morris Cooke.[73] Again, these sources (and others used for Brown's first chapter, which dealt with rural electrification history before the creation of the REA) seemed to demonstrate a partiality for the administration and a hostility toward utility companies. At least one reviewer also noted that favoritism, suggesting that the book might have too much of a "saint-devil bias" in which the author drew on the analogy of the "forces of light" versus the "forces of darkness."[74] Brown further observed, without offering evidence, that "[a]t the industry's rate of extending service, electrification of the remaining 5,000,000 farms would take 100 years." In fact, an extrapolation of the growth rate of farm electrification from soon after establishment of the CREA in 1923 to the end of 1929 (just as the Depression began) shows that the number of energized farms would have reached more than six million (that is, almost all farms) by 1941.[75]

In an article in the *Encyclopedia of the Great Depression*, published in 2004, Brown reprised his interpretation. Accurately, he noted, "During the 1920s[,] the privately owned electrical companies recognized the importance of serving the countryside and created the Committee on the Relation of Electricity to Agriculture (CREA)." But he seemed to contradict himself later in the same piece. "For the *first time*," he asserted after describing the passage of the Rural Electrification Act of 1936, "the privately owned power companies showed an interest in the rural market."[76] Moreover, though he observed that only about 10 percent of farms received electricity in 1935, he failed to acknowledge that this percentage had increased by a factor of more than four since the end of 1923.[77]

Brown's work gains importance because it served as the basis for other assessments of private industry's rural electrification efforts. In his biography of Lyndon Johnson, Robert Caro drew on Brown's monograph to portray electric utilities as evil actors. While depending on a host of other sources, such as the writings of REA administrators, NRECA's executive manager, and interview subjects, the Pulitzer Prize–winning author claimed that private utilities in Johnson's state of Texas "denied" farmers the electricity they eagerly sought. He began one chapter by describing the misery and low agricultural productivity of residents in Johnson's congressional district—all "because there was no electricity."[78] He started another by using Leuchtenburg's assessment (quoted above), and then noted that Johnson triumphantly obtained REA funds for his constituents. According to a cited interview subject, the success—achieved after convincing skeptical REA officials that farmers would use enough electricity to make creation of a co-op feasible—so elated people that they "began to name their kids for Lyndon Johnson."[79] Like several authors, he cited other REA-friendly sources to suggest that utilities quoted expenses of $5,000 per mile—a very high cost at the time—to build distribution lines.[80] Overall, the chapter portrayed government efforts to bring power to desolate farmers as awe-inspiring while characterizing private utilities' concern for the rural market as indifferent at best.[81]

References to Brown's book as an important authority also appeared in a wide variety of academic treatments. In an article about the REA on the Economic History Association's website, Laurence J. Malone of Hartwick College drew heavily on Brown's treatise (along with Cooke's 1934 "National Plan" and 1948 article) to argue that power companies "*ignored* the rural market due to its high network construction costs and the prospect of meager immediate profits."[82] A dissertation dealing with Louisiana's co-ops depended on Brown's book and other REA-friendly sources to conclude that lack of service to farms resulted from utilities' "greed and social irresponsibility." Moreover, after the success of the REA, the CREA "accepted failure and disbanded without a whimper."[83] Similarly, a law journal article observed that rural electric co-ops "generally formed as a response to the unwillingness of private companies to extend service into rural areas," citing Brown as the source.[84]

A detailed history of the REA at Encyclopedia.com continued this trope. Written for students, the piece emphatically dismissed rural electrification activities pursued by utilities before the Great Depression. "Early efforts by private power companies to encourage rural electrification failed," the piece read, "because they did not address the main obstacle, which was the cost of the service. By the 1920s, some European countries, Canada, and New Zealand had made much better progress in electrifying rural areas through public cooperatives and government assistance." The author also noted (better than most other pieces), however, the context of the times. "[B]ut until the Depression," the article explained, "most people in the United States believed that such approaches were un-American and threatened time-honored values of self-reliance and private enterprise."[85] Predictably, the article listed standard references of traditional REA historiography, such as those written by Brown, Cooke, and Slattery.

Though not necessarily using Brown as their primary source, other authors seemed unknowledgeable about efforts pursued by utility companies before creation of the REA. In her 1988 essay on women's use of technology on Great Plains farms, Katherine Jellison perhaps inadvertently gave the impression that rural folk needed to wait for the REA before they could enjoy electricity. She cited a Kansas woman's letter to the US Department of Agriculture in 1915 that tells of the critical need for electricity on the farm. Solely with that magical form of energy can chores be performed easily and lighting provided in a safe manner, allowing "for a social life and the improvement of [one's] mind."[86] The missive ended with a plea: "The only way I can see is for the Government to furnish, at a reasonable price, electricity to every farm."[87] According to Jellison, as noted in the next paragraph of the paper, "[t]his woman's proposed solution was the goal of the New Deal's Rural Electrification Administration some twenty years later." In other words, the author gave full credit for rural electrification to the New Deal program, apparently disregarding efforts by utilities (which, perhaps for the geographical area of her interest, may not have served farmers much) and by companies that sold isolated power plants. This last dismissal seems odd, because Jellison also displayed (without discussion) a 1920 advertisement from the Delco-Light Company in her article.[88] As discussed in chapter 5, the manufacturer produced a generator-and-battery system that enabled farmers to obtain electric power without depending on central station electricity sent over an expensive distribution network.

Authors of popular histories have also adopted the standard narrative of the REA's glorious work. In his highly readable 2012 volume, *The Men Who United the States*, journalist and history writer Simon Winchester provided an emotional description of how sinister capitalists prevented farmers (especially in the American Midwest, his apparent focus) from getting electricity. "The electric power was there," he observed, "ready and waiting and straining at the leash, to give them [farmers] relief and hope, but in the 1930s, the chiefs of the utility giants judged it as being too costly to bring to their doorsteps. So," he continued, "their hardscrabble lives were to remain that way for much longer than seemed the right of every other American."[89] Thankfully, FDR campaigned in 1932 "against what he saw as an inequity," attacking Insull and his utility magnate collaborators, who "had established policy of not doing business with the faraway farms."[90] Once elected, the new president acted "swiftly to apply government right, as he saw it, to a monstrous capitalist wrong."[91]

To further dramatize the desolate situation faced by farmers who endured without electricity, Winchester quoted from what he claimed was a lengthy government publication to describe the "woes of the powerless."[92] The five-paragraph block quotation, beginning and ending with "Because there was no electricity," in fact, did not come from an official document. Rather, it originated in Pence's book (noted above) published in 1985 by the National Rural Electric Cooperative Association—a trade organization.[93] Moreover, the quotation, dramatic as it was, actually appeared in Robert Caro's book on Lyndon Johnson, noted earlier, which the NRECA publication properly cited, but which Winchester did not. (Winchester's book contained no footnotes or endnotes, though it included a good-size bibliography. Pence's book—but not Brown's—was listed in it.) Overall, Winchester—like others—perpetuated the view that utility companies and their leaders malevolently denied farmers a privilege that city folk enjoyed.

More recently, another accessible popular history, Doris Kearns Goodwin's *Leadership: In Turbulent Times*, repeated the conventional narrative. Though not a major element in a chapter describing Lyndon Johnson's work as a Texas congressman during the late 1930s, Pulitzer Prize–winner Goodwin nevertheless drew on historian William Leuchtenburg's study, observing that "private utility companies had refused to install power lines in rural areas, maintaining that the rate of return in thinly populated areas precluded a profit."[94] New Deal agencies, she noted, "brought electricity to *millions* of farm families, but the needs of the people of the Hill Country [in central and southwest Texas] had been ignored" because even the REA felt that farms remained too thinly spaced to be economically feasible. With President Roosevelt's intervention, though, Johnson obtained the REA's willingness to provide loans for co-ops in the region in September 1938.[95]

While explicitly describing the heroic efforts of the REA, most accounts of rural electrification also implicitly represented farmers who lacked electricity as backward and without resources—material or otherwise. The characterization made for a more dramatic story in which the money-grubbing utility managers maintained the agriculturalists in a subservient position, robbing them of the benefits of modernity that arrived through electric power lines. As noted above, authors such as Schlesinger, Leuchtenburg, Caro, and Winchester suggested that ruralites had been struggling to keep up with their urban counterparts who delighted in the technological wonders of the twentieth century. Indeed, farm life was then (and remains today) more arduous and less comfortable than urban existence in many ways. Nevertheless, as historians and others have demonstrated, rural folk quickly adopted new technologies—especially the automobile, tractor, telephone, and radio—suggesting that they may have embraced innovative technologies and behaviors more aggressively than imagined. And as I illustrate in chapter 5, some farmers proved remarkably creative in embracing sophisticated technologies that produced electricity outside the realm of the central station power system.

The view that selfish private utilities shunned efforts to energize farms, requiring the federal government to pursue the activity as a matter of social justice, constitutes the traditional narrative of rural electrification. The account emerged with the creation of the Tennessee Valley Authority and REA during the 1930s, a period of increasing distrust of free-market institutions such as banks and large corporations that seemed to fail the public during the Great Depression. It continued with histories written by several academics, such as Schlesinger, Leuchtenburg, and Brown, which have become sources for many subsequent accounts.[96]

Undoubtedly, the standard account thrived because electric utility companies in the 1930s obtained a deservedly bad reputation for their excesses. Having emerged as small enterprises operating independently in cities during the 1880s, power firms accelerated their consolidation efforts after the creation of the holding company, a managerial invention that bundled securities from small operating companies into offerings of larger enterprises. The practice, first pursued on a large scale by the Electric Bond and Share Company in 1905, enabled financing of individual operating utilities (companies that generated and sold electricity to ultimate customers). It also offered greater security and financial leverage to share owners of the more diversified holding companies, higher up in the financial pyramid. Additionally, the new firms provided centralized technical and management support to the operating companies and helped them perform more efficiently, interconnecting with other firms via high-voltage transmission wires and exchanging power during emergencies and when they found economic advantages in doing so.[97] Throughout the 1910s and 1920s, holding companies acquired other firms in various geographic locations and at often-exaggerated prices. Especially during the go-go 1920s, industry leaders continued expanding the scope of these business entities, benefiting from little state or federal regulation, such that, by 1932, the eight largest holding companies controlled almost three-quarters of the private utility industry.[98]

The financial leverage that made holding companies so popular when operating companies made money also doomed them when the economy deteriorated during the Depression. Many of them (and their operating affiliates) defaulted on bond payments and went into bankruptcy or receivership, while share prices of utility firms lost more than 80 percent of their value (on average) from September 1929 to June 1932.[99] The Federal Trade Commission, which had already started investigating the industry's concentration, issued voluminous reports in the 1930s that described financial manipulation and the use of propaganda to influence educational institutions, newspaper editors, politicians, and regulators.[100] The high-profile collapses of forty-one companies operated by Samuel Insull, once a high-flying celebrity, reportedly produced losses of more than $638 million among investors, many of whom came from the middle and lower classes.[101]

Largely as a result of these events, the commercial electric utility industry earned public scorn as a symbol of corporate excess and a cause of the Depression. (Of course, this business was joined in infamy by the banking and securities industries.) Franklin Roosevelt took aim at utilities in his 1932 presidential campaign, and Insull became the personification of the power industry's abusive behaviors. In his January 1935 address to Congress, the president reminded the nation of the financial harm caused by the electric firms and called for "the restoration of sound conditions in the public utilities field through abolition of the evil features of holding companies." In August of that year, he signed the Public Utility Holding

Company Act, which broke up the multilevel structure of power organizations except for those that managed interconnected operating companies serving contiguous geographical regions.[102]

In such a hostile, anti-utility environment, REA promoters crafted a narrative that served their own needs. They did so by using an approach later advanced by theorist Michel Foucault, who argued that historical writing often reflects the aims and goals of "winners" of social struggles.[103] Using this interpretive framework, one can see that REA supporters took advantage of their dominant position in the 1930s and later to emphasize a positive account linking rural electrification to social equity and democracy, while discounting alternative stories perpetuated by the disgraced electric power industry. In addition, REA managers and employees undoubtedly realized that glowing accounts would publicize the organization's mission and help it earn popular and political support.

To be absolutely clear, I am not critical of the REA and its affiliates for producing such a narrative through promotional publications and various art forms. My only concern, for the purpose of this book, consists of the fact that many commentators and academics have uncritically accepted much of the conventional storyline. That school of history reflects, in part, the interpretations of esteemed scholars such as Schlesinger and Leuchtenburg, who clearly admired the liberal policies of President Roosevelt and his efforts to rescue the nation during the Depression.[104] At the same time, the account benefited from an understanding of the apparent success of government agencies in effecting rural electrification. When one looked back from the 1960s and later, it seemed obvious that huge progress in stringing lines to farms had occurred, and it appeared logical to accept the dominant account (even though private companies built many of those lines). And of course, the traditional narrative simply makes for a wonderfully exciting story, in which the good guys in government defeated the predatory capitalists who had thrown the country into financial and social chaos.[105]

By reading history from the perspective of the winners, however, we discount too much of the context and nuance of important events. As upcoming chapters illustrate, one can imagine an alternative history of events in the 1920s and early 1930s, describing the creation of institutions and well-organized efforts pursued by utility companies to power America's farms. At a time when the general public did not expect government to involve itself heavily in the business realm, the firms made genuine efforts to stimulate rural electrification in ways that made eminent sense, resulting in what also appeared to be significant—but unheralded—successes.

Chapter Two

Unattractive Economics in the Rural Electricity Market

One big obstacle that has limited the extension of electrical service to rural districts is the fact that usually there are only two or three farms that can be served per mile of electric line, whereas in the city there are anywhere from 50 to 200 customers on a mile of line.

—ELECTRIC UTILITY HANDBOOK, 1927

H aving just devoted a chapter to arguing that the traditional narrative of rural electrification needs revision, I now observe that the standard interpretation had at least one thing correct: the economic rationale for limiting service to rural districts had merit. While sometimes exaggerated for rhetorical and political reasons, most accounts rightly point to the unfavorable financial calculus of supplying electricity to farms. Utility companies began producing power in the 1880s for urban areas, where densely packed customers enabled firms to earn a quick return on their investment in generating plants, transformers, and distribution lines. By contrast, these businesses rarely extended lines to sparsely settled areas, where higher per-customer costs diminished opportunities to make respectable profits.

Power company managers understood that electrification of farms would enhance the lives of ruralites. As early as 1911, an industry report acknowledged that the "practical value of electric service to the farmer includes both the saving of his time and the reduction in the amount of hired help he needs."[1] Nevertheless, utility leaders concluded two years later that "[t]he 'farm business,' as developed at present, does not furnish the central-station companies with returns commensurate with the necessary capital investment."[2] Moreover, the appalling financial plight of ruralites in the otherwise prosperous 1920s and the existence of more lucrative customers militated against reaching out to farmers. The rural market simply appeared unattractive—at least to the majority of power company executives.

Basic Economics

The electrification of cities began in the late 1870s, with the Brush Electric and Thomson-Houston Electric companies stringing wires to a few arc lamps in cities that provided bright lights for street intersections. The Brush firm, which first illuminated thoroughfares in Columbus, Ohio, in 1879, began experiments in New York City in December 1880 and lit thirty streets by 1886.[3] But the illumination of homes and businesses emerged only after Thomas Edison invented incandescent bulbs, which created less intense and more palatable light than arc lamps, and had developed an entire system for producing and distributing electricity.[4] With construction in 1882 of a power plant in New York City's financial district and wires radiating from it, Edison offered a paradigm for electricity supply: a central station generated electricity and conveyed it to nearby customers through wires strung above (or concealed below) streets. Edison's direct current system operated at around 110 volts, such that the resistance in wires only allowed electricity to travel within about a half-mile from the generator. Competitors Thomson-Houston and Westinghouse developed alternating current systems, which facilitated the transmission of electricity over greater distances (by raising the current's voltage with transformers and reducing resistance). The latter firm demonstrated the value of such transformed electricity in 1896 when it built a transmission line, operating at about 11,000 volts, from a generating station at Niagara Falls to the city of Buffalo, a distance of more than twenty miles.[5] By 1911, numerous companies had erected transmission lines of more than one hundred miles at "tensions" of up to 110,000 volts. California's Pacific Gas and Electric Company became a technical leader at the time by constructing about one thousand miles of 60,000-volt lines.[6] Companies reduced the voltage at substations, with distribution lines taking power, usually at a few thousand volts, to ultimate customers. Transformers brought the power down to a few hundred volts for industrial equipment and to about 110 volts for operation of most residential lights and appliances.

During the first decades of electrification, pioneers such as Samuel Insull exploited several evolving managerial and technological innovations. President of the Chicago Edison Company, one of many firms vying for business in his city, Insull realized the worth of consolidation and purchased many of his competitors, creating the Commonwealth Edison Company.[7] Earlier than other utility leaders (in 1898), he also promoted state regulation as a means to legitimate natural monopoly status of electric supply companies.[8] Becoming the only commercial producer and distributor of power in the area, Insull's company took advantage of increasingly large generating technologies, such as newly invented steam turbines, whose economies of scale helped bring down the unit cost of electricity.[9]

Insull further understood the value of enhancing his company's "load factor," the mathematical ratio of the average demand of electricity divided by the maximum demand sustained during a period of time. The concept has great significance in the electricity business because power cannot be efficiently stored; it must be generated and used at almost the same instant because batteries then (and now) can only hold a limited amount of energy, and usually at high cost. Since power companies make large capital expenditures for equipment, they sought to use it as much as possible and in a way that would push up the load factor. Insull learned that he could boost the metric and earn greater profits by selling electricity—even at lower-than-average rates—to customers who used it during times when overall demand slackened. These periods consisted of the middle of the day (when electric street railroads and residential customers did not use much power) and late at night (when most people slept and had turned off electric lights). In speeches to lay and professional audiences, Insull took pleasure in noting how his firm's technological policies and management practices brought down costs and prices. He observed that a typical residential customer saw his or her bill drop by 69 percent between 1892 and 1912 for the same amount of electricity consumption.[10]

Insull's proud claim of declining prices, while true, hides some of the complexity of how consumers paid for power. To stimulate demand for electricity among manufacturers using huge amounts of electricity, for example, Insull's company offered a rate schedule (also known as a rate structure) specific to such customers. That schedule in 1916 included a monthly charge of $2.00 per kilowatt (kW), compensating the company for the user's highest incurred demand. Large consumers of power required utilities to invest substantial capital in generating capacity to meet their needs, and this "demand charge" reflected their portion of the investment. Beyond this amount, manufacturers paid for the actual energy used at a rate of 3 cents per kilowatt-hour (kWh) for the first 5,000 kWh consumed. The rate dropped to 1.1 cents per kWh for the next 25,000 kWh and to 0.65 cents per kWh for use above 100,000 kWh.[11] Other rate structures helped increase the load factor by offering lower demand charges, but only for customers employing electricity during off-peak times.[12]

For residential users, who drew considerably less power than businesses and factories, the company offered simpler rate schedules. In 1916, Commonwealth Edison charged 10 cents each for the first 30 kWh of monthly consumption, then 6 cents for the next 30 kWh, and 3 cents for subsequent use.[13] This "declining-block" structure did not include a separate demand charge, but it incorporated the pro-rated cost for the generating equipment into the first and most expensive block.[14] As customers used increasing amounts of power, they paid less per unit, since their share of the supplier's fixed costs (expenses that existed regardless of electricity consumption, such as for meter reading and billing) had largely been reimbursed. Once residents used more than 60 kWh per month in this case, they essentially paid just for the fuel to generate electricity.[15] And to ensure that customers paid their share of costs, even if they used little power, utilities often imposed an additional charge. Commonwealth Edison required a payment of 50 cents per month for customers who used electricity for lighting and other domestic purposes.[16]

Because of the mathematical nature of declining block rates, consumers saw their average cost of energy fall as they used more of it. In the 1916 Commonwealth Edison example above, a residential customer using just 20 kWh of electricity per month paid $2.00 plus a fixed charge of 50 cents; on average, each kWh cost 12.5 cents. But if that same customer consumed 100 kWh, he or she would have paid $3.00 for the first 30 kWh, $1.80 for the next 30 kWh, and $1.20 for the last 40 kWh, plus the 50-cent fixed charge, for a total of $6.50. The average price then dropped to 6.5 cents per kWh. The rate structure, in other words, encouraged greater energy consumption. And despite the lower average unit price, utility companies still recovered all their costs while earning more revenue from high-use consumers. This economic logic motivated power companies to promote electricity consumption throughout much of the twentieth century.[17]

Most electric utility companies established themselves in large population centers because they could offer service with relatively small investments (on a per-customer basis) in distribution lines, transformers, and associated equipment.[18] In a 1914 talk, Insull gave an example of a set of apartment buildings on a downtown block, approximately 0.2 mile long, which contained 193 apartment dwellers and 34 other customers. The customers used a total of about 50,000 kWh during the year, generating income of $18.34 each. If this block can be considered typical, then a full mile of distribution line would produce a load of about 250,000 kWh and yield income of almost $21,000. To be sure, the mile would require more transformers and other equipment than would a mile of rural line.[19] Still, this example will be useful later.

High population density, in other words, constituted the bywords of the nascent electric power business. That condition, however, did not apply in most rural settings. Unlike the Chicago neighborhood, which boasted the equivalent of more than 1,100 customers in a mile, the number of farmsteads in New Hampshire stood at about 3 to 5 per mile of distribution wire in the early to mid-1920s.[20] In 1923 Wisconsin, each mile of rural line that had already been built served 2.7 customers, while the density rose to 35 or 40 per mile even in small villages with populations of 500 or 600 people. (In the state's larger cities, the density jumped to 300 customers per mile, which further drove down the per-user cost of lines.)[21] Utilities serving Iowa's cities and towns in the 1920s typically could share costs among 10 to 75 customers per mile of distribution line. But in rural areas, only 2.5 customers drew power from each mile of wire.[22] Reports from central station power companies around the country in the 1910s and 1920s offered similar information: one mile of rural line provided power for three or four customers, with stated costs running from about $1,200 to $2,000 per mile.[23]

Distribution expenses did not constitute the only factor that raised the amount of investment in each potential farm customer. Transformer costs also played a big role. As noted in a 1922 analysis by an Iowa utility, each device could usually serve only one rural customer. By contrast, the close spacing of urban homes and businesses meant thirty-three consumers shared one transformer.[24] Consequently, the Iowa firm anticipated investment costs for transformers at about $5 per urban customer and $164 for each rural user.[25] Furthermore, while transformers designed for single customers were smaller than those serving several users, energy losses from them rose disproportionately. Quoting an electricity textbook of the time, Idaho researchers noted, "It is estimated that transformer losses are ten times as great per customer on rural as on urban lines."[26] Long distribution lines to farms also suffered large energy losses, which naturally raised costs. Wisconsin agricultural engineers commented in 1923 that about 45 percent of the energy put into the lines dissipated as it moved through wires and transformers. The equivalent loss on urban lines came to just 18 percent.[27]

The high fixed costs to distribute electricity to dispersed rural customers startled researchers. The University of Minnesota agricultural physicist, Earl A. Stewart, observed in 1924 that distribution costs to farmsteads amounted to $8.75 per month, exclusive of the cost of energy. He concluded, "Even if electricity could be generated by water power for nothing, which is not the case, it would have little bearing on the cost of electricity to the farmer, for . . . the big cost in rural electrification is not in generation but in distribution."[28] Iowa State researchers concurred after a 1928 study: "It is safe to say that under present average conditions, from 80 to 95 per cent of the cost of electricity at a farm can be charged to distribution."[29] The authors noted that utility companies could not just give away power;

they needed to obtain a "return commensurate with the expense involved and the risks taken" so they could maintain solvency.[30]

The National Electric Lighting Association (NELA), the major trade group of private utilities, summarized the economic rationale for limiting service to farms in a 1924 study. It noted that "the investment necessary for a distribution system to serve 100 rural customers is many times the investment necessary for a distribution system to serve city customers."[31] Beyond the higher up-front expenditure, however, are the costs for taxes, depreciation, and interest paid annually for the distribution line. In one example, rural customers cost a utility $56 per year in these expenses compared with about $10.50 for urbanites.[32]

Of course, electrification could be profitable if rural customers used considerably more power than did their urban counterparts. In most cases (with exceptions especially in western states, as noted in chapter 6), however, farmers did not seem likely to consume much power outside of what they needed for illumination. As almost everyone recognized early in the history of electrification, farmers—like city residents—truly enjoyed electric lights. In a 1909 thesis, Oregon Agricultural College electrical engineering student Walter Baker observed, "Farmers and residents of small country towns . . . appreciate the safety, cleanliness, and convenience of this method of illumination and would gladly adopt it in their homes and about their farms if possible."[33] But electric lights did not require much power, especially as light bulb manufacturers continued to improve their product. (Commonwealth Edison's Samuel Insull noted that incandescent lamps had increased in efficiency by more than 260 percent from 1886 to 1912.)[34] David Weaver, the head of the North Carolina State College agricultural engineering program, observed that even when farmers obtained high-line electricity, they often did not realize the value it could provide if used for more than lighting. "We know," he commented in 1935, "that a farmer with three or four 25-watt lights thinks he is on Broadway after years of kerosene lamps and . . . he cannot see why he will have to use more current if he is to get electric service."[35]

To be fair, farmers showed justifiable reluctance in using electricity for anything beyond the obvious application of lighting. Although a 1913 NELA report described thirty applications of electricity outside the farm home, most of these could not guarantee favorable economic returns to most farmers.[36] Highly touted by their manufacturers, electrical devices had not yet been used much by rural customers, and they often had no track record of providing financial benefits. One utility executive remarked in 1913 about the difficulty in selling to the hard-nosed agrarian customer: "You must show the farmer where he will get back $6.00 for every $5.00 he spends."[37] Unfortunately, power companies and manufacturers could not yet make that demonstration with most electrical machines. It required another decade before the utility industry established the Committee on the Relation of Electricity to Agriculture to address this concern directly, namely, to identify clearly productive equipment that would yield increased energy consumption and sufficient revenues to central station companies.

Farmers' Economic Woes and Tenancy

The straightforward economic rationale for unenthusiastic service to farms— that utilities generally earned poor returns on investments in them—seems compelling enough. Yet another element of rural life reinforced power company managers' reluctance to invest in rural lines: the decayed agricultural circumstances of the 1920s that followed several years of well-being. That deterioration countered the experience in other parts of the American economy, especially manufacturing, which saw its efficiency and production increase dramatically, making it more appealing to serve.[38] Put simply, utility leaders realized that a business sector's economic output correlated with electricity consumption and revenue. If a sector's prospects appeared poor, service to it would not likely yield profits.

The agricultural realm enjoyed a period of prosperity in the 1910s. Demand for crops exploded, and commodity prices rose as war erupted in Europe in 1914. The farm economy continued to flourish as the United States sent troops to fight in 1917. Even after the armistice began in 1918, demand for foodstuffs in Europe remained high, causing prices to double from the 1914 level.[39] By the 1919 crop year, gross income for farmers peaked at $16.9 billion, 2.4 times the amount of 1914.[40]

But the good times ended in spring 1920, when the nation entered a serious recession—one that hit farmers especially hard.[41] Wheat prices, which had reached about $2.56 per bushel (as a national average price) in June 1920, fell by more than half ten months later and sank to 93 cents in December 1921,[42] remaining well below its peak for many years thereafter.[43] Though the farm recession bottomed out in 1921, with gross income falling to 53 percent of its 1919 height, the rest of the decade remained bleak. In 1929, gross income reached $11.9 billion, still about 70 percent of its highest level a decade earlier.[44] Moreover, the value of agricultural land and buildings declined throughout the decade, such that 53 percent of farm homes had values below $1,000 (with 33 percent worth less than $500) in 1930.[45] A Department of Agriculture report summarized the "farm problem" or "farm crisis" of the 1920s as a multifaceted event consisting of war-induced inflation and acquisition of debt, resulting in farmers' income falling far behind that of

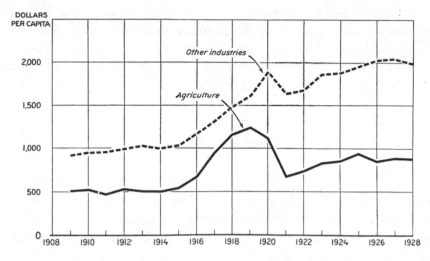

Fig. 2.1. Income per person engaged in agriculture and in non-agricultural industries, 1909–28. *Source*: L. C. Gray and O. E. Baker, *Land Utilization and the Farm Problem*, USDA Misc. Pub. no. 97 (Washington, DC: GPO, 1930), 2.

nonagricultural workers (figure 2.1). Related problems consisted of farm product price deflation, high taxes, greater costs for transporting agricultural goods, and changing consumer demand.[46] Utility managers, in other words, had good reason to question investments in the agricultural sector, seeing that it remained in dire straits throughout the 1920s.

The agonizing financial situation intensified during the Great Depression, which followed the stock market crash in 1929. Gross farm income slid from $11.9 billion in 1929 to $5.3 billion three years later.[47] At the same time, land prices plummeted, meaning that people often could not sell their properties for more than they owed on loans.[48] The value of farmers' real estate, which had reached an index of 173 at its peak in 1920 (with 100 as the index's base for the 1912–14 period) had fallen to 116 in 1929 and tumbled to 70 in 1933. And as banks struggled during the financial crisis, they cut lending to all customers by 54.2 percent in mid-1933 (compared with the amount loaned four years earlier).[49] Worse, many bank loans to farmers went unpaid. At the nadir of the Depression in 1933, the foreclosure rate reached 38.8 per thousand farms, which compared to 3.2 per thousand farms in the period from 1913 to 1920 and 10.7 from 1921 to 1926.[50]

The unusual structure of the farm economy, such that many agrarians did not own the land they cultivated, reinforced the view that rural electrification would not provide reasonable returns to power companies. As tenants, these workers often paid rent in a portion of the crops they produced. While the arrangement allowed cash-poor families to subsist off the land, it did not provide the means for investing in electrified equipment that would yield higher farm productivity. An enthusiast of farm electrification in Texas acknowledged this fact in 1928, noting that a large number of owners might electrify their farmsteads, but only a small percentage of tenants would do so.[51] Two years later, an agricultural engineering graduate student noted that even if the tenant farmer wanted power, "wiring for electricity is a permanent improvement that he is not justified in making, and often the landlord is not sufficiently interested to meet the necessary cost."[52] The 1930 census confirmed this phenomenon: among all farmers who had electric service on their properties (either from central stations or from isolated power plants), 85 percent owned or managed their farmsteads; only 15 percent of tenants worked on electrified farms.[53]

An increasingly recognized problem in the United States, tenancy reached 25 percent in 1880. That number surged—to more than 35 percent in 1900—and had severe consequences. Ohio State University professor Homer Price observed in 1908 that the growing number of tenant farmers threatened American economic and social welfare. Holding relatively short leases, these ruralites had little incentive to pursue investments that would raise productivity. "Such a tenant is not interested in improving the farm unless immediate results can be realized," Price wrote.[54] Tenancy held sway especially in southern states, largely because (according to experts) of the large number of African Americans who could not afford to purchase land.[55] About 45 percent of the nation's farmers did not own the land they cultivated in 1935, a great concern to Secretary of Agriculture Henry Wallace, who noted that these workers usually occupied property for slightly more than four years before moving on.[56]

Tenancy and its associated problems continued to linger the most in the South, where both Blacks and whites suffered due to the agricultural collapse that began in 1920 and accelerated during the Depression.[57] The situation called for dramatic government action since "[t]he present conditions, particularly in the South, provide fertile soil for Communist and Socialist agitators," which should be met, Wallace asserted, not with "violence or oppressive legislation to curb these activities but rather to give these dispossessed people a stake in the social system."[58] A report produced by a special committee authorized by President Roosevelt in 1937 (and chaired by Wallace) noted that tenancy in the South had reached 54 percent in 1935, well above the national average (figure 2.2).[59]

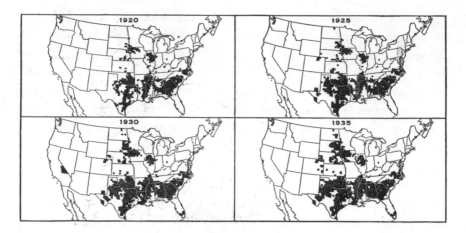

Fig. 2.2. Extent of farm tenancy, 1920–35, showing counties in which at least half of the farms were operated by tenants or sharecroppers. *Source*: H. A. Turner, *A Graphic Summary of Farm Tenure*, USDA Misc. Pub. no. 261 (Washington, DC: GPO, 1936), 13.

In short, high rates of tenancy appeared to correlate with low levels of electrification. Put differently, the plight of the poor tenant farmer helps explain why rural electrification lagged so badly in southern states. With tenancy running at about 70 percent in Mississippi, only 1 percent of the state's farmers received central station service in 1935.[60] By contrast, tenancy in New England states (such as New Hampshire, Rhode Island, Massachusetts, and Connecticut) remained considerably lower, and they exhibited higher electrification rates before the REA began its efforts.[61]

Focus on Other Markets and Activities

The dismissal of the farm market also occurred because power company managers saw better opportunities in other areas. Though Thomas Edison's pioneering 1882 station sold direct current to businesses and wealthy homeowners primarily for electric lighting, utility companies in subsequent decades found more lucrative uses of power in, for example, transportation networks. The electric streetcar, first showing its technical and commercial superiority over horse-drawn carriages in Richmond, Virginia, in 1887, became familiar sights in many cities.[62] In fact, many urban central station companies, such as the Milwaukee Electric Railway and Light Company (formed in 1896) initially supplied power primarily for streetcars, but they benefited from selling electricity for lighting to businesses and homes at off-peak times.[63] In other words, serving residential customers for lighting loads initially constituted a secondary interest. In 1917, the capitalization of American electric railway companies exceeded that of utilities providing telegraph, telephone, and power services.[64] The prospect for further electrification of main railway lines (and associated profitability) seemed bright in the 1920s.[65]

Outside of the transportation sector, the manufacturing realm appeared attractive to power companies. As factory owners and managers learned that electricity offered huge productivity gains, they replaced waterwheels and steam engines with in-house electricity generators.[66] These isolated power plants required investment in equipment and manpower, but they enabled manufacturers to reduce costs by employing low- (and fractional-) horsepower electric motors. Self-generation became so popular, in fact, that nonutility companies in 1912 operated more than half of the nation's electricity production capacity (5,800 out of 11,000 megawatts, abbreviated MW; 1 MW equals 1,000 kilowatts). As consumption of electric power grew fortyfold in the years between 1899 and 1919, however, central station operators replaced local electricity production with power generated in plants exhibiting increasing economies of scale.[67] And as Samuel Insull realized earlier than other utility managers, selling power to manufacturers at times when businesses, electric streetcar companies, and homes did not need much electricity helped increase load factors, yielding improved financial results. Consequently, power firms sought and captured greater amounts of the overall electricity market. By 1922, electric companies owned more than twice the capacity of businesses that generated power for themselves (13,400 MW to 6,300 MW).[68]

During and after World War I, central station electric companies also devoted much of their capital to construction of transmission lines between distant power plants. They did so as electricity demand increased among manufacturers of wartime supplies, with orders pouring in from England, France, and Russia. At the same time, energy shortages in parts of the country sometimes forced rationing and government-mandated industry shutdowns. By erecting transmission lines between large, efficient generators, companies could draw on excess power capacity in one region for use in another. As historian Julie Cohn observed, transmission lines provided the key to conserving fuel and guaranteeing sufficient power for manufacturers and businesses.[69] To extend those benefits into the postwar era, consulting engineer William Murray proposed in 1918 a "superpower" system connecting large-scale power plants in northeastern states for use, among other things, by an increasingly electrified railroad network.[70] Governor Gifford Pinchot of Pennsylvania proposed in 1923 a similar system relying heavily on transmission links between power plants. Though neither plan came to fruition, largely for political reasons, interest in transmission line construction did not wane. Economist Martin Glaeser observed in 1927 that the war experience had taught how interconnection lessens fuel consumption and capital investment per unit of capacity and enables the profitable use of surplus power. It also "safeguards the service from interruptions, and secures the advantages of long hour use of investment by building up the diversity of use."[71] Though utility holding companies would obtain a bad reputation for their financial machinations in the late 1920s, they often spent available capital to purchase small companies, consolidate their assets, and tie them together with transmission lines.[72]

As firms expanded their web of interconnected systems and promoted electricity sales to manufacturers, they often neglected residential customers. In 1917, only 24 percent of homes had been wired, such that electric service remained an extravagance for the vast majority of city dwellers, no less for ruralites. With increased business activity and growing urban affluence in the 1920s, along with the greater availability of electrical appliances, managers saw improved markets emerging in cities. Owners of homes and apartments wired their structures, and power companies extended service in 1922 to 40 percent of them. By 1928, that percentage jumped to 63. To encourage residential customers to purchase increasing amounts of electricity, utility companies offered promotional rate structures that promised lower prices for enhanced consumption.[73]

To summarize, marketing to manufacturers and urban customers, while also spending heavily on an increasingly interconnected transmission network, took on greater importance to the utility industry than serving rural districts. When power company leaders considered how to deploy financial resources, these areas of concern took precedence—as long as they only evaluated business considerations.

During the first three decades of the twentieth century, utility managers generally viewed farm electrification as a laudable—but challenging-to-achieve—goal. A major impediment consisted simply of the high cost incurred to serve sparsely distributed customers who used electricity as their urban cousins did. While farmers appreciated electric lights, it had become clear that, according to a 1930 agricultural engineering thesis, "electricity for its convenience alone will never make rural electrification practical and profitable." Farmers needed to consume greater amounts of power, such that increased revenue would offset the higher costs of providing service.[74] (Of course, companies could levy extremely high prices for each unit of electricity as a means of recovering expenses, but classical economic theory suggested that those rates would depress consumption and revenue still more.)

This strict, profit-seeking attitude may seem indifferent and callous today, given the experience of government agencies and regulatory bodies since the 1930s that took on goals of equity. In recent decades, government has moved to ensure fair consumer and housing practices, safety of workers, environmental protection, comparable workplace pay for men and women, and many other social functions.[75] Electricity eventually became viewed as a necessity of life, and its provision became embodied in law, such as in California's Warren-Miller Energy Lifeline Act of 1975. (The legislation declares that "[l]ight and heat are basic human rights," and it directs the state regulatory commission to establish cheap gas and electric rate structures for low-income consumers.)[76] But before the 1930s, state and federal regulatory commissions dealing with industrial enterprises had narrower mandates, focusing predominantly on economic—not social—considerations. They sought, for example, to ensure that railroads and other industries did not gouge customers and that companies rendered their services without undue discrimination. Such regulation sought to impose a substitute for competition in the marketplace, but it remained economic regulation.

Put simply, private power company managers felt no obligation to provide service to customers who could not pay for it. Unless farm customers found ways to increase their power consumption enough to become profitable customers, utility firms overlooked them and sought more rewarding markets in the transportation, manufacturing, and urban residential markets. Doing so appeared rational, and it comported with attitudes of most business and government policymakers of the day.

Chapter Three

Business Attitudes toward Farmers in the 1920s

> The more important sources of difficulty in agriculture . . . are in part due to the peculiar deficiencies of organization and management which are characteristic of the agricultural industry. The trend toward organization and collective action, which has been so marked a characteristic of non-agricultural industry in the last half century, has made relatively little headway in agriculture.
> —BUSINESS MEN'S COMMISSION ON AGRICULTURE, 1927

The economic and market realities of the 1910s and 1920s likely played the largest role in utility managers' decision to decline service to most farmers. Even so, power company leaders' condescending attitudes about rural folk probably reinforced those hard-number truths. Big-city executives generally thought of agrarians as backcountry rubes with little technical and economic savvy; unlike urban customers, farmers would never develop the know-how to use electricity in ways that would benefit themselves or yield sufficient profits for power companies. Like other businessmen of the day, most—but not all—utility leaders adopted the belief that progress first occurred in cities, where people experienced modernity in the form of automobiles, radios, telephones, moving pictures, and electric motors. Though necessary for sustenance of urban activity through the food they produced, farms carried connotations as places where life may have been simpler (in a rustic, sentimental sense) but not necessarily as up-to-date nor economically as productive as cities.

Businessmen's Condescending Views about Farmers

The twentieth century began with enthusiasm for new technology-based services that depended on physical infrastructures within densely populated cities. Indoor plumbing required mains that drew water from miles away and pumps to deliver it to homes and businesses. Telephone connections and electricity delivery depended on wires strung overhead or beneath streets, linked to switching exchanges and substations scattered throughout cities. Heating (and some lighting) necessitated underground pipes, while streetcar conveyance needed rails running through public thoroughfares and another set of overhead cables.

This urban technological system inspired awe. "Wonderful, indeed!" exclaimed the New York City commissioner of franchises in 1910 as he proudly described the "bundles of tubes" running beneath the streets, bringing gas and electricity to thousands of appreciative customers. "Light[,] heat and power distributed through-out a city!" But the existence of these networks of power also spurred feelings of pity for the poor ruralite: "The city is ablaze at night. How strange and impene-trable is the darkness of the country, with only here and there a flickering ray from some farm-house window! How short the winter's day! How long and dark the winter's night! How meager the opportunity for social life and gaiety, if one must be at home when night falls."[1]

Taking note of the growing divide between urban and rural life, characterized in part by differences in fundamental infrastructural systems, a group of reformers and progressives created the "Country Life" movement.[2] Seeking to improve farm-ers' efficiency and stanch the migration of intelligent people from farms to cities, the initiative won a sympathetic ear from President Theodore Roosevelt, who in 1907 established a Country Life Commission to investigate problems of rural life. The body's report, published in 1909 with a preface written by the president, re-flected the already condescending attitude about farmers who needed to employ business techniques to keep up with their urban counterparts.[3] More specifically, the document highlighted a series of problems, including farmers' "lack of knowl-edge . . . of the exact agricultural conditions and possibilities of their regions," poor training, inadequate highways, and a "general need of new and active leadership."[4] To overcome deficiencies, ruralites already had support from major institutions such as the US Department of Agriculture (USDA), land-grant colleges, states' agri-cultural experiment stations, and national farmers' organizations.[5]

Farmers often resented "book learning" and advice offered by experts at the USDA and colleges, however, as historian David Danbom has detailed.[6] And they certainly had little interest, it appeared, in approaches promoted by these institu-tions to grow the best-suited crops for specific locales or to employ state-of-the-art fertilization practices. Secretary of Agriculture David Houston noted in a 1913 report, for example, that his agency had been "giving special attention to the subject of farm management with the view of rendering to the farmer service similar to that rendered to the business man and the manufacturer by efficiency experts and engi-neers."[7] But he also lamented the fact that only about one-eighth all farmland had been used efficiently.[8]

Businesspeople reflected a similar view that farmers needed to boost their productivity. Frank Vanderlip, president of the National City Bank, argued in 1914 that the press and public should stop complaining about the inefficiency of big business enterprises, such as railroad companies. (The transportation firms had been in the news as the "people's attorney," Louis Brandeis, demonized them, claiming since 1910 that their poor management resulted in excessively high freight charges to customers.)[9] But farmers operated even less productively and in a way that should make people upset with them: "If a railroad manager is culpable and is answerable to society for anything less than a hundred per cent of efficiency, what of the great farmer and the planter holding the great agency of production—land—and utilizing it with but 40 per cent of efficiency? That is the indictment that stands against no small part of the agricultural community."[10]

The editors of the *Banker-Farmer* magazine continued this line of criticism and urged ruralites to wake up at "the dawn of a new agriculture." A scathing editorial cartoon appeared on the cover of the May 1914 issue, portraying a county agent who offers information about business methods and science, pounding at the door of an unmoved, sleeping farmer (figure 3.1.) Better methods for raising crops and animals await, but the farmer continues to employ poor agricultural practices. Inside the issue, an author observed how the average farm of the day "is running on 'low speed.'" By contrast, the "business man, transplanted to the soil, has shown himself a remarkably efficient chauffeur and able to get speed out of a farm," using "distinctive methods" that are "worth watching and studying."[11] Another contributor wrote a few months later that the farmer, being individualistic by nature, does not eagerly work with his country or in-town neighbors. That lack of cooperation needs to change if farmers seek to make progress in the future.[12]

Concerns for rural welfare and efficient production dissipated somewhat in the following years, as farmers improved their economic fortunes during the run-up to and years following World War I. The recession of 1920 to 1921 ended that prosperity and (as noted in the previous chapter) agriculturists struggled in subsequent years. By contrast, other segments of the economy—especially manufacturing and commerce in cities—improved dramatically after 1921: by 1929, the nation's gross private domestic product for nonfarm activities had grown 48.1 percent (in inflation-adjusted terms) over its 1920 amount. The same index for farm activity advanced only 12.6 percent.[13]

To some people, the success of the overall business sector resulted, in large part, from the growth of the engineering profession and its partnership with corporations to solve social problems. In this view, which evolved in the Progressive Era

The BANKER◆FARMER

Reviewing the Banker's Activities for a Better Agriculture and Rural Life
Conducted by the Agricultural Commission of the American Bankers Association

| VOL. I | CHAMPAIGN, ILLINOIS, MAY, 1914 | NO. 6 |

Fig. 3.1. Time for the farmer to wake up to modern agricultural techniques.
Source: Banker-Farmer 1, no. 6 (May 1914): cover.

before World War I and until the Great Depression, engineers within companies spurred civilization's advancement. They did so by adopting the notion of efficiency as a professional mantra, drawing on Frederick W. Taylor's scientific management principles as the means to eliminate wasteful industrial practices.[14] During the Great War, corporations and their managers—many of whom had training as engineers—helped mobilize resources to produce the weapons needed to win the conflict and the food used to feed starving masses in Europe and elsewhere.[15] In this last effort, former mining engineer Herbert Hoover played a key role, using his expertise in such a way that he became an exemplar of the professional expert who resolved humanitarian challenges.[16] After the war, several companies—with General Electric serving as a preeminent model—undertook efforts to continue exploiting technical innovations while providing benefits to laborers in the form of profit-sharing schemes, life insurance, pension plans, and improved working conditions.[17] Managers of these corporations also felt that they should sometimes collaborate with the federal government to solve large-scale problems that individual firms or states could not manage individually.[18] Allied with engineers and professing to understand "scientific" means of organizing factories and laborers, academically trained business managers made these corporations more efficient, profitable, and (supposedly) valued by citizens.[19] Together, the businessmen and engineers invented assembly lines in factories, which spewed out Model T Fords and other consumer goods in prodigious numbers and at decreasing cost, enabling material bounties for a growing population.[20] The business community in the 1920s, according to historian David Goldberg, "had become the dominant if not the hegemonic force in American society."[21]

The success of business managers and engineers—in contrast to the apparent failure of farmers who proved unable to recover quickly from the 1920–21 recession—suggested the wisdom of the modern approach of producing goods in a free market, laissez-faire environment. To be sure, farmers still represented the "paragons of virtue" (in Thomas Jefferson's words), and they formed the mythical, individualistic, moral core of the country, embodying romantic values of probity, spirituality, and individual responsibility. The noble agrarians conquered the wild frontier, turned it into gardens of Eden, and produced the necessities for everyday life.[22] However, to many urban, elitist managers and their allies, farmers appeared stupid, backward, and lacking in knowledge of basic economics and the latest business principles. They urged farmers to adopt techniques of the industrial world to overcome their failings. In the words of historian David Danbom, agriculturalists in the early twentieth century "had been transformed from paragon to problem," and rural America had devolved "from backbone to backwater." He further observed that by the 1920s, the general urban population viewed farmers "as retrograde elements in an increasingly sophisticated society."[23]

Business leaders frequently discussed the primitive tendencies of farmers, who acted so differently from their enlightened city cousins, especially after the recession of 1920. US Chamber of Commerce president Julius H. Barnes, for example, urged farmers to take advantage of the new knowledge created by state land-grant colleges and disseminated by the Department of Agriculture. And as farmers competed with cheap-labor farmers in other countries, they should exploit "American resourcefulness" and "mechanical processes" to improve productivity and reduce costs.[24] A few years later, Barnes took issue with farmers who complained about their plight and sought government subsidies to raise prices of agricultural products (as provided for in the McNary-Haugen Farm Relief bills, vetoed twice by President Coolidge in 1927 and 1928).[25] Discounting such grievances, Barnes noted, "The voice of business speaks in sober, sound terms and in serious modulations, instead of sounding the paeans of quick and easy panaceas," such as federal assistance.[26] If only farmers would act more like urban businessmen, the farm problem would be mitigated.

A similarly disparaging characterization of ruralites as country bumpkins appeared in a 1930 book describing the origins of the radio. Writing about the already profound impact of the new technology in *This Thing Called Broadcasting*, Alfred Goldsmith (vice president of RCA) and Austin Lescarboura (former managing editor of *Scientific American*) observed that farmers who visited cities easily made themselves known by their unsophisticated attire and awkward manner. They seemed astonished by the urban environment and the excitement it offered in terms of culture, entertainment, and comforts. No wonder the cities tempted farmers—so much so that increasing numbers left their rural homesteads for the vibrant metropolises. (The 1920 census indicated, for the first time, that more people lived in cities than in rural areas.)[27] The remaining agrarians clearly lacked the smarts to make it in the city, leading to "the conclusion that by staying on the farm[,] a man showed himself to be a dullard, lazy, stupid, or all three."[28] Of course, the advent of the radio would help remedy the problem of rural flight. A farmer could live by the fields and still obtain valuable market information and education (especially from broadcasts produced by the Department of Agriculture). By listening to the radio, he would acquire the "business acumen and knowledge" that will "make him no longer a Rube but a man of the world."[29] Perhaps correctly observing the likely impact of the new technology on rural life, the authors nevertheless depicted the average agrarian as a naive citizen, one who needed enlightenment by more urbane and urban business leaders.

Farm journal authors sometimes agreed with such assertions made about their readers. A 1919 article in *System on the Farm* magazine implicitly acknowledged ruralites' deficiencies by exhorting them to get up to speed with new technologies and to modernize their farmsteads. It told the story of an Ohio family that installed its own on-site power plant, enabling "modern water systems" and many other conveniences that enhanced productivity and daily life. The electrical amenities also played no small part in keeping the almost twenty-one-year-old son content and reluctant to leave for the city.[30] And in a 1920 piece titled "Is the Farmer Behind the Times?," a writer conceded that farmers often have "little respect for their own reasoning powers" and take pride in their independence. They exult in the knowledge that their forebears sustained themselves on remote homesteads and produced food and other necessities by themselves. "But the demands of modern life have robbed the farmer of this independence[,] until today he is almost as dependent upon grocers, miners, carpenters, tailors and mechanics as is his city brother." To improve their lot, the article added, farmers should take advantage of modern business principles and concentrate on only the most profitable activities instead of trying to diversify too broadly. Beyond taking such measures, farmers needed to exploit the resources of agricultural experts and adopt "modern business methods."[31]

Criticism from the business sector sometimes caused bitter feelings among ruralites accused of retaining premodern and preindustrial mindsets. Farm magazine editor E. R. McIntyre expressed resentment about efforts in 1926 to force his readers to become more disciplined and efficient, sure that they would destroy the culture of the agrarian worker. "Woe to the conscientious objector in agriculture who persists in growing whiskers or wearing overalls to the bank," wrote McIntyre. "Open derision is the share of him who feels an inherited hankering for the sentimental things of the soil, or who, perhaps, has held on to a few legends and customs that were cherished by his misguided sires." The author celebrated these quaint, backward-appearing pretenses, rejecting the perception propagated by reformers that farming ought to become a "highly serious business."[32]

City-based electric utility managers appear to have adopted the general businessperson's view of farmers as resistant to learning modern methods. In 1913, a company engineer looked forward to governmental efforts (through a proposed Bureau of Farm Engineering) that taught current accounting techniques so agrarians could more effectively manage the costs of doing business.[33] NELA president

Walter Johnson expressed a similar concern in 1924 about farmers' inability to become properly educated. On one hand, he readily conceded that the nation's prosperity depended on agriculture and noted that ruralites had been victimized by international economic circumstances that resulted from the Great War. Nevertheless, the farmer had exhibited little "flexibility in adapting himself to changed conditions," he noted unsympathetically. Worse, the farmer "has shown little ingenuity in working out his own salvation and has been inclined to rely upon the government and special legislation for relief."[34]

To overcome rural problems, power industry leaders urged farmers to become more receptive to experts. In a 1925 speech to representatives of farm organizations, electric utilities, and universities, Owen D. Young, chairman of the General Electric Company, observed that farmers must become more interested in their own business and accept the lessons, taught by agricultural colleges, that demonstrate the value of electrical tools. Instead of resenting outsiders' suggestions, they should welcome the efforts of power and manufacturing companies and "co-operate with them to work out practical plans to get effective results." By doing so, farmers would learn that electricity could yield increased profits so long as they used enough power to make service worthwhile to utilities.[35]

Farmers' independent tendencies also hindered their reception of electric power. While rural citizens showed enthusiasm for other modern technologies, such as the radio and the automobile, they expressed less gusto for electricity, largely, argued Young, because farmers did not like to cooperate with others. Purchasing radios and cars incurred a personal, individual decision, one that did not require much concern for the actions of others. But to obtain electric power, farmers needed to collaborate, presumably by agreeing with neighbors that they would create enough electricity demand so utility companies would string lines to their homesteads. Young asked the farmer to "be open-minded as to new methods and to show readiness to abandon the old. I ask him to learn to co-operate with his neighbors for their mutual benefit as well as to act individually on his own account."[36] In short, ruralites needed to give up their old-fashioned independence along with their ambivalence to learning new methods. Otherwise, utility companies would not see value in offering electricity to them.[37]

Social Responsibility to Serve Farmers?

Even if utility moguls viewed farmers as backwater hicks, some rural advocates thought they should be provided electricity as a matter of social justice. Lesher S. Wing, a California Farm Bureau Federation engineer, opined in 1926 that utilities had a moral obligation to offer power to agriculturists. Farmers constituted the "real backbone of our nation," he argued, and they should "be given an opportunity to maintain themselves upon a plane of living comparable with workers in other industries."[38] Acknowledging that providing power to rural customers proved more expensive than selling to city residents, Wing did not seek to cause financial harm to utilities. Instead, he suggested subsidizing farmers with part of the savings accrued as the industry saw its urban business costs (and prices for power) decline owing to increasing overall consumption of electricity and improving load factors.[39] In other words, utilities could make up the extra cost of serving rural customers by slowing the pace of rate reductions to urban customers. City consumers would continue to see their rates drop, but not as quickly as in the past, and the companies would not lose revenue. Moreover, as they began increasing their demand because of their adoption of electrified technologies (because "soon what was deemed a luxury will become necessity just as have the telephone, the automobile and the radio"), farmers would ultimately use enough power to become profitable customers who did not require subsidies.[40]

Efforts to force one class of customers to pay for another, even for noble reasons, held little traction in the utility community. A 1925 editorial in the prominent trade journal *Electrical World* urged power companies to help the farm market evolve, but it warned that the industry should avoid riding "the social hobby . . . to the point of financial bankruptcy."[41] At the same time, utility consultant William Murray rebuffed arguments calling for subsidies by noting that state regulatory policies generally required each class of customer to "pay rates proportionate to the cost of such service." In other words, existing policy discouraged Wing's plan for cross-class assistance. "[F]armers should have electricity wherever it is possible to furnish it economically to them," Murray argued, "but they should not get it at the cost of others."[42]

Utility managers moderated their views toward achieving social goals such as rural electrification while the Depression deepened, but not by much. In May 1932, an *Electrical World* article observed: "The primary interest of the electric utility in rural electrification is revenue. Social responsibility is a factor, a strong one, but electric utilities are not eleemosynary institutions and they cannot undertake to serve any class of customers on any narrower base than that the revenue will pay at least the cost. Therefore, conspicuous advances in farm electrification must wait until the converging efforts in reduction of cost of service and in persuading the farmer actually to *use* electricity have met and merged into a single stream of progress."[43] Utility officials who testified in hearings about the REA in 1935 continued to note that the "urge for rural electrification [is] a social rather than an economic problem."[44] They declared that electric utilities would serve rural customers who prove economically viable for profit-making companies—a sentiment that seemed perfectly rational at the time.

Decentralizing Industry and Rural Electrification

According to some utility managers, the emergence of the postwar movement toward decentralization of industry would gradually bring electricity to farms. Becoming an important concept among city planners, engineers, and architects, the redesign of urban manufacturing on a regional scale took on new meaning as many cities suffered fuel and food shortages during the winter of 1917–18.[45] Perhaps such problems could be mitigated by establishing communities in rural districts, where industry and agriculture would more easily find integration.[46] Among those who felt this way, Clarence Stein, a New York architect, helped organize in 1923 the Regional Planning Association of America (RPAA), which sought to create (among other things) "garden cities." Outside of established industrial urban areas, these communities would be surrounded by protective agricultural belts and would balance "industry and healthy living."[47]

As the organization's secretary, Lewis Mumford became a major proponent of establishing outlying industrial towns, observing that new infrastructural technologies would help stem the tide of what he called the "third migration" of citizens from farms to cities. Instead, people would move to rural areas, as part of a "fourth migration," where they would produce goods and services without the problems experienced in increasingly populated cities.[48] According to cultural historian Mark Luccarelli, nonurban districts would become homes to planned manufacturing sites that did away with the disorganized extension of factories in cities. Expansion of the benefits of modern civilization to less urban areas would depend on new technologies, such as telephones, radios, and electric transmission lines.[49] Robert Bruère, the editor of *Survey Graphic*, a publication of social and sociological research, observed that electricity, in particular, would offer "the opportunities for a balanced city and country life, of diversified industry and indigenous culture."[50] And as these decentralized communities obtained electric power, so would it spread to the even-less-densely populated farms, assisting in another way to stem the tide of migration from rural areas to cities. The combination of electric power and regional planning of garden cities would reinvigorate America's social structure.[51]

The notion of industrialized and electrified rural areas co-evolved with Henry Ford's notion of "village industries." As early as 1916, the industrialist who produced the wildly popular Model T cars worked to create manufacturing towns distant from his Dearborn factories. Between 1918 and 1944, Ford established nineteen such centers in southern Michigan, where rural residents could both farm and produce components for his companies' products. Concerned that nonurban Americans suffered from, as historian Howard Segal put it, "poverty, loneliness and cultural deprivation," the business magnate experimented with ways to enhance rural communities.[52] The future of America, Ford apparently told a reporter, would be one with "no mammoth collections of skyscrapers and teeming tenements in which millions of people are cooped up within a few square miles of territory."[53] Instead, he foresaw and helped establish small towns clustered around factories. Adopting this notion, Westinghouse's board chairman, Guy Tripp, observed in 1926 that such decentralization would require high-voltage, interconnected transmission networks for the industrial sites. And once these came in, nearby farms could obtain reasonably priced electric power.[54] Decentralization of industries and rural electrification, in other words, would go hand in hand.

Though not explicitly referring to the RPAA's or Ford's work, the Middle West Utilities Company, one of the Insull holding companies, appeared to support the concept of electrified garden cities and nearby farms. In a 1928 company-produced publication, *America's New Frontier*, utility leaders observed that many trends until about 1910 pushed industry to concentrate in cities, where large-scale machines, railroad nodes, and labor supplies joined to make manufacturing efficient for corporations. But centralization also meant social and technical inefficiencies, including the energy wasted by people who traveled within cities and the capital spent on transportation infrastructure. (In Chicago, 29 percent of the electricity generated at the time went for railway and streetcar use.)[55] But as the railroads continued to broaden their networks, and as highways and electric transmission networks expanded, it became easier to imagine moving people and industries outside of cities. In a self-congratulatory manner, the book told the story of one of the company's predecessor firms that began in 1910 to string high-voltage transmission lines from Chicago to Lake County, Illinois, replacing small generating technologies in towns that only supplied power at night. Quoting Samuel Insull, the book observed, "Within two years[,] all but two of those little villages and towns, and 125 farms, were getting full 24-hour electric service."[56] More utility companies would likely emulate such practices, the publication explained, partly by installing power lines along roads to smaller communities, a practice that also improved load factors by serving a more diverse customer base. In doing so, the firms would bring high lines closer to farmers, who would then receive extensions to their farmsteads.[57]

Decentralization of industry therefore appeared—at least argumentatively—to constitute a means for rural electrification to occur spontaneously and without dedicated efforts by the utility industry.[58] It allowed electric companies to continue their already existing efforts to deliver power to small communities, where population densities allowed for reasonable financial returns, and then radiate to less densely inhabited regions from the towns or from the power lines built along roads near farms. Rural electrification would therefore continue increasing, without the need of government intervention or requiring companies to subsidize a class of customers (farmers) at the expense of others.

The attitudes of urban businesspeople—and utility executives in particular—suggest an often-unarticulated reason why farmers did not easily get service from the central station companies. In the years since Edison introduced electricity on a grand scale for lighting and power, utility managers had brought the marvelous energy form to millions of customers, improving their material standard of living and raising industrial productivity at declining unit costs. Managers viewed themselves as stewards of technological progress, in line with the work of innovators and business executives in other realms, such as in automobile manufacturing and telecommunications. With people such as Henry Ford, George Westinghouse, Alexander Graham Bell (inventor of the telephone), Thomas Edison, and Herbert Hoover as exemplars, businessmen saw the modern world leaving farmers in the dust. As summarized by historian James Shideler, the city person imagined the rural countryside in a condescending manner—as "a dull foreign country" and "an anachronistic brake upon progress whenever agriculture exercised power, or as a source of amusement."[59]

Of course, utility managers shunned farmers' business because they felt it would not prove profitable. But perhaps as important (and in a way that historians have largely neglected), they discounted the rural market because they regarded farmers with scorn—as backward, antimodern, and unsophisticated. Moreover, in an era that extolled the virtues of free enterprise and individual responsibility, the executives felt no great pressure to achieve goals of social improvement, nor did they think it reasonable to force one group of customers to subsidize another. In the days before Franklin Roosevelt's administration altered perceptions of the roles of government and business, such practices appeared un-American, socialistic, and unworthy of serious consideration.[60] The decentralization and regional planning movement also gave utility leaders rhetorical ammunition to deflect critics of their practices by suggesting that farmers would obtain electricity through a gradual process that accompanied industrial readjustment.

Utility managers' focus on economic realities and their lack of interest in social responsibility did not appear unusual at the time. Even Harry Slattery, REA's third administrator, seems to have remained sympathetic to company managers' behavior at a time before the government became involved in rural electrification. In a 1940 book, he stated that neglect of farmers by utility companies

> must not be ascribed to any exceptional hard-heartedness on the part of the youthful "power trust." The men directing the destinies of this new but essential industry were personally just as good and bad as the rest of us. Doubtless, they would have glowed with satisfaction at the sight of a light in very farm house and a motor in every barn. But the first reaction to any proposal whatever had to be "Will it pay us?" not "Will it promote the general welfare?" . . . Utility leaders, like most men, were out to make money and charged all the traffic would bear, or franchises would permit. The idea of social responsibility was a new one to Americans.[61]

Slattery further noted that investors in utility companies justifiably sought to earn as much money as possible. Though power companies may have been able to extend service after 1915 to more farmers and secure a "reasonable profit," he wondered why they should have worked with difficult-to-serve farmers when they could realize more income from densely packed customers in cities. Should investors be interested in simply earning about 6 percent returns from farmers, he asked, "when ten to fifty percent could be made in urban business[?]" Utility managers should therefore not be held solely responsible for the slowness of rural electrification before the late 1930s; rather, "[t]he financial system then must take its share of the blame."[62] Writing at almost the same time, Ernest Abrams, a utility bond salesman and critic of federal electrification programs, explained the apparently meager performance of private utilities to power up rural America in similar terms: ". . . it is well to consider that rural electrification by private enterprise is primarily an economic, and not a philanthropic activity."[63]

The disdain for farmers by mainstream utility managers originated in contemporary thinking and a commonly held view of modern industrial life. Clearly, ruralites did not appear to be ideal potential customers of central station companies, largely because they would not immediately offer generous financial returns. But from the standpoint of company executives, farmers also constituted a class of people who had been left behind by modern society, not smart enough or too independent to exploit the many benefits offered by electricity. Even if they realized that the energy form could improve the lives of their rural cousins, utility officials largely did not feel an obligation to provide electricity unless they could earn a reasonable profit from doing so. Such an attitude may strike today's readers as insensitive, especially seeing how electrification eventually helped rural citizens improve productivity, health, and comfort. But to provide a product or service at below cost did not harmonize with business people's and government officials' sense of duty at the time. The onus to obtain electricity seemed to fall on farmers—not utility companies—who should prove that service to them would be profitable enough to attract the interest of private enterprise.

Chapter Four

The Lure and Lore of Rural Electrification

Today the actual amount of current consumed on the farm is still very small; but the rate at which this amount is increased in connection with agriculture and its different branches, bids [sic] well for the future. The farmer is beginning to realize what the electric current can do for him.

—ARMIN KARL NEUBERT,
ELECTRICAL ENGINEERING THESIS, 1916

Whether or not power company managers wanted to sell electricity to rural residents, a latent demand for the energy source began emerging in the early years of the twentieth century. A promoter observed in 1910, "Electricity on the farm sounds like a luxury for the rich, but it is now perfectly feasible for everyone. In most cases it means real economy, and in all cases it means convenience and comfort. The comforts of the city with the delights of the country is no longer a mere dream."[1] Though certainly an exaggeration, this claim of affordable and useful electricity on the farm became a common refrain in early-twentieth-century literature.

The lure of rural electrification drew from accounts of successful uses of electric power and from often-embellished assertions made by popularizers, engineers, and manufacturers. Amplified by publishers and overly enthusiastic authors, these pronouncements helped establish electricity as a component of widely held views concerning progress and modernity.[2] While the realization of these expectations first accrued to only a small number of farmers, such early experiences—along with knowledge of urbanites' uses of electricity—intensified people's hopes that rural electrification would become commonplace in the near future.

Early Electrification of Farms

From the start, electric illumination excited the mind and spirit. When Thomas Edison demonstrated electric lights in 1879 at his Menlo Park, New Jersey, laboratory, reporters raved about the "wizard's" success: "More glorious than any triumphs that were ever won on fields of martial glory are the victories of Edison with the electric light," wrote the *Omaha Daily Herald*.[3] Among the first customers of Edison's New York City company, small businesses found they could attract people to shops whose display windows exhibited electrically illuminated wares.[4] Demonstrators of the new technology wowed audiences by holding electric lights upside down and showing that the lamps continued to "burn" without igniting fires.[5] On a symbolic level, electric illumination held great meaning, signifying contemporary and highly evolved lifestyles. As late as 1914, an author observed that when God created the world, He only declared it "good" after commanding "let there be light." Going beyond biblical descriptions, the writer asserted, "One of the great differences between civilization and savagery is the abundance of light which marks the dwelling places of those living on the higher plane[,] and the lack of it among those who live in mental darkness."[6]

On the farm, electric lights had great appeal. In August 1882—a month before Edison demonstrated a successful central power system in New York City—the *Electrician* magazine enthusiastically asserted that electric illumination would extend the working day and "become one of the farmer's best friends," enabling the harvesting of hay at night.[7] An author in 1910 extolled similar benefits: "The average barn is a gloomy, cavernous sort of place, where time may be lost in semi-groping about, but the availability of electricity enables the interior to be flooded with light and renders it just as easy to work in as any city workshop."[8] Writing in 1913, another enthusiast noted that electric lighting offered wondrous benefits: "Good lighting . . . is of value to every one [sic] in the country as well as in the town. Better light secures greater efficiency and cleanliness[,] while fire risks are diminished and insurance rates are reduced. Electric lamps require no matches, burn without flame, consume no oxygen, and therefore do not vitiate the air of a room, and are unaffected by any change in weather conditions. Electric lighting is particularly of great service for stables and barns, where the use of lanterns has caused numerous fires and destroyed millions of dollars' worth of property."[9] Differentiating rural areas from the cities, electric lighting also acquired significant social consequences. One author noted that "there is more truth than fiction in the saying that the boys and girls are wont to leave the country for the brightly lighted streets of the cities."[10]

After lighting, electricity had potential for improving the delivery of motive power. Traditionally, energy for farm tasks came from humans, horses, and other animals. They provided traction for plowing, hauling produce, and turning machines that threshed grains, for example.[11] The use of new power sources began soon after the Civil War ended, as some farmers employed steam engine–driven tractors and, starting in the 1890s, gasoline-fueled internal combustion engines.[12] Lightweight gasoline-powered tractors became popular in the 1920s because they allowed for quicker work while also speeding the shift from cultivation of crops grown for animal fodder to foods sold at market.[13] As farmers took advantage of these new power sources, productivity soared: between 1850 and 1924, the average farmer nearly tripled his output.[14]

But some promoters saw the use of electrically operated traction machines— for plowing and harvesting in particular—as the way to extend this progress. Since the 1890s, test farms in Germany and elsewhere replaced steam-powered tractors (and horses) with those that used electric motors to transport cable-connected plows across fields.[15] Because an electric motor demonstrated greater reliability than other forms of tractive power, advocates asserted its superiority, including lower first and operating costs. Reflecting the optimism created by these experiments, an 1892 editorial in *Engineering* magazine suggested that "the day is not distant when the entire labor of preparing and tilling the ground, as well as that of seeding, harvesting, threshing, and transporting the crops to the nearest railway station, wherever done on a large scale, will be performed by electric motors." Such a transformation of tractive sources, the editorial continued, "will constitute an industrial revolution of almost inconceivable magnitude."[16] Seeing that American agricultural equipment "is far superior to that of any foreign make," one booster argued in 1912, it will only be a matter of "a short time when the farmer would recognize the advantages of electric plowing."[17] An engineering student reiterated this sentiment in his 1912 bachelor's thesis, lamenting the fact that Americans still relied on "more cumbersome and less efficient" steam and gasoline tractors and paid little attention to the clearly superior electric machines.[18]

It also appeared that electric motors would power a host of smaller machines, previously operated by gasoline engines. Unlike these prime movers, which required a good amount of maintenance and remained valuable mostly for high-horsepower needs, electric motors started and stopped easily with a flick of the switch and did not require combustible fuels. Electric motors promised other advantages, such as being able to be attached, via leather belts, to equipment already used on the farm, such as grindstones.[19] The expectation of electric motor–operated appliances led a *Farm Journal* author to exclaim in 1906 that the "churn, the chopper, the thresher, the pump, the saw, the forge—everything that makes an up-to-date farm, may be operated by electricity, and the experts contend that there will be economy in doing the farm work with a power that never stands in the stable eating its head off and doing nothing for the farmer. . . . Speed the day!"[20]

Overall, "the wondrous agent" of electricity offered great potential benefits, as described by the *Ohio Farmer* magazine in 1896.[21] Admitting that electricity remained an undefined and poorly understood natural force, a newspaper writer in the same year nevertheless expressed confidence that it would reduce production costs.[22] In 1903, a farmer wrote in awe of seeing the various applications of electricity in the big city. But he also contemplated how farmers could employ the energy form. While the rural citizen "may not have electric cars to supply current for, . . . a farmer could use electricity every day, if it were only at hand." He further noted that "[w]hile there is practically no limit to its uses, it would be very convenient to touch a button in your sitting room and have your pump working as by magic, or your lamp burning in house or barn, or your cream separator running."[23]

A few lucky farmers began realizing these heretofore imagined benefits, and because they received journalists' attention, they helped perpetuate visions of more widespread rural electrification. A visitor to an Illinois farm stated in 1911 that electricity enhanced its prosperity and comfort by operating a separator and churn in the creamery as well as a vacuum pump for milking cows, who "seemed to enjoy the process." Inside the farmhouse, an electric radiator heated rooms, electric lights illuminated spaces, and electric appliances (such as a vacuum cleaner and iron) made life easier for the homemaker.[24] At about the same time, an article observed that electrically operated machines would increase productivity and help farm owners overcome problems caused by the growing expense of hiring farmhands. With small electrical motors operating an assortment of equipment, farm work "can be done quicker and better."[25]

Proponents of farm electrification often pointed to the benefits women and children would receive from the use of electrical appliances.[26] One power company official noted in 1910 that, before electricity's appearance, "life of the women on the farm has been one of constant drudgery and maddening monotony." But electrically operated lights "mean comfort in the farmhouse" along with a reduction of fire perils.[27] An author of a trade journal explained that a "small [electrical] motor attached to the sewing machine takes away much of the drudgery of this work; a small buffing wheel polishes the silver; the washing machine will eliminate much of the work at the tubs; [and] the vacuum cleaner sweeps better than a broom with less labor in one-tenth the time and leaves no dust." Beyond these advantages, women will find that "[e]lectric flat irons make Tuesdays a pleasure; a small electric stove in the chamber will warm baby's milk or give heat immediately for any nocturnal emergency."[28] And as another author noted, by reducing the backbreaking grind of farm work, electricity "would certainly help to solve the problem of keeping the young folks on the farm."[29]

Electroculture

For lighting, motors, and home appliances, electricity seemed to offer endless possibilities. Beyond these uses, European and American experimenters worked on stimulating soil or nearby air with electricity, with the hope that plants would grow faster, stronger, or more resistant to disease. This line of work, often known as "electroculture," had an apparently strong logic. As a meteorologist noted in 1892, the atmosphere serves as a "great reservoir for all electricity"—the source of lightning and static charges that enter the earth through trees and buildings.[30] It appeared reasonable to consider the possible impact of these electrical phenomena on plant life.

Research on the connection between agriculture and electricity went back centuries. The natural philosopher Joseph Priestly recounted work done in 1746 by "Mr. Maimbray of Edinburgh," who electrified myrtle trees and claimed to have observed faster branching and blossoming than on unelectrified shrubs.[31] Historian Clark Spence described blacksmiths and inventors who claimed (as early as the 1830s) to have created electrical machines for stimulating plant growth and for enhancing agricultural productivity.[32] The English *Electrical Magazine* in 1845, meanwhile, reported on experiments demonstrating "that the electro-vital principle in vegetation is capable of being greatly promoted, both by means of free electricity skilfully [sic] drawn from the atmosphere, and also by well-directed voltaic currents through the soil."[33] In the same year, the *British Farmer's Magazine* published a letter that reviewed various experiments in which seeds of several plants germinated faster and grew better when exposed to low-voltage electricity.[34]

As scientists gained more understanding of electricity in the late nineteenth century, experimenters and academics renewed their interest in electroculture. European investigators in 1891 tried using electrical illumination to force plants in greenhouses to grow fruits and vegetables during the off-seasons.[35] A year later, entrepreneur Elihu Thomson (one of the founders of the General Electric Company), imagined that passing current through the soil might stimulate plant growth, yielding such edible wonders as "pomme de terre à la dynamo" and "asperges électriques."[36] Going beyond assertion and into practice, a Frenchman constructed a "geomagnetifer" that collected atmospheric electricity and distributed it through wires in a field, which supposedly yielded great gains in potato crops, suggesting the wisdom of Thomson's speculation.[37] Physics professor Karl Selim Lemström at Finland's University of Helsingfors conducted well-respected experiments motivated by his research on the aurora borealis. When subjecting plants to static electricity in the soil and in the air above them, the output of potatoes, carrots, and celery allegedly increased between 30 to 70 percent.[38] In the production of strawberries and raspberries, electricity spurred yield improvements of up to 75 percent and shortened their ripening period by one-third, he reported.[39]

Further electroculture experiments in England, France, and elsewhere in the early 1900s often appeared to demonstrate positive results. One researcher in 1913 observed that the "action of the electrical current . . . seems to be somewhat analogous to that of a tonic in the human body," perhaps as the electricity produces ozone and nitrates that fertilize the soil.[40] Likewise, a University of Wisconsin electrical engineering student noted that electrified air currents near plants might boost productivity.[41] Another investigator commented in 1917 that "reliable data" demonstrate "that electro-culture opens up a vast field for serious work, for not only hastening growth, but superior and much greater yield."[42] Though some reviews properly noted that not all electroculture investigations proved successful, the efforts illustrate that researchers eagerly sought practical agricultural applications of the magical electric force.[43]

Other Advocates for Rural Electrification

The chorus of advocates of rural electrification in the years before and soon after World War I included engineers. Among the most prolific, Putnam A. Bates, a New York City–based consulting engineer and operator of a New Jersey farm, penned a series of magazine articles and spoke before various audiences, arguing that electrical farming lay on the horizon. In one presentation given at a meeting of the American Institute of Electrical Engineers in 1912, Bates described the use of electric pumps to irrigate fields in western states, which yielded bountiful fruits and vegetables.[44] These devices demanded prodigious amounts of power and operated twenty-four hours per day for five to six months a year.[45] Central station companies gladly sold power to these farmers at low rates, realizing how much energy they consumed, and the agriculturalists also benefited from employing electric motors to grind and cut grains for ensilage.[46] (See chapter 6 for more on western electrification.)

Beyond anticipating the wider use of electrically operated irrigation pumps as a means of advancing agricultural productivity, Bates provided examples of already-electrified tasks performed on dairy farms that boosted output: electric lighting and refrigeration devices made facilities cleaner and more sanitary, while electrically operated cream separators, churns, butter workers, and milking machines generated productivity gains. Most impressive, "electric energy is greatly cheaper than man or horse power" and can be useful when human labor proves difficult to acquire. He observed that "when it seems impossible to secure men on the farm, the turn of a switch brings electric energy, begetting production and wealth."[47] In a set of articles published in *Scientific American*, Bates offered more details about irrigation pumps, watertight electric light fixtures (enabling farmers to spray the walls of their dairies to achieve high levels of cleanliness), and electric motors for harvesting ice from frozen lakes.[48]

Engineer Frank Koester similarly argued that the "greatest agent of agricultural progress is electricity." Serving as the "emancipator of the toiler," electricity "does the work of a man at far less expense." But even while praising the virtues of electricity in a 1913 book, Koester, who received training in Germany and held credentials as a member of professional engineering societies, hinted at a condescending attitude toward the farmer, describing his work as largely unskilled. With the use of electricity and other modern knowledge, "[a]griculture is no longer to remain a practice of yokels but is to become an applied science." Such advances will lead to "a happier and a fuller life" for the farmer, enabling an "escape from the drudgery and the grinding toil that have made the farm a place to be abandoned to the less enterprising."[49]

Businesses Spreading the Promise of Farm Electrification

Despite the relative difficulties of electrifying rural districts, some companies saw niche markets there, and they contributed to the notion that farm electrification would bring considerable benefits. General Electric managers realized that agriculturists potentially could consume large amounts of power, and they could become good customers of pumps, motors, and other production equipment. The company began experimenting with electrically operated appliances and electric plows as early as 1900, no doubt as part of its already existing strategy of manufacturing equipment for power producers and consumers alike.[50] A GE electrical engineer announced in 1910 that his firm had begun investigating the farm market and sought to take advantage of agricultural engineers to help educate farmers about the value of using electricity to increase productivity and living standards.[51]

Proving its interest in the agricultural market, GE published a lavish primer, *Electricity on the Farm*, in 1913. As much a piece of advocacy as a catalog of products, the seventy-two-page illustrated book listed fifty beneficial uses of electricity in the home alone. Using a GE sewing machine motor, for example, the farm wife obtains "30,000 Stitches for One Cent," while that same penny would energize an electric vacuum cleaner over a 150-square-foot carpet.[52] The most productive device for use outside the home appeared to be the electric motor, which could replace gasoline- or kerosene-operated engines or, more important, human energy. In all, the book listed thirty applications of electric drives for barns and in the field, including their use to power water pumps, feed grinders, grain threshers, hay cutters, and wood splitters; it added another twelve applications of electricity in the farm shop.[53] Beyond these devices, the company highlighted an electric (battery-containing) truck, which could gather hay and wheat during harvest season and deliver it to markets. Owing to its speed and structural strength, an electric vehicle could carry almost 2.4 times the amount of produce drawn by a two-horse wagon.[54] Such a truck also started easier, climbed steeper grades, and required less maintenance than gasoline engine–driven vehicles having the same horsepower rating. Demonstrating the machine's superiority, the book noted that "the operating and controlling mechanism is so simple that either pleasure or working vehicles of this type can be safely handled by women."[55] The company argued further that the use of electric motors and pumps as replacements for steam- and gasoline-powered equivalents reduced the cost of irrigation activities. When considering all expenses (including those for equipment purchases and for maintenance, repair, fuel, and depreciation), electrically operated irrigation pumps saved sizable sums on 40- and 160-acre test farms.[56]

General Electric also publicized agricultural uses of electricity through its lecture service and with lantern slides and films lent to schools and agrarian associations. In 1916, for example, it offered the movie *Back to the Farm*, described as a "two-reel picture full of human interest, depicting a farmer involved in labor difficulties." Staged in California, the film showed how the farmer resolved his problems through the use of electricity.[57] Near the end of the film, the script read: "By Electricity on the Farm[,] toil is turned into pleasure, frowns into smiles[,] and happiness is brought to the home."[58]

As the farm market grew, General Electric continued its publicity campaign, helping to stoke expectations of the wonders of electricity in rural settings. Many of the company's products, such as small GE generators (with capacities of 0.75 to 1.25 kW), appeared as components of isolated power plants and were listed in publications such as the *Farm Light and Power Year Book* in 1922.[59] (See chapter 5 for more on isolated plants.) *The G-E Farm Book* appeared in early 1925 and saw a second edition in November 1926.[60] Like earlier forms of promotion, the publication illustrated astonishing possibilities, many of which could be substantiated by the fortunate few farmers who already employed electricity. Aside from descriptions of improved productivity and a happier life in the home, barn, and field, the edition boasted about the newest and perhaps most extraordinary electrical technology to appear—the radio. "Farm isolation is a thing of the past," an article reported, since people living far from cities could now tune in to clergymen and statesmen and—perhaps best of all—to baseball contests, following "the progress of the games, play by play." The technology also enabled farmers to listen to courses that taught scientific methods for improving production of crops. Not going unsaid, General Electric owned three broadcasting stations (in Schenectady, NY; Oakland, CA; and Denver, CO), one of which offered a weekly farm program. "It may be aptly said today," noted the piece, "that the farmer is as near to town as he is to his radio set."[61]

Exuberant portrayals of the benefits of electricity multiplied as it saw growing farm use starting in the 1880s. Even if promoters of rural electrification exaggerated their claims or expressed their hopes too vigorously, they nevertheless established an expectation that most farmers would ultimately thrive as they adopted the new energy form. Put differently, the accounts that everyday farmers read in the popular literature and in manufacturers' catalogs may have created a cultural imperative that emboldened people to seek—either explicitly or subconsciously— increased access to electricity. As historian Kenneth Lipartito observed when describing the technical failure of the Picturephone, a video telephone system released in the 1960s, engineers and consumers often create representations of future technologies, despite the hardware's failure in the market when initially introduced.[62] In other words, even when a technology remains elusive, it still sparks the imagination and inspires people to develop similar forms of it, while also creating a market when it emerges later. (Though the Picturephone never achieved commercial success, the idea of televised personal communications now lives in the form of numerous computer-based and smartphone applications.) Like the proponents of the Picturephone, technically trained people and other rural electrification advocates glowingly described the possibilities and actual uses of electricity on farms, and they may have legitimated expectations of its greater rural consumption.[63]

One can also view the evolving discourse of farm electrification as the creation of a "sociotechnical imaginary." Developed by social scientists Sheila Jasanoff and Sang-Hyun Kim, the concept describes the "collectively imagined forms of social life and social order reflected in the design and fulfillment of . . . technological projects."[64] Other scholars have adopted a similar view of "imagined use," in which "[b]eliefs about technologies and their potential capabilities for a technological future . . . influenced innovation and use," while also shaping "cultural narratives about technology as harbingers of modernity."[65] In the case of rural electrification, a sociotechnical imaginary (or sense of imagined uses) emerged in which various parties considered the extensive use of electricity on farms as a desirable goal. That imaginary served as an unarticulated rationale and motivation for efforts to pursue the extension of electrical power to rural areas, even in the 1910s, when electricity largely remained an urban and luxury commodity.

Turning the imaginary of electrification into reality required, as with similar imaginaries, a combination of social constructions and technical innovations. It also needed concrete demonstrations that the highly touted electrical equipment truly improved efficiency and reduced farmers' overall expenses. As documented later in this book, new players in the 1920s established institutions for creating productive electrical technologies and for discounting overblown claims made about others. Concurrently, private companies organized divisions of agricultural experts to work with farmers, and regulators reduced barriers to erecting distribution lines. Additionally, the ultimate success of the rural electrification effort benefited from early adopters of electrical technology—often those with above-average incomes and with technical savvy—who showed that electricity on the farm could become practical and beneficial.

Chapter Five

Farmers on Their Own

A farm equipped with a good electric light and power plant can be made even
more modern than a city home and by far a more delightful place to live.
—*FARM MECHANICS* MAGAZINE, 1922

T he actual implementation of the rural electrification imaginary did not occur
quickly. As described earlier, central station power companies generally did
not string distribution lines outside of densely populated regions. But in a way
that displays their creativity, sophistication, and willingness to take risks, many
rural citizens obtained electricity by producing it themselves using isolated power
plants.[1] While several firms offered hardware to generate electricity from water
turbines and windmills, the Delco-Light Company, as a prime example, manufactured thousands of generators powered by internal combustion engines, enabling
customers to enjoy amenities such as electrically pumped indoor water supplies,
washing machines, refrigerators, electric irons, and lights.

An examination of farmers' adoption of isolated power systems highlights several elements of this book's alternative narrative of rural electrification. First, it
demonstrates that agriculturalists did not play the role, as often imagined by urban elites, of technological naïfs. Through their use of isolated plants, they exhibited an eagerness and ability to embrace complex technical systems. Second, the use
of the plants may have suggested to central station managers that, given the existence of nontraditional generation technologies, they did not need to supply
power themselves. After all, if farmers could obtain electricity for the limited, yet
desirable, electrical applications within the home and nearby barn at a reasonable
cost, then utility leaders could honestly claim they had no obligation to serve such
customers. Third, the use of isolated plants contributed to the sociotechnical imag-

inary and helped accustom rural denizens to the notion that electrification remained both possible and imminent. In fact, some utility managers suggested that
experience with isolated power plants would help ruralites become better central
station customers because they would already know how to employ electricity to
accomplish various tasks.

Electricity from Flowing Water

From the earliest years of electrification, farmers comprehended they would
benefit from many of the same applications of the new energy form as urban customers, especially for illumination. Sadly for most ruralites, access to a power
line did not occur frequently. Many electrification advocates therefore expected
farmers to fend for themselves, such that power would come not from a utility's
generating plant but from a local source of energy that could be converted into
electricity. The choices for self-generation depended on the natural environment,
with the harnessing of rivers and streams constituting an early approach.

Commentators often remarked that farmers had foolishly squandered a wonderful water resource that could literally empower so many people. In describing
the use of streams to electrify a New York State farm, a *Saturday Evening Post* writer
in 1910 lamented the earlier frittering away of power, observing that the water
"had been doing little more than make a merry noise over the rocks and pebbles."[2]
The New York State Water Supply Commission concurred, suggesting in 1911 that
farmers should exploit the "valuable power which is now running to waste in thousands of small creeks and brooks in all sections of the State"—power that could
furnish light and operate feed grinders, churns, cream separators, and other machines.[3] Frederick Irving Anderson, a journalist and popular magazine author,
poetically compared the use of water power in 1915 to observations of horses
running freely on the plains. While the farmer might lament seeing the animals'
"horsepower running to waste," he rarely becomes "inspired by any similar desire
of possession and mastery by the sight of a brook, or a rivulet that waters his mead-
ows." But placing a small water-turbine generator in a stream would yield electricity in a way that comes close to obtaining "something out of nothing." Even
better, "the task of harnessing and breaking this water-horsepower is much simpler and less dangerous than the task of breaking a colt to harness."[4]

All that water power could be put to good use. An *Electrical World* writer noted
in 1906 that by building a small powerhouse, the size of a "modest corn barn," a
farmer could employ a 15 kW dynamo to yield enough power for "sawing, grinding, pumping and light motor work of the neighborhood." The equipment would

require little maintenance and would provide electricity for about one hundred incandescent lights, claimed the author.[5] Another idea for exploiting water power emerged in 1911: groups of farmers could work cooperatively to convert abandoned mills into power plants, generating electricity at up to 10,000 volts, thus enabling (owing to the high tension and small line losses) transmission to distant farms. While the investment in such a converted facility might run $30,000, the cost would be borne by twelve neighbors, bringing the initial expense down to $2,500 each. The annual running cost might total $5,000 (shared by all subscribers), an amount that included interest, depreciation, and salaries of plant operators.[6]

To further illustrate how farmers could electrify their properties, *Popular Electricity* magazine highlighted in 1911 the work of childhood friends Eli Crosiar and Henry Grove, who transformed a decrepit mill into a small generation station. After designing a channel that directed water to a turbine, the men installed a generator, transmission line, and associated equipment that provided abundant power for a 320-acre dairy farm. Designated the "chief engineer," Mr. Crosiar's wife controlled the switchboard and regulated the unit's voltage. Besides describing the benefits of using electrical equipment for milking cows and other functions, the article included a financial calculation showing that the plant and equipment cost about $2,500.[7]

As experience with them mounted, water generation systems gained further proponents. The State College of Washington extension service in 1920 extolled the virtues of a farmer-installed hydroelectric generator that powered sixteen light bulbs, a washing machine, and wringer.[8] The same organization offered advice and examples of several water wheels in its 1924 publication, *Let the Creek Light Your Home*, which illustrated the use of small streams to produce "cheap and serviceable power."[9] And in 1925, the US Department of Agriculture partnered with agricultural engineers at Virginia Polytechnic Institute to print a thirty-five-page manual on designing and operating similar hydroelectric plants that promised to enhance productivity and life on the farm (figure 5.1).[10]

Of course, the use of water-powered generators required farmers to have sufficient capital to invest in them. In the years before World War I, it appears that only the richest agriculturalist could afford them. The typical farmer in 1913 earned an annual income of about $657.[11] And while the average value of American farms (measured by adding together the worth of land, buildings, implements, machinery and livestock) grew from $6,444 in 1910 to $12,084 in 1920, few farmers had access to enough disposable income or set-aside capital to purchase generation devices, such as those mentioned costing $2,500 or more.[12] Additionally,

Fig. 5.1. Government advice on how to generate electric power from streams. *Source*: Cover of A. M. Daniels, C. E. Seitz, and J. C. Glenn, *Power for the Farm from Small Streams*, Farmers' Bulletin no. 1430 (Washington, DC: GPO, 1925).

because of the huge variation in local conditions, manufacturers could not create standardized equipment at reasonable costs that could be easily attached to a water wheel or turbine.[13]

Wind Power

Though not always blessed with good water resources, ruralites could sometimes take advantage of wind, which had already become common as a motive force for pumping water and performing various mechanical tasks.[14] In some cases, farmers crafted wind units to generate electricity in the same manner that they designed water-powered generation hardware. As discussed in a 1912 article, a Wisconsinite constructed a system for $1,250 that yielded 210 watts, enough to charge fourteen batteries and illuminate twenty-four electric lamps.[15] According to a 1922 catalog, the idea of using wind to power an electric generator, with the current stored in batteries, already had a half-century history (though, "like many other valuable ideas it was first presented many years in advance of its time").[16] Often, "kitchen mechanics" constructed these machines for individual use and did so with some success.[17]

Along with other companies, the Perkins Corporation of Mishawaka, Indiana, eliminated the need for homemade designs by producing commercially successful wind generators. The firm's "Aeroelectric" device, advertised in 1922 as "simple, silent, [and] harmless,"[18] incorporated well-tested Westinghouse hardware and components from other established manufacturers.[19] Placed at least fifty feet above the ground, the generator produced up to 1 kilowatt of power for a battery consisting of sixteen cells.[20] A switchboard contained fuses and a circuit that automatically allowed the batteries to draw power whenever the wind speed exceeded six miles per hour.[21] Winning praise from *Current Opinion* magazine as a "practical plant" that "should need no attention on the part of the man in the family," the technology would put "electric light within the reach of many who live where no electric lighting plant exists."[22]

Wind-electric plants provided good service under proper conditions. Agricultural engineers at Iowa State College in 1925 tested a generator that yielded 842 kWh during a year, which compared more than favorably with engine-driven lighting plants used on four farms to produce about half as much energy.[23] In subsequent years, the researchers studied wind plants already employed by the state's farmers, finding that each family connected its isolated system to about thirty lights and a few electrical appliances: of the sixty-six surveyed households, fifty-two also used irons and washing machines; thirty-six had water pumps; thirty-five owned vacuum cleaners; twenty-five enjoyed radios; and sixteen operated cream separators.[24] The plants had high first costs—an average of $720—though the fuel cost was zero. When including depreciation of the plant and batteries, along with other expenses, the annual cost of a wind generator ran between $47.50 and $125.38.[25]

But the technology did not seem appropriate for everyone. When responding to a handwritten inquiry from a farmer who wanted to pump water and produce electricity with a wind-powered system, Alabama Polytechnic Institute agricultural engineer M. L. Nichols in 1928 urged caution, likely because of different meteorological conditions in his state. He noted that colleagues at his institution had not tested wind generators, but his impression, based on work done by other agricultural engineers, has "not been particularly encouraging." Given the farmer's situation and needs, Nichols suggested the purchase of an internal combustion–powered lighting system (see next section) and a dedicated windmill just for pumping water. Interestingly, Nichols made no mention of ways to obtain power from a utility line that ran 2.5 miles away from the letter writer's home.[26]

For those who desired (or could only afford) less energy, a Wincharger wind generator often fit the bill. Producing electricity to charge a 6-volt battery, the device (initially offered in 1927) powered the newly popular radio. In 1935, the Zenith company bought a controlling interest in the manufacturer and promoted (in the following year) a 6-volt radio and a wind generator for only $15.[27] An article in *Popular Mechanics* indicated that the outfit would not only provide enough power for the radio, but that it "has enough extra juice to run two lights, one in the kitchen and one in the living room."[28] The company expanded its offerings, marketing in 1936 a 32-volt "Giant Wincharger" for the "unheard price of $69.95." Claiming to be the "[w]orld's largest makers of wind driven generating machinery," the firm advertised a cost of only "50c a year for farm electric power."[29] The company publicized its product more aggressively in the following year, proclaiming "Rural Electrification [I]s Here," and telling readers that they did not need to "wait for the high line" so they could enjoy "the same comforts that city folks have!" Unlike gasoline-operated electric plants, the wind generator saved $30 to $60 annually on gas and oil and, better yet, "There's no tax on the wind—IT'S FREE." The 32-volt, 650-watt machine came from a company that served (according to its advertisement) "[m]ore than 500,000 farm folks all over the world."[30]

Of course, even with batteries, electrical energy would eventually run out if the wind failed to blow enough. And in most cases, the generators could provide only limited amounts of electricity—usually just enough for a radio and a few lights.

Nevertheless, people must have viewed wind-generated power as a welcomed extravagance, especially if they lacked any other access to electricity.

Internal Combustion Units

While water and wind served some farmers, many more depended on isolated plants that employed internal combustion motors to drive electric generators. In a special 1910 issue devoted to farm electrification, a trade journal observed that small units on farms could power many useful devices, such as cream separators, mechanical refrigeration devices, and (of course) lights.[31] Making these plants more popular, General Electric had recently introduced its "Mazda" tungsten filament lamps, which produced about three times more illumination per unit of electricity than older, carbon filament bulbs. The lower power consumption meant that small generators could energize a larger number of lights.[32] Meanwhile, the lamps operated at 25 to 32 volts—making the system "entirely harmless to the operator," claimed one article, and enabling the system to depend on smaller batteries.[33]

Various companies sold this hardware. The Fairbanks-Morse firm advertised in 1908 an engine-generator set that "operates incandescent or arc lights, electric fans, pumps, motors, etc."[34] The Dean Electric Company sold a "Home-Lighting and Power System" in 1911 that one could set up "without the help of an electrician and have perfect illumination at any point in your home or in any of the buildings on your farm."[35] In what appeared to be an increasingly lucrative market, more than 160 manufacturers produced internal combustion generating units in 1917.[36] Several publications, such as *Home-Farm Power and Lighting*, produced by the editorial staff of the *American Automobile Digest*, provided instructions and tips in 1920 on operating the devices. In general, these power-production units enabled the use of "labor-saving devices" that offered "comforts which were formerly only enjoyed by the residents of the large cities."[37]

Delco-Light

Among the companies that produced power generation sets, the Delco-Light Company stands out. Using modern manufacturing and selling techniques, the firm assumed a large portion of the isolated plant market by the 1920s. Of 1,786 electrified farms in New York State surveyed in 1926, 605 had isolated units; Delco-Light made 374 of them, and twenty-four manufacturers supplied the balance.[38] Reflecting the company's success, the product's name became a generic term, as indicated by its use in a 1934 report authored by Morris Cooke. In providing the context for the need of a government-managed electrification effort, he observed that of the 800,000 energized farms at the time, only 650,000 of them received power from central plants. "The balance," he commented, "have individual Delco plants," even though, as just noted, many other firms produced similar electrical generators.[39]

The company entered the farm electrification business as a sideline activity pursued by Charles Kettering and colleague Edward Deeds, who created the Dayton Engineering Laboratories Company (known as Delco) in 1909. The two engineers had previously worked for the National Cash Register Company, and Kettering had designed small motors that powered the workings of the firm's major product. Employing a similar high-torque motor that operated only for a short period of time, he and Henry Leland, one of the founders of the General Motors Corporation, introduced a successful electric starter for the company's Cadillac vehicles.[40] The device drew power from a lead-acid battery, which sent electricity through an ignition system (also made by Delco) to the spark plugs, exploding gasoline vapor in the engine's cylinders. Once turning, the engine powered a small generator that produced electricity to charge the battery. The automobile essentially included a small power plant, one that continuously charged the battery and energized the spark plugs. Because of these significant innovations, Delco won attention from auto manufacturers, with the United Motors company buying the firm in 1916 and General Motors purchasing United in 1918.[41]

According to a story repeated in a popular history of electricity and in Delco-Light advertisements, the idea for an isolated, farm-based power plant using the automobile generator principle came from a customer.[42] A Floridian wrote to Delco to obtain electrical parts for his car. The letter motivated a Delco salesperson to visit the man, thinking that he needed the components to repair an inoperable vehicle, but he discovered that the customer wanted the equipment to produce electricity for his summer cottage. When Kettering and Deeds heard this account, they began experimenting with a device that provided power to a remote home or farm in a more efficient manner.[43] The Delco managers certainly knew that various companies had already marketed internal combustion motors to generate electricity, with some selling lead-acid batteries to store power. But they also understood that many of these setups remained difficult to operate and maintain.[44]

The Delco patent for a "system of electrical generation," filed in 1915, describes an internal combustion engine that directly powers a generator. A set of lead-acid batteries stores the electricity for use in the home and other structures. So far, the

Delco product offered nothing original. As described in the patent, however, the device proved innovative by stopping the engine and generator automatically when the batteries reached full charge; the motor reengaged when detecting nearly discharged batteries, requiring no intervention by the user except to supply fuel and oil. Moreover, the engine and generator fit together with a direct connection, such that it did not need a belt, which could slip, wear out, and break.[45] To minimize the need for maintenance further, the engine did not use a water cooling system (as did most automobile engines). Rather, the engine dissipated heat with large-surface-area fins attached to the cylinder head (that is, it employed air cooling), thus eliminating concerns associated with water freezing in cold climates.[46]

With a few modifications (such as the ability to burn kerosene—a fuel widely used by farmers already), the system enjoyed commercial success. Incorporated in 1916 as the Domestic Engineering Company to sell the generation sets,[47] the Delco-Light firm advertised the hardware as "a complete electric power plant for churches, stores, farms, etc."[48] The system cost between $350 and $425 and had sold (according to company advertisements) more than thirty thousand units by 1917.[49] Aside from illuminating lamps, which naturally attracted most customers' attention initially, the units offered enough power to energize a variety of appliances, such as "electric irons . . . seed cleaners, milking machines, water pumps, vacuum cleaners, . . . and a hundred other pieces of light machinery around the farm and home."[50] (In the 1918 advertisement shown in figure 5.2, the firm claims its equipment "improves living conditions" in many ways.) The company itself sold many of these accessories, including well water pumps, light fixtures, irons, and washing machines,[51] the latter of which included a light-sensitive photometer for determining when clothes had been properly cleaned (plate 4).[52] The generating set could also power Frigidaire refrigerators—a product line acquired by General Motors in 1919.[53] (The auto manufacturer had already absorbed the lighting company and changed its name from Domestic Engineering to Delco-Light.)[54] In 1921, GM reorganized management such that the Delco-Light Company manufactured the Frigidaire machines.[55] The company often sold the power source and cooler together; a 1927 advertisement recommended a 750-watt Delco-Light unit for use with "Frigidaire electric refrigerators and pumps."[56]

Customers appreciated many features of the Delco-Light generator, especially its operation only when the batteries needed charging. One happy user wrote that, beyond working at lower cost than a previously owned competitor's model and with easily available kerosene, the Delco-Light plant "does not run while it is lighting so there is *no noise*." Another buyer complimented the product because "when

Fig. 5.2. Advertisement for Delco-Light isolated electric plant. *Source: Country Gentleman* 83, no. 22 (1 June 1918): 1.

a person *is sick you don[']t have the noise from the constant* running of the motor"
while also observing that he only needed to operate his motor "*once or twice* a
week" to recharge the batteries.[57] A magazine description of the hardware in 1917
observed, "The slogan 'So simple a child can operate it,' may well be believed when
we see the simplicity and safety of the system." With air cooling and few necessary adjustments, the device could endure "wear and tear in all kinds of weather."[58]
Well-designed and sturdy, the machines also could handle other tasks besides making electricity. The model 1286, for example, consisted of the standard engine
and generator, but it included an extended shaft from the engine "[e]quipped with
[a] pulley to operate line shafting or heavy power appliances directly from [the]
flywheel."[59] In other words, the device produced electricity and direct mechanical
power (figure 5.3).

Aside from use on farms, the isolated power sets seemed popular in other ven-
ues. The company claimed that it sold almost five thousand units to the US government during the World War, with about one thousand employed in hospitals to
energize X-ray machines. Some also found applications on ships to pump water from
bilges, to provide light, and to charge batteries.[60] Church leaders seemed to think
that electric lights helped attract parishioners to services. A 1917 news piece noted
that the Crystal Springs, Mississippi, New Zion house of worship had recently installed "a beautiful Delco light plant for the church."[61] Meanwhile, a religious leader
in White Stone, Virginia, raised enough money ($802) in 1925 to finish paying for
the building's land, church house, and for "installation of its Delco lights."[62] Such
statements suggest that contemporaries considered the Delco plants as advanced
and modern—particularly in rural settings.[63]

The company sold its products through a sophisticated network of distribution
and service offices around the country and with representatives who dealt directly
with customers.[64] Agents came to the Dayton headquarters in March 1916 to witness ground breaking for a new factory that produced the light sets, returning for
annual sales meetings thereafter.[65] At the March 1918 convention, select salesmen
heard speeches from top corporate leaders (such as the general manager, Richard H. Grant, and vice president, Charles Kettering). They also met salespeople
from companies offering other electrical equipment (from air compressors to
X-ray machines) and enjoyed tours of the manufacturing facilities. To motivate salesmen, the company publicly regaled those who sold twenty power plants in two
months and one hundred or more annually.[66] In 1919, dealers who moved at
least nine units every two months could attend the meeting with all expenses
paid. Corporate literature taught the latest in marketing techniques and offered

1¼ K. W. Delco-Light Pulley Plant
(32 Volt)

Model No. 1286

Model No. 1278—Generating Unit Only. Model No. 1286—with 160 A. H. Battery

FOR LIGHTING AND POWER PURPOSES

For installations where it is desired to cut out the generator and have a 2½ H. P. engine with pulley available for power purposes such as driving line shafting, feed grinders, buzz-saws, milking machines, deep well pumps, and other machinery, requiring up to 2½ H. P. Will produce 1250 watts as a standard Delco-Light Plant.

Specifications

ENGINE: Air-cooled, valve-in-head, 4 cycle, single cylinder, 2½ H. P., Delco-Light Battery ignition, fuel pump and fly-ball governor. Operates on kerosene or gasoline. Self-cranking. Ball and Roller Bearings. Plant stops when fuel tank becomes empty or can be stopped by hand. 3½" diameter pulley for 3" flat belt on flywheel end of crank shaft. Speed, 1250 R. P. M. (Approximately). Splash system of lubrication—one place

to oil. Height, 32"; length, 27"; width, 20".

GENERATOR: Direct connected to engine. For 32-volt service. Full load output, 1250 watts.

SWITCHBOARD: Mounted on generator frame.

SHIPPING WEIGHT: With 160 A. H. Battery (glass jar), five crates, 1298 lbs. Model No. 1278, generator only, one crate, 450 lbs.

Fig. 5.3. Delco-Light multipurpose unit. *Source: The Delco-Light Story* (Dayton, OH: Delco-Light Co., 1922), 45. Used with permission of General Motors Media Archive. Photograph courtesy of the Richard P. Scharchburg Archives at Kettering University.

inspirational messages: "Some men have a wishbone where their backbone should be; Sales are not made by wishing for them. Canvass and demonstrate."[67] Agents often visited potential customers in the evening, after daily chores had been completed, and demonstrated light sets carried in their vehicles. Cleverly, they handed the farmers sales literature, which could only be read in the dark after the salesman turned on a Delco machine. "When the farmer was reading by the light of the 'miracle' bulb," observed an internal history, "the salesman would turn off the plant, leaving the prospect to squint at the print by the light of his gas lamp. If the farmer were a ready prospect, this dramatic demonstration would end resistance."[68]

Sales of Delco-Light units expanded over the years. In September 1918, the company claimed to have put "more than 50,000" units in use.[69] By 1921, the company sold several versions of the systems, each with different capacities: a 16-volt, 300-watt system for $250; a 32-volt, 600-watt system for $295; and another twenty-three styles and sizes with prices up to $1,675.[70] Advertisements proclaimed "over 100,000 satisfied users" in 1920,[71] and 160,000 users by 1922, when the company cut prices for its most popular units.[72] Benefiting from its ties with the parent company, the firm offered the electrical plants on credit through the General Motors Acceptance Corporation.[73] The model 620 (a 600-watt unit) in 1923 cost $404.50 in a one-time payment, or it could be purchased with a down payment of $80.50 and twelve monthly remittances of $27.[74] Improving farm conditions and the introduction of new models of water pumps pushed sales higher in 1926, such that shipments of power plants in the first seven months of the year soared 40 percent above the previous year's corresponding period.[75] The firm expanded production to meet demand, and in 1928, its advertising proclaimed "more than 300,000 satisfied users."[76]

Delco-Light advertisements and sales literature emphasized the many benefits of electrification. Because of its obvious appeal, electric lighting received much attention in publicity campaigns as a replacement for dangerous kerosene lamps. In a dramatic 1921 ad, the company noted that if Mrs. O'Leary owned a Delco-Light system instead of a lantern that offered only "a sickly light," she never would have kindled the Great Chicago Fire of 1871.[77] As important, the company argued that on-site production of electricity replaced human-power and saved huge amounts of time and effort. The firm asserted that its equipment "increases farm efficiency" by allowing workers to toil with little effort, even at night, with clean and safe illumination.[78] From a purely economic perspective, the Delco-Light system "pays for itself." The amount of time saved doing various activities on the farm—by the husband in the barn and the wife in the house—easily accounted for labor reduction of about fourteen hours per week, the company suggested. At 30 cents per hour for the work, savings accrued to more than $218 per year, about half the first cost of some units.[79] In 1928, the company promised, quite simply, "shorter hours" and "bigger profits" by reducing "work in a hundred ways."[80] As late as 1935, Delco-Light advertised its units as enabling farmers "to increase the profits of your farm, to assist you in your chores, [and] to save labor and increase productivity."[81]

Electricity also made life better outside of the work realm, claimed the company. Electric lights reduced eyestrain, caused by reading in the faint glow produced from kerosene lamps (plate 5). Promotional literature harped on the value of the electric lamps for making "the girls and boys contented on the farm. Let them enjoy the cheerfulness of bright electric light while growing up."[82] Mothers also appreciated electricity in the home, the company averred. An electric sweeper "quickly picks up the dirt with little labor and no dust," observed a sales brochure. Electricity also powered the electric iron, "which will give your wife great comfort and satisfaction and save her much time."[83] Of course, an "electric washer cleans the dirtiest clothes quickly and without drudgery."[84] Showing a proud father watching his son play with an electric train set, a 1922 advertisement urged farmers to "put Delco-Light in your Home for Christmas," since the power plant will "[b]ring greater happiness into the lives of everybody on the farm."[85]

Many of these advertisements tapped into the farmers' eagerness to be considered as modern, despite what big-city people may have believed about the supposedly rustic bumpkins. Modernity likely had a different meaning for rural folk, who (as historian Ronald Kline has documented) often adapted "urban" technologies, such as telephones, automobiles, and radios, to meet their specific needs. Other historians, such as Deborah Fitzgerald and Joshua Brinkman, have argued that farmers in the early twentieth century took on sophisticated management roles and developed new hardware to increase efficiency in production and in the home.[86] Delco sales literature contributed to and drew from an emerging discourse of rural modernity by noting that the company's product enabled electric appliances to make work easier and faster (figure 5.4 and plate 6).[87] Customers' letters abound in sales literature, attesting to these benefits and pointing to the advantages that accrued from living on state-of-the-art, electrified farmsteads. Impressed with the hardware, a farmer wrote in 1921 that his family members "would rather part with our car than the [electric] plant."[88] Even a complaint letter addressed to "Mr. Delco" suggested a craving to use electricity in a way that yielded an up-to-date

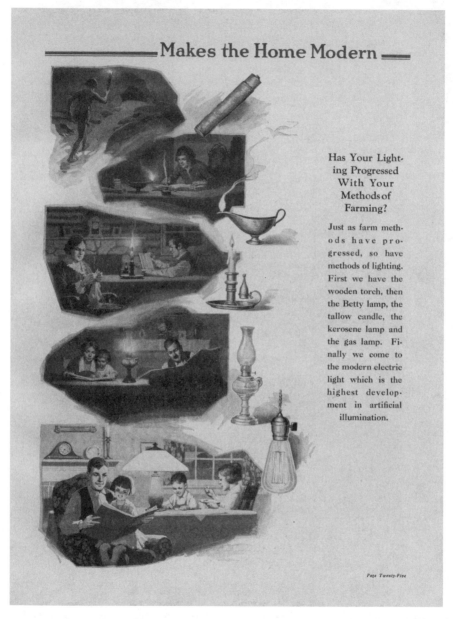

Makes the Home Modern

Has Your Lighting Progressed With Your Methods of Farming?

Just as farm methods have progressed, so have methods of lighting. First we have the wooden torch, then the Betty lamp, the tallow candle, the kerosene lamp and the gas lamp. Finally we come to the modern electric light which is the highest development in artificial illumination.

Page Twenty-Five

Fig. 5.4. Delco-Light modernizes the home. *Source: The Delco-Light Story* (Dayton, OH: Delco-Light Co., 1922), 25. Used with permission of General Motors Media Archive. Photograph courtesy of the Richard P. Scharchburg Archives at Kettering University.

farm: in search of parts so he could fix his machine, a customer wrote to the Dayton headquarters that, without electric lights "the possums are ketching my chickens so bad all on acct my Delco plant gone bad." Sorrowfully, he continued, "I hates to go back to dat coal oil age[.] I'se trying to be progressif and run my place modern way[,] so I askes you again pursonally to help me out [*sic*]."[89]

The company specifically urged salesmen to emphasize modernity as an attribute of its product. "What do you sell?" asked a company publication in 1919:

> Do you sell an electric light plant—just a piece of machinery? Or do you sell electric light and power—modern conveniences? Farmers buy Delco-Light primarily because they desire the things that Delco-Light will do. A man seldom has any desire for a piece of machinery, *just* because it is a *good* piece of machinery; but he may long for clean, bright, safe, electric lights in his barns when he is milking, or in his living room when he sits down all tired out to read his paper. . . . Sell electric light and power, sell modern conveniences, comfort, happiness![90]

Shrewdly, the company also associated its machines with the modern factories that produced automobiles and appliances at higher speeds and lower costs. In a *Greensboro Daily News* advertising supplement published in October 1919 (in anticipation of the Central Carolina Fair and the State Fair in Raleigh), Delco-Light asserted that "[e]very farm becomes a food factory and every farmer is a factory manager," on whom the world depends. "Your farm must be an efficient factory," the ad continued, since "[e]very little scheme of invention that saves work and utilizes a mechanical device means greater production, less labor expense and more profit to you."[91] In a not-too-subtle fashion, the company implied that a Delco-Light plant epitomized such a beneficial invention.

Ruralites could read similar statements about the equivalence of electricity and modernity in farm magazines. A 1917 article observed that "the time will come when practically every farm home in the country will appreciate the benefits of the use of electricity." The same piece described the results of a survey of ruralites using isolated systems, with one farmer noting that electric lighting in the henhouse increased egg production significantly. It concluded by stating that the delayed purchase of a small power plant overlooked "one of the greatest advantages of living in the present age."[92] In case readers did not get the message, they could read a Delco-Light advertisement on the same page, sold by an agent's business, the *Modern* Appliance Company of Seattle, Washington.[93]

Promotional literature also addressed the differences between life in cities and on farms, highlighting the notion that "Delco-Light brings city conveniences to the country."[94] Among the most treasured of such comforts were modern bathrooms, made possible by electrically powered water pumps.[95] A *Farm Mechanics* author in 1922 explained how electricity (in this case generated from isolated plants, though he did not specify a manufacturer) would change the attitudes of those intending to run off to the cities as soon as they came of legal age: "The world is too small any more for us to expect keen, live, intelligent youth to be contented with inconveniences on the farm while other young folks are having the joy and brightness of modern surroundings in the city. When isolation began to be broken down and the boundaries of the world were drawn in, with the advent of the interurban [railroad], the telephone and then the automobile, it was at first assumed that the only way to be rid of the old-fashioned farm life was to leave the farm and go to the city." Happily, the writer concluded, "It is the introduction of electricity to the farm that has smashed that idea all to smithereens."[96]

Costs and Deficiencies of Isolated Power Plants

While various companies produced isolated power plants, offering good service for those who could not get electric service otherwise, they still required substantial investments from farmers.[97] Beyond putting up the initial cost of the units, isolated power plant owners needed to wire their farmsteads and purchase lights, appliances, and electrical equipment, another strain that only the most successful farmers could endure.

To how much did these costs accumulate—beyond the initial cost of the plants—and what were the actual economic benefits? A 1929 master's thesis written by a Virginia Polytechnic Institute agricultural engineering student provides answers. In a detailed study of an entire Delco-Light system under true farm conditions, Lawrence Koontz recorded performance data of a unit installed near the college's home in Blacksburg.[98] The 1,500-watt power plant and sixteen batteries cost $866, and associated expenses brought the total price of the generating and storage system to almost $900. Wiring the farm and purchasing lights cost about $291. Taking careful note of the cost of fuel, oil, and labor, along with consideration of interest charges and equipment depreciation, Koontz studied the system when it generated 19 kWh and 60 kWh in different months. (The greater consumption during the second test month resulted from the addition of a Frigidaire refrigerator to the load.) During the first period, the per-kWh cost came to a little over 86 cents; when the farm used more electricity, the cost dropped to about 31 cents.[99] (For comparison, urban customers of electric power paid much less—

generally no more than about 8 cents per kWh in 1927.)[100] Since a large part of the cost consisted of power plant and battery depreciation, which did not change much regardless of usage, per-kWh costs declined as farmers consumed more electricity, leading Koontz to observe the "necessity of building up the kwhr load."[101] Despite these relatively high costs, Koontz demonstrated that the isolated sets still provided financial gains; estimating labor cost at 25 cents per hour, he concluded that use of an electric iron alone yielded savings of $42.90 per year (compared with employment of old-fashioned flat irons that required slow and frequent reheating on stoves).[102]

Another analysis of operating isolated electric plants—not necessarily Delco Light sets, but likely including them—illustrated similar costs for producing electricity. Researchers at the University of Illinois Agricultural Experiment Station reported in 1929 that after considering expenses for depreciation, repair, and other items, the cost per kilowatt-hour of electricity ranged from 64 cents for farmers who used a small amount of power (10 kWh monthly) to 22 cents for those who consumed seven times as much.[103]

Central station power clearly held great advantages over home-generated electricity, as had been expressed as early as 1913 in the General Electric Company's book *Electricity on the Farm*. Unlike the jack-of-all-trades agrarian, managers and operators of utility-owned plants "make the generation and distribution of electricity their specialty." Moreover, they had reduced the cost of electric power equipment and electricity in ways that could not be duplicated by smaller, isolated plants. Consequently, "the isolated plant can no more compete with the central-station plant than an individual making his own shoes can compete with a large shoe manufacturer."[104]

Isolated electric plants had other downsides. As noted in a 1923 Wisconsin Extension Service circular, 32-volt systems required nonstandard bulbs and motors that could not always be found in electrical supply stores. The low voltage also meant that more expensive wires had to be used than those employed on utility system circuits operating at 110 volts, especially if farmers sought to string lines farther than five hundred feet from batteries. These storage devices, meanwhile, demanded unusual care and attention to ensure they would endure "a good life" of five years.[105]

Users of isolated plants also needed to be careful about which appliances to employ concurrently. Generating units that produced a few hundred watts could only supply enough power for lighting (which, of course, appealed initially to most people) and a few other small loads at the same time. If too many devices overburdened the system, the lights would dim and battery life would decline substantially.[106] While Delco-Light manufactured and advertised several electrically operated farm products, not many became commonly used, likely because of the capacity limitations of their generation systems. A 1928 survey of more than 13,000 electrified farms in Nebraska—of which 10,229 had isolated plants—found that 100 percent contained electric lights. About 82 percent used electricity to operate washing machines, and 77 percent had electric irons. But only 44 percent pumped water with electricity, and just a few employed electric power for field or barn operations.[107]

Farmers who first bought isolated plants and then obtained central station power reflected ambivalent feelings about their earlier purchases in a 1923 survey conducted by Wisconsin extension specialists. Though appreciating the value of self-generated electricity, they much preferred high-line service from private companies. In fact, 71 percent of customers favored utility-provided power, and only 2.4 percent opted for individual light plants. (The remaining 27 percent did not respond to the survey question.) Among the reasons for this satisfaction were "less bother" (51 percent) and "more power" (29 percent) from central stations.[108]

Of course, if utility-provided electricity power remained unavailable, "then the next cheapest method is the isolated electric plant."[109] And despite high costs, Nebraska researchers in 1929 concluded that, because they proved so productive, the devices made for "very efficient and inexpensive hired men or chore boys."[110] Illinois investigators concurred, observing in the same year that farmers obviously desired high-line power. Even so, individual power plants "can render great service where central station service is not available, as it provides the power for most of the conveniences found in the city home."[111] As late as 1935, an author of a rural electrification book written for high school vocational instructors noted that, despite "construction of many miles of rural distribution line, it is still necessary, if electricity is desired, to use farm lighting plants on many isolated farms or in sparsely settled farming districts." Because it may take several years "before the majority of farms will be reached [with central station power] . . . it will be a good investment to install a farm electric-lighting plant."[112] Three years later, an agricultural journalist made the same point, even though the REA had begun operations amid great fanfare. He observed that the economics of central station service militated against the idea of universal rural electrification; he therefore suggested that remotely located farmers purchase a wind- or engine-powered isolated set so they could immediately begin enjoying "the ease, comfort, convenience and safety that electricity only can bring. They need not wait for the high line to begin to enjoy its many benefits."[113]

While viewed today as a nostalgically quaint technology that circumvented the problems of being unable to receive central station electricity, isolated generation sets, which powered 4 percent of rural American farms in 1930, provided a viable option.[114] Just as important, perhaps, they fit in well with some farmers' self-identity as technically savvy and hardy individuals.

Even some government officials who worked with farmers realized that isolated systems had greater value than many contemporaries imagined. In a 1924 report, for example, US Department of Agriculture economist Emily Hoag Sawtelle observed that rural women often resented the efforts of city-based reformers, such as the members of the Country Life Commission, who acted in a condescending manner. Citing a farm woman from Iowa, Sawtelle relayed the commonly held view that farm life had improved dramatically in recent years and that "the march of progress" had yielded many new conveniences on the farm, some of which were powered by isolated electric power plants. She described how life now included enjoying the use of a washing machine, a mangle (a mechanical clothes wringer), and a flat iron. But the small generating plant also signified the independence that farmers relished. The Iowan observed that, because of the nonexistent connection to a central station company,

> [w]e farm women never have to watch a meter as we have our own electric plants. Within three miles of my home there are only three out of 14 farmers that haven't electric plants on their own. Eleven of us farm women have the use of electricity and we don't have someone always sending us a light bill either. . . . I have the dustless mop which lightens one's work so much and which is in use every day on my hardwood polished floors. My sewing machine is run by an electric motor[,] and while I am busy sewing[,] I have an electric fan to keep me cool.[115]

In other words, owning isolated electric power plants did not appear as a sign of backwardness. On the contrary, rural citizens often viewed such devices as symbols of rugged modernity, progress, and self-sufficiency—all virtues within the farmers' culture.

Perhaps ironically, the isolated plant market served interests of central station operators because the existence of an alternative technology reduced pressure on them to supply power to farmers. Using electricity largely for lighting, typically consuming only about 30 kWh per month, ruralites simply could not be served economically (from the power company's perspective) with high lines.[116] Such little use would not yield enough income to utilities unless customers paid extremely high rates. But farmers could employ Delco-Light or other manufacturers' sets to serve themselves. As early as 1922, an *Electrical World* editorialist expressed a positive view of the isolated systems. He urged utility managers to avoid a short-sighted perspective that viewed such generating units as competition. After all, he noted, "it is believed that the individual plant has a very definite part to play in the development of electric service on the farm and that the central station cannot afford to assume other than a friendly attitude toward this development as one of the means of giving the farm the benefits of electrical energy."[117] The western editor of *Electrical World* agreed, commenting in 1923 "that the individual plant[s] must not be ignored. They have a field to fill that the central station cannot fill and should be helped, not hindered."[118]

Highlighting this point, NELA's Rural Electric Service Committee published a policy statement in January 1924 noting that, because power companies could not economically supply electricity to all farmers, the "individual plant is a practical method of securing electric service where central station service is not available." The statement further implored utility men to avoid taking "an attitude of criticism or fault-finding toward such plants" because it would "tend to deprive farmers of electric service where central station service . . . may not be available until some future time."[119] In other words, utility leaders should view the isolated plant business as an ally in the effort to bring electricity to farmers. The statement, meanwhile, harmonized with the decision to include a representative from the individual plant industry on its recently created (in September 1923) Committee on the Relation of Electricity to Agriculture.[120]

The use of isolated generation plants would also benefit the utility industry by giving farmers more experience with electricity and help them build demand for the form of energy. Most ruralites wanted electricity first to provide safe and convenient lighting for their homes and barns. Unfortunately for private power companies, as noted, that use yielded relatively low consumption and could not justify the extension of power lines to them. But once farmers discovered the benefits of electricity, using it for appliances such as Delco-Light fans, water pumps, and Frigidaire refrigerators, for example, they would eventually exceed the capacity of their self-generation units.[121] They would then appeal to the central station operators, who would feel more inclined to do business with them, seeing that they already created high loads. Continuing the line of thought, a 1935 report written a few weeks before the creation of the REA suggested that the "isolated plant not only will serve sections which otherwise would be unserved[,] but will build up gradually such a load that connection with a central station later will be justified."[122]

In short, generation of electricity using water, wind, or isolated power plants aided the movement of rural electrification in at least two ways: it contributed to the growing notion (and the sociotechnical imaginary) supporting the desirability and economic value of electricity on the farm. Moreover, it helped reinforce the view, held by farmers and their advocates, that at least some rural citizens were up-to-date and that they could become valued customers within the expanding central station network at some future time.

This chapter's discursion on farmers' use of isolated power systems—rather than how rural citizens won connections to what we now refer to as the grid—serves another purpose: it highlights the fact that people had alternatives to what we today view as the modern and "correct" way of obtaining electricity.[123] Conceivably, if private utility companies had not continued their pre-Depression pace of supplying electricity to farmers or if the federal government had not intervened, a larger number of farmers would have purchased independent means to generate power. As ruralites came out of the Depression and World War II with greater prosperity (farmers' net cash income rose from about $3.5 billion in 1940 to $15.3 billion in 1947), more of them would have had the wherewithal to purchase products such as Delco-Light sets, which likely would have improved owing to new manufacturing techniques and materials that evolved during the war.[124] Consequently, rural America would have become electrified through a hodgepodge of interconnected and isolated power systems.

Such hypothesizing constitutes more than just playing a game of "what if"—an entertaining approach used by popular fiction writers who imagine an America in which the Confederacy had won the Civil War, for example.[125] In fact, it has a sound basis among academics who do work in contingency and counterfactual history. Contingency refers to the notion that events might have played out in unrecognizable ways over time, the result of people choosing different options or because of accidental, serendipitous, or apparently "irrational" occurrences. Proponents of counterfactual history explicitly consider contingency and urge practitioners to focus on alternative decisions that stakeholders could have made without being influenced by knowledge of events that actually occurred. Through an examination of the real choices that actors confronted, scholars can "compensate for the hindsight bias that distorts historical vision," according to historian John K. Brown.[126] By examining viable choices at moments of indeterminacy in the history of rural electrification, such as the increasingly popular and sophisticated isolated power plants of the 1920s, one learns that the electrical landscape of nonurban Americans might have evolved differently, but also as "logically" as it actually occurred.[127] The approach forces historians to focus on how and why people selected among options, and it encourages them to write history from the past to the present—not from the present to the past—so we gain more nuanced perspectives of significant events. As I have argued in this book, maintaining a sense of the context in which actors made choices during the 1920s and 1930s, and disregarding how events played out later, contributes to a more accurate and sensitive narrative of the rural electrification process.

The focus on counterfactuals and alternatives to central station electric power on farms has unexpected significance in the realm of energy policy. While rural electrification in the United States and most industrialized countries now remains largely of historical interest, it constitutes an important goal in several developing countries, where (in 2019) about 13 percent of the world's population—940 million mostly rural inhabitants—do not have access to electric power.[128]

For many years, the World Bank and other international aid agencies have provided resources to build massive generating plants such as hydroelectric dams. Moreover, they provided support to continue the practice, begun in some African and Asian countries during colonial periods, to construct central stations and distribution systems in cities and surrounding regions—similar to the networks that evolved in the United States.[129] But perhaps policymakers should pay more attention to the early history of American rural electrification in which many farmers took advantage of isolated generation technologies, such as wind-turbine dynamos and Delco-Light generators and batteries. While farmers using such power systems may not have enjoyed the maintenance-free operation that the advertising literature promised, they nevertheless obtained several productivity- and life-enhancing benefits that electricity offered. Taking advantage of today's small-scale and renewable-energy generators—those benefiting from eighty to one hundred years of continued innovation that improved their diversity, efficiency, and affordability—rural electrification advocates in developing nations could gain greater success by pursuing a course of action that Americans abandoned in the 1930s.

Of course, this suggestion is not original, and many academic researchers, nonprofit organizations, and corporations have employed the approach with positive outcomes.[130] Today's version of isolated power systems, known as distributed generation (DG), can often provide reasonably priced electricity and higher reliability than can traditional generators linked to transmission and distribution lines.

And just like American companies one hundred years ago, private and state-run utilities remain unwilling to invest in expensive infrastructure. DG appears as an appealing alternative.[131]

In experiments and pilot projects in Africa, communities have already begun establishing small-scale networks (often called microgrids or minigrids) that exploit renewable and nonrenewable distributed generation sources.[132] For example, in one village in Kenya (a country in which 93 percent of rural residents lack access to electricity), more than sixty homes and businesses obtain power from an 8.5 kW solar photovoltaic cell system.[133] Using "smart" (computer- or Internet-connected) technologies that manage demand and distribute power, similar networks provide an electrified infrastructure that enables substantive economic and social activities in outlying districts.[134] In Sokoto State, Nigeria, a 60 kW photovoltaic array provides power for a battery bank, which is supplemented by a diesel generator, sufficient for a community of 350 households and 20 small businesses.[135]

As with all situations of technology transfer and adoption, consideration of DG systems requires more than a simple analysis of costs. Large-scale grids managed by governments or other large organizations remain easier to subsidize than myriad small- and community-scale systems; this fact explains, in large part, why about 70 percent of all funding goes to grid approaches.[136] Interconnected networks also can be supervised more efficiently (on a national scale) than disaggregated systems. But the cost for grid components (generators, transmission lines, and associated equipment) has remained constant for several years, while the cost for DG elements, such as wind- and photovoltaic electricity–generators, has declined. And since they require less time to deploy than large-scale grid technologies, DG systems can more quickly begin serving communities that have modest energy needs.[137] In theory—and at times in practice—distributed generation competes favorably with grid-connected networks on certain technical and economic criteria, though they sometimes meet with administrative resistance.

In short, while central station service and interconnected power networks have long been viewed as the optimal way to deliver electricity to all residents, the approach may not satisfy almost a billion people who currently lack access to electric power. The future may actually lie in the past—employing a century-old model in which American farmers exploited small-scale and off-grid energy systems to achieve what they could not obtain through conventional means.

PART II / Alignment of Rural Stakeholders

As the sociotechnical imaginary of rural electrification emerged in the early twentieth century, so did examples of successful farm use of high-line electric power. Though economic considerations seemed to rule out extension of distribution wires beyond cities, innovative power company managers began demonstrating the occasional merit of serving agricultural communities in the East and Midwest, with the most positive experiences occurring in western states where irrigation activities consumed enormous amounts of electricity. Utility leaders also recognized that, for largely political reasons, they would benefit by devoting resources to expansion of the farm market.

In its pursuit of increased rural electrification, the industry began establishing the institutions and collaborations for translating the imaginary into a more widely experienced reality. Most notably, company executives organized research groups within the National Electric Light Association, and they initiated partnerships with agricultural engineers employed at land-grant colleges. Working with the Committee on the Relation of Electricity to Agriculture—but also acting independently—the engineers demonstrated new, productive farm applications of electricity while educating the manpower to serve as rural experts within power companies. At the same time, many firms engaged with stakeholders to streamline the regulatory process and facilitate more liberal financial arrangements with farmers who desired high-line power. By doing so, the industry and its partners created substantial momentum in the rural electrification subsystem during the decade before the New Deal began.

Chapter Six

Utility Interest in Rural Electrification Awakens

> To visualize the blessings of rural electrification, one needs only to picture the California farm with electric service. There the farmer's wife may cook meals in a cool kitchen on an electric stove with electric fans relieving the valley heat; there she cleans house with a vacuum cleaner; there electric lights await only a touch of the button; there an electric refrigerator keeps foods fresh and supplies ice; there small motors do countless chores and much hard work, with a saving of time and effort; there the farmer and his wife live in the country under city conditions.
>
> —WIGGINTON CREED, PRESIDENT,
> PACIFIC GAS AND ELECTRIC COMPANY, 1926

Though most power company leaders viewed rural electrification with disdain, some nevertheless advocated for it. Starting in the late nineteenth century, the case emerged for bringing electricity to scattered farmers as part of an attempt to reduce the increasingly apparent chasm between city dwellers and ruralites. A few utility executives experimented with rural electrification by stringing distribution lines from existing power lines to sparsely populated regions and found that, in certain circumstances, these extensions earned good economic returns. And they discovered that, especially in western states, electrification could become immensely profitable when farmers used large amounts of power for irrigation purposes. These experiences provided suggestive evidence that farmers might eventually make good customers for utility companies and that they deserved more attention from managers.

Emerging Undercurrent Movement of Rural Electrification

In the years after the end of the Civil War, differences between the nation's agricultural and manufacturing sectors grew more noticeable. Industrialization gained momentum, with cities becoming the dominant places of economic activity, and the percentage of people living in rural areas declined. The former Confederacy remained exceptionally drained, with tenancy and out-migration common, particularly on land that once grew cotton (the so-called Cotton South).[1] Nationally, farm productivity increased because of the growing use of mechanized equipment, such as McCormick reapers and Deere plows, often manufactured in urban factories. Too frequently, however, overproduction drove down commodity prices. And while a majority of Americans still worked in rural areas as late as 1910, providing the necessary foodstuffs to support the populations in burgeoning metropolises, farmers did not appear to benefit as much by modern technology as did city folk.[2]

Efforts to improve the lot of farmers included creation of a cabinet-level Department of Agriculture in 1862 (during the Civil War), empowered to "acquire and to diffuse . . . useful information on subjects connected with agriculture."[3] That same year, Congress passed the Morrill Land-Grant Act, which established a mechanism for founding and supporting colleges in each state devoted to educating working-class students in the methods of modern agriculture and engineering.[4] The schools benefited from passage in 1887 of the Hatch Act, which provided funding for experimental research stations, and by the 1914 Smith-Lever Act, which afforded a means (through county agents working with extension services) to disseminate knowledge about agricultural innovations.[5]

Some political leaders and activists even saw a role for electricity to play in the revitalization of the agricultural realm. As early as 1892, Kansas senator William A. Peffer pressed for a congressional resolution to obtain information from other countries on "the application of electricity to the propulsion of farm machinery and to the propagation and growth of plants."[6] Two years later, the senator unsuccessfully sought to expand on this work by establishing "an electrical experiment station for farming."[7] The authors of the 1909 Country Life Commission report also observed that farmers could boost productivity through the use of streams to generate electricity, though they highlighted potential impediments to gaining legal rights to use the water.[8] Progress in extending electric lines could occur more quickly, the document noted, if farmers copied the approach of ruralites who created cooperative associations to acquire telephone service.[9]

Reflecting the views of those seeking improvement of rural life, Illinois congressman Henry T. Rainey made a well-regarded effort to increase knowledge about agricultural uses of nonanimal energy sources. In July 1912, he introduced legislation to create a Bureau of Farm Power within the Department of Agriculture. The bill directed the agency "to investigate and report . . . upon all matters pertaining to methods of furnishing power on farms and all labor-saving machinery adapted for use on farms, and the use of electricity, gasoline, and steam in propelling farm vehicles."[10] Dealing with more than just electrification, the bill sought to help farmers employ machinery instead of human and animal power at a time when obtaining rural labor proved difficult. Receiving a good amount of publicity, the bill constituted "a measure well worth of encouragement and support," according to the *Farm Journal*.[11] The American Society of Agricultural Engineers concurred, endorsing the idea of the bureau and looking forward to the bill's passage.[12] Likewise, *Scientific American* magazine noted that the federal government pursued valuable programs to help farmers in a variety of realms, from understanding plant diseases to predicting the weather, but not how to maximize value from power sources. "The American farmer has outgrown the age of hand labor, and even animal labor is now on the wane," the magazine reported. While some farms had begun using steam power, more recently "the internal combustion engine has been playing a very important part, and now electricity is beginning to be used."[13] The bill won positive notice from utility industry leaders too: at a 1913 convention, NELA secretary T. Commerford Martin observed that the legislation enjoyed the organization's official support.[14] According to one news account, the bill would help farmers find information required to satisfy production needs in a way that would prove the wisdom of the article's title, "U.S. Government to Solve Farm Power Problems."[15] High hopes and broad support notwithstanding, the bill languished in House committees, never to gain passage even after being reintroduced a few times.[16] Its existence, however, suggests that farmers and their advocates sought more support to raise productivity through the use of new mechanical devices, some of which could be operated by electricity.

Even without a new government program, the Department of Agriculture pursued efforts to gauge farmers' needs and wants. In 1913, the department secretary sent questionnaires to 55,000 farmers' wives, asking about ways to improve rural life for women. When describing labor-saving devices, many respondents commented that they would enjoy converting the power of water rushing past their homes into electricity to "relieve the housewife of her most laborious and distasteful work." The correspondents clearly understood that electricity could make "the women in the rural districts healthy, happy, and contented," as one Ohioan wrote. "My neighbor thinks electricity is one of the most needed conveniences on the farm to lessen the labor of the housewife," observed a Marylander. "Hasten the day when all rural districts could enjoy or have the benefit of either a company's plant or a simple plant that could be established at a reasonable figure in each home that might be applied or harnessed to the washing machine, milk separator, churn, [and] vacuum cleaner."[17]

Early Utility Efforts to Expand Rural Electrification

Some commercial firms had already begun modest efforts to serve the rural market. Surely, these companies (and their managers) constituted exceptions to the generally accepted "wisdom" that farm customers would not make worthwhile customers. Nevertheless, they contributed to an undercurrent movement that ultimately provided the manpower and knowledge to pursue more strenuous efforts in the 1920s and later.

Perhaps the earliest rural electrification efforts materialized as interurban railroads became popular toward the end of the nineteenth century. Speedy, cheap, and clean, these electrified railway lines used equipment that could stop frequently and more efficiently than steam-powered trains.[18] Their success relied on overhead cables that provided electricity to the rail cars—and potentially to ruralites near them. An Ohio author observed in 1894 that extensions from the increasingly common railway power lines could provide electricity to "any wide-awake enterprising farmer."[19] Employing such an approach, the Milwaukee Light, Heat and Traction Company first sold power to a farmer in 1897.[20] A decade later, an Indiana interurban firm sought to increase revenue by convincing "the farmer how much more economically he can thus plow, sow, harrow, run his feed-cutter, churn and separator, and light his house and barn" with electricity.[21] Other railroads developed comparable plans, encouraging one executive in 1907 to suggest that the "time will come when on farms near electric railroads, fields will be plowed and harvested, machinery operated, even cows milked by the same power which takes the farmer's family to town and his products to market."[22]

Herman Russell, a manager working for the Rochester (New York) Railway and Light Company, reported in 1910 that his firm had begun selling electricity to farms from lines that paralleled train routes in the countryside.[23] He observed that the power could be used profitably for irrigation, a high-demand application that helped utilities recover their investments quickly, especially when farmers drew power during off-peak hours. Russell indicated that his company had invested in generating equipment yielding 38,000 horsepower (about 28,337 kW), but 10,000 horsepower remained unused from April to October each year owing to lower urban demand during that time. Clearly, this manager understood the value of raising his firm's load factor. The farm customer, who uses power off-peak—mostly in the summer—would produce welcome income to the company from this unused capacity. Russell hoped that the time would soon arrive when "distribution lines for supplying electric energy will be as common throughout the country as the telegraph and telephone lines are to-day, and when the present enormous idle investment of central-station companies will be employed in making farms more productive."[24]

Managers of the Edison Company of Boston held similar ideas about the potential value of rural lines. Observing that the utility had installed transmission and distribution lines across eastern Massachusetts, many of which spanned farming districts, they realized that they could easily connect new rural customers who might yield profitable business. In 1910, the firm began an extensive advertising campaign employing a rapidly erected wooden demonstration house that it transported throughout the company's service area.[25] Complementing that publicity approach, the firm displayed a "Farm of Edison Light and Power" in a circuslike tent containing electrical appliances.[26] Carried by two electric vehicles, the exhibit highlighted electrically operated pumps for irrigation use and for land reclamation.[27] It also contained "other money saving devices" such as feed grinders, milking machines, cream separators, butter churns, and bottle washers. Employees showed visitors how a small electric motor, mounted on a portable truck, could be moved where needed on a farm for driving an ensilage cutter, wood saw, wood splitter, and hay hoist. And while the company did not exhibit equipment that stimulated plant growth with electricity, it offered information on the subject.[28]

To be sure, Boston Edison's managers had self-serving motives. A company engineer in 1912 remarked that while electricity use on farms would greatly increase productivity, rural distribution of power enabled the firm to help "mankind as well as ourselves."[29] It did so because the company could make good money from the investment in transmission and distribution lines, especially as farmers consumed power at different times from those favored by urbanites.[30]

Because of his personal interest in life on farms, Samuel Insull, the former assistant to Thomas Edison and president of Chicago's Commonwealth Edison Company, demonstrated the value of modest efforts to distribute power to rural users. He did so in 1910 by extending power lines to Libertyville, Illinois, a small town north of Chicago (in Lake County) and near a farm estate he purchased in 1907. He continued the experiment by electrifying adjacent villages. Eventually tying together twenty-two towns and several farms, Insull found that the peak seasonal demand in rural areas came in midsummer, when farmers used power for agricultural machines. That maximum occurred as the cities' use for power declined, only to rise again in the winter. By serving both sets of customers, the company's load factor improved, and costs per unit of energy and capacity decreased.[31]

Elsewhere, motivations for rural electrification differed. In Wisconsin, connecting power plants to farms resulted from efforts to exploit water resources early in the twentieth century. Though encountering technical and financial challenges, entrepreneurs sought to emulate the experience at Niagara Falls by using rivers to generate electricity.[32] The Wisconsin River Power Company, for example, built a hydroelectric plant near Prairie du Sac and arranged to sell power to a Milwaukee utility and a few towns, but it still had excess power available. In 1917, Samuel Insull's Middle West Utilities Company, a holding company formed in 1912, bought the financially troubled Wisconsin River firm, combining it with several other small power businesses in the state.[33]

Possibly influenced by Insull's work on stringing power lines to rural customers, Grover C. Neff, a manager of Wisconsin River Power Company and a Purdue University–trained civil engineer, developed plans to interconnect small cities and farms in south-central Wisconsin (figure 6.1).[34] According to historian Forrest McDonald, Neff established a "radical" and "visionary" plan to serve a nonurban customer base, one that harmonized with Insull's own achievement in electrifying rural areas outside Chicago.[35] Undoubtedly, Insull bought the Wisconsin utilities for investment reasons, not because of his interest in rural electrification. But according to McDonald, the utility mogul often asked managers of newly acquired companies to propose novel ideas as a way to gain insight into the strengths of his new employees. "Sometimes, as in Neff's case, the plans were adopted," he wrote. In his book on the creation of power policy, Philip Funigiello further noted that Insull and Neff deviated from traditional utility company practice, "experimenting with rural electrification long before it became fashionable."[36]

As early as 1920, some of these efforts won accolades. Madison's *Wisconsin State Journal* newspaper published a glowing account of the benefits provided by the hydroelectric plant owned by Neff's company. Not only did it save almost 290,000 tons of coal annually by producing electricity from a natural source rather than from a fossil fuel; it also was a "boon to farmers" who took advantage of nearby electric wires to secure power.[37] Upcoming construction of a 66,000-volt transmission interconnection line between Insull's companies in Wisconsin foretold of even more opportunities to supply electricity to residents of the state's "best manufacturing, agricultural and dairy sections."[38]

Perhaps because of his innovative and well-regarded work with Insull's growing empire, Neff gained stature within the utility industry. In a 1928 article on midwestern utilities, Don Sterns, vice president of the Iowa Public Service Company, gave credit to Neff as the executive "who aroused the utility industry to a consciousness of its obligations and opportunities in serving rural districts."[39] Neff even won respectful recognition from REA administrators Morris Cooke and

Fig. 6.1. Grover C. Neff. *Source: Edison Electric Institute Bulletin* 3, no. 1 (1 January 1935): 13. Used with permission of the Edison Electric Institute.

Harry Slattery in the 1940s, who described the utility manager as "progressive."[40] Hence, when Neff urged his colleagues within NELA to create a study group and then pursue more substantive research on rural electrification, his suggestions won a positive response—especially when the work fit into a larger program of the industry's public relations efforts.

California and the Fulfilled Promise of Rural Electrification

While some eastern and midwestern power companies occasionally obtained positive outcomes when serving farms, the case for rural electrification was decisively made in the far western states. In California especially, exceptionalism (in the realm of electrification) stemmed from an unusual energy situation in which large population centers near the Pacific Ocean lacked easy access to traditional fossil fuels. Urban utility planners overcame the problem by exploiting the bountiful hydroelectric resources in the distant Sierra Nevada mountains, which offered power that could be transmitted efficiently via high-voltage lines.[41] As the wires traversed large agricultural expanses, companies sometimes allowed rural folk to draw power from them, with the first farm energized in 1898 near Yuba City. Electrified farms sprang up a few years later near Visalia, in California's agricultural San Joaquin Valley (the southern end of the state's Central Valley).[42]

Utility companies felt disposed to serve new customers because farmers consumed large amounts of power for crop-watering purposes. As historian David Nye has written, electrically operated irrigation pumps served as an important precondition for farming in much of California, as well as in Utah, Nevada, and Arizona.[43] Hydraulic pumps existed before the use of electricity, driven by windmills and gasoline-, oil-, or distillate-fueled internal combustion engines. But farmers quickly learned that electrically driven devices proved cheaper to install and demonstrated greater efficiency, reliability, and safety.[44] A 1908 study noted that gasoline engines generally delivered only about one-fourth their rated capacity, while electric motors offered up to 47 percent and allowed for quicker service and simpler maintenance.[45] An editorial in the *Journal of Electricity, Power and Gas*, published in San Francisco, described the electrically operated pump in 1912 as a "joy forever to the farmer."[46] General Electric pointed out in a 1917 advertisement that with "'electric' irrigation," farmers "merely throw the motor switch—a few cents' worth of electricity per hour will do the rest."[47]

Electrically powered irrigation proved a boon to utility companies, and not only because it provided a new revenue stream. Rather, the use of the relatively large electric pump motors (from 2.5 to 100 horsepower in 1916) generated great demand, especially in the spring and summer, which complemented the higher urban demand (largely for lighting) in the fall and winter.[48] Such diversity therefore helped elevate the companies' load factors. As an example, the Pacific Gas and Electric Company, which served the San Francisco area as well as rural customers to the east, witnessed load factors of about 60 percent in the 1910s, which compared impressively to eastern companies' factors running in the 30 percent range.[49] The San Joaquin Light and Power Corporation, which provided electricity to many farmers using irrigation pumps, recorded annual load factors in 1915 and 1916 of 61 percent.[50]

Positive early experiences motivated managers to develop promotional campaigns to spur further rural electrification. In 1912, Pacific Gas and Electric assem-

Fig. 6.2. Pacific Gas and Electric Company railroad demonstration car.
Source: "Building the Agricultural Load," *Journal of Electricity* 45, no. 10
(15 November 1920): 458.

bled a demonstration railroad car that carried electrical equipment for farm use
(figure 6.2).[51] Irrigation and pumping information consumed nearly half the
car's space, with much of the balance used to exhibit refrigeration devices, dairy
machines, and household appliances.[52] Extremely popular, the car hosted about
six thousand people during the year, some of whom picked up a booklet that
praised electrical irrigation with lists such as this:

Choose Electricity!
Irrigation by electrical pumping is superior to any other method because:
The water is under complete control of the land owner.
The irrigator is not dependent on his neighbors.
The water is flowing in an instant, day or night.
The machinery has no complicated parts to keep in repair.
There are no boilers to explode.
There is no gasoline tank to catch on fire.
It is cheaper to install.
It costs less to operate.
The cost of electricity is growing less every year.

It is easily and quickly started.

It is on the job day and night.[53]

Once farmers obtained electric wires for irrigation, they expanded their use of the new energy form for other purposes: electric lights illuminated homes, barns, and henhouses; electric heaters staved off frost in citrus fields; and electric flat-irons, washing machines, and vacuum cleaners reduced menial labor in the farmhouse. The *Journal of Electricity, Power and Gas* explained in 1912 how irriga-tion customers used electricity to add a "thousand comforts . . . to the daily life on the farm."[54] Many of these electrical delights, by the way, were manufactured by companies such as General Electric and Westinghouse, which had begun rec-ognizing the value of the farm market. GE ran an advertisement for electric mo-tors in a western publication in February 1915, for example, exclaiming that their use saves labor costs and does "practically all your chores."[55]

During the early years of rural electrification in California, power companies required farmers to advance the cost of distribution lines (operating at 2,300 or 11,000 volts) and transformers. But as companies became convinced of the prof-itability of irrigation loads, they paid for the lines and transformers themselves, such that consumers only needed to purchase irrigation motors. (See chapter 10 for more discussion of such arrangements.) A 1920 article noted that utility financ-ing of the rural line extensions had been "exceptionally generous," contributing to the "especially large agricultural load."[56]

A growing number of farmers produced that load, with 18 percent of California farms receiving central station power in 1917.[57] Applauding the virtues of electri-cally operated irrigation pumps and the use of lights and motors, the *Journal of Elec-tricity* headlined a photo montage in a 1920 issue focusing on rural electrification with the words: "Electrifying the West's Most Important Industry" (figure 6.3). More significant, perhaps, the piece's subtitle highlighted the value of the farmer to the utility: "Where the farmer is the power company's best customer."[58] By the end of 1923, 23.5 percent of California's farms had high lines coming to them. That fig-ure stood almost an order of magnitude greater than the national percentage: only 2.8 percent of American farms obtained central station service at the same time.[59]

The extensive farm use of electricity, largely for irrigation, and the increased load factor it provided, enabled companies to offer relatively cheap power. In a Central Valley community in 1915, seven hundred ranchers had connected an irrigation load of almost 2,000 horsepower and paid 3 cents per kWh for the first 1,000 kWh, 2.5 cents per kWh for the next 1,000 kWh, 2 cents per kWh for the next 1,000 kWh,

Electrifying the West's Most Important Industry

WHERE THE FARMER IS THE POWER
COMPANY'S BEST CUSTOMER

The use of a single motor to drive a number of different machines is characteristic of farm installations. In this barn the motor operates a feed cutter, and the idle belt just to the left of it goes down a pit behind the cutter where it can be connected with a pump for filling the tank on the roof, or supplying water for irrigation. Note that the barn is also lighted by electricity as it is too dark even during the daytime for work to be carried on easily and safely.

Electric lights in the farm buildings are among the greatest benefits which electricity brings to the farmer. Not only are they clean, convenient and safe, but they enable the farmer who must use the last daylight hours in the fields, to do certain indoor work after dark, which would be impossible by the light of a lantern or kerosene lamp. The elimination of the fire hazard is also very important.

Where electric milking machines are used one person can attend to the milking of several cows simultaneously, thus saving a great deal of time on this otherwise laborious task. Milking done this way is not only quicker and easier but far more cleanly and satisfactory.

Only those who have drawn and carried bucket after bucket of water from a well for watering stock, for washing, cooking and a thousand other uses can fully realize the inconveniences of the old-fashioned farm, and the savings of an electrically-pumped water supply.

Fig. 6.3. Electrifying farms in the West, "where the farmer is the power company's best customer." *Source:* "Electrifying the West's Most Important Industry," *Journal of Electricity* 45, no. 10 (15 November 1920): 462.

and 1.5 cents per kWh for subsequent use. Comparable to rates paid by big businesses in cities, these prices undercut those paid by most eastern farmers lucky enough to have power by a factor of two or three.[60] By 1925, 40 percent of California's farms enjoyed central station service. The average farmer paid 1.55 cents per kWh and used 17,678 kWh per year for power alone (in other words, *not* for lighting, heating, or cooking).[61] The average American farmer who did not use irrigation purchased less than 5 percent of this amount—only 707 kWh per year for all purposes, paying 6.93 cents per unit.[62]

Rural use of electricity continued to increase during the next few years such that more than 63 percent of California farms received central station power by the end of 1931.[63] Looking backward from 1937, Ben Moses, an agricultural engineer at the University of California, Davis, observed that the electric pump constituted the "Daddy of Rural Electrification" in the state, since it provided the economic rationale for utility companies to serve farmers. In that year, California agriculturalists operated more than sixty thousand electric irrigation pumps (averaging 15 horsepower each).[64]

Farmers in other western states also benefited from the extensive use of electrified irrigation. In eastern Washington, ruralites in 1906 obtained a power line emanating from Spokane, which built its first hydroelectric plant in 1887 and expanded generation capacity thereafter.[65] As early as 1913, the trend toward increasing agricultural consumption of electricity caused a vice president of Pacific Power and Light, based in Portland, Oregon, to shower praise on progressive western farmers who helped themselves and his company. "Prepare for the new era of the electric light and power industry," he concluded in a message to his managerial brethren, "which has unparalleled possibilities for the territory you serve now and hereafter."[66] Another utility executive remarked on the growing fondness for farm electrification. "Those of us who have been in the business a number of years know that up to a short time ago the farmer made us tired," he remarked. But his opinion of the agricultural customer, based on the use of electricity primarily for irrigation, "is reversed now."[67] In the following years, the managers' expectations were fulfilled: at the end of 1923, 18.3 percent of Washington's farms enjoyed central station service;[68] by 1931, that number had risen to 51.0 percent.[69] Likewise, largely owing to heavy irrigation demands, Utah farms reached a 56.8 percent electrification rate in 1931, up from 12.9 percent in 1923. Oregon's rural electrification saturation remained lower, but still impressive at 29.4 percent in 1932, rising from 9.4 percent in 1923.[70]

Early efforts at rural electrification—and especially the work done in western states—provided a glimpse of what could be accomplished elsewhere. After supplying lines that energized the high-demand (and high-load-factor-generating) irrigation pumps, utility companies supplied transformers to provide lower-voltage power for homes and barns. In such a way, Californians enjoyed the nation's highest percentage of electrified farms, the greatest consumption per farm, and among the lowest electricity prices.[71] Evaluating the situation from a longer perspective in 1988, historian James Williams remarked that, even before the federal government became involved in rural electrification with the creation of the Tennessee Valley Authority in 1933, "California's irrigated and reclaimed agricultural valleys already illustrated the power of electrical redemption. And when the Rural Electrification Administration set out to light the rest of the nation's farms, despite some fears that farmers would not use enough electricity to make it worthwhile, California provided a continuing example of success and profit."[72]

To be sure, the western states constituted an anomaly for central station service to rural citizens.[73] The huge irrigation demand served as the primary stimulus for electrification of farms, and that same source of demand only existed in select parts of the nation. Statistical reports (such as NELA's compendium of rural electrification data for the years up to 1932) made that distinction explicitly, often by categorizing data on power use as coming from irrigation or non-irrigation states. On average, farm customers in 1931 consumed 775 kWh annually in non-irrigation states; those in irrigation states used nearly ten times as much—7,487 kWh. Total consumption of farms in the fifteen irrigation states exceeded by more than a factor of four the combined total amount of the other thirty-three states and the District of Columbia.[74] In a similar fashion, a 1936 Federal Power Commission report deliberately omitted California's data when describing average electricity consumption and prices paid by the nation's rural customers; the numbers from the Golden State simply skewed average information too greatly.[75]

This regional divergence in power consumption made a great difference to company managers. While the West may have demonstrated how rural electrification could benefit farmers and utility firms alike, the example also reinforced views that farmers elsewhere needed to consume large amounts of power. The challenge, in other words, consisted of encouraging farmers to use increasing numbers of kilowatt-hours in such a way as to make service profitable to all parties. And it wouldn't hurt if power companies felt external pressures to help farmers achieve this goal.

Chapter Seven

The Unexpected Public Relations Value of Rural Electrification

> We must recognize that the farmer is going to have his service. It must be recognized by [private utility] companies generally throughout this country that unless we are ahead of the farmer and point the way out by which he can get service, he is going to get it through some other means. He will have the power through the [regulatory] commission, and politically through his legislature, and we would be compelled to do something in a way that may not be as advantageous for the company or consumer.
> —UTILITY OFFICIAL M. T. D. CROCKER, 1922

The farm market officially received institutional recognition when the National Electric Light Association created a group in 1910 to investigate the supply of power to nonurban areas. The effort began for two reasons. Founded in 1885, NELA regularly established bodies to examine various technical and policy issues that arose. As the previous chapter suggests, some companies, especially those in western states, embraced rural customers, and it therefore seemed reasonable to investigate the potentials of (and inhibitions to) the farm business in a more systematic fashion.

Perhaps as important, the industry's modest exploration of the rural market stemmed from its leaders' perception that such work could help deflect political challenges. Like other rapidly growing enterprises in the late nineteenth century, the commercial power business witnessed regular attacks from elected officials and populist activists who sought to make electricity supply a function of government rather than that of private enterprise. Some of the public rebuke focused on the apparent neglect of private companies to serve farmers, and industry leaders likely felt a need to respond by demonstrating an interest in rural electrification. A 1926 report noted, "Whether from fear of government ownership, from a desire to expand markets, or for some other motive, the public utility companies have recently become aware of the great importance of rural electrification."[1] Put simply by historian Abby Spinak, "the American power industry was deep in politics," and farm electrification played a role in the conflict between for-profit power companies and other electricity stakeholders.[2]

NELA's Efforts to Study Rural Electrification before 1923

Less than three years after Thomas Edison illuminated parts of New York City in 1882, representatives of sixty-five power companies and electric-equipment manufacturing firms met for the first convention of the National Electric Light Association.[3] The association's revised constitution, taking effect in 1891, stated its goal "to foster and protect the interests of those engaged in the commercial production of electricity, for conversion into light, heat, or power."[4] Over the years, members altered the governing document to account for new classes of members and creation of state organizations. In the further amended 1921 constitution, the organization established a more profound objective, namely, "to advance the art and science of the production, distribution and use of electrical energy for light, heat and power for public service."[5]

Electrification of farms did not appear important to NELA members for several years after the organization's creation. A computer search for the words "farm" and "rural" in digitized convention proceedings resulted in only a few hits.[6] Serious attention to rural electrification appeared to grow at the May 1910 convention, where several attendees expressed interest in Herman Russell's presentation on using high-voltage lines, designed for electric railways, to power nearby farms (described in the previous chapter). That concern appeared to prompt creation of a Committee on Electricity in Rural Districts in November 1910.[7] The group (which included Russell as a member) began collecting information on farmers who already received electricity from central station operators, and it inaugurated efforts to educate farmers through the technical and popular press.[8] In its first published report, the committee in 1911 described beneficial uses of electricity on the farm—such as lighting, incubation of chicks, and charging batteries of electric vehicles. But significant impediments to widespread rural electrification remained, especially since the farmer still constituted part of the "'show me' class" that needed to "be convinced of the economical and practical value of electricity in performing his work before he will spend his money" on electrical wiring and equipment.[9] To do that persuading, the committee recommended advertising directed toward farmers, including the use of cartoons published in newspapers (some of which exaggerated the value of electrification [figures 7.1, 7.2]) along with utility men learning more about the needs and desires of farmers.[10] Much work remained to be done, of course, since "the majority of Central Station companies are not active in getting business in the rural districts."[11]

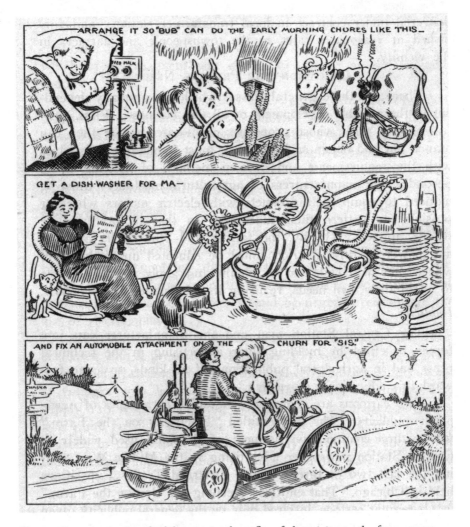

Fig. 7.1. Cartoon touting the labor-saving benefits of electricity on the farm, 1911. *Source:* John G. Learned, "Report of Committee on Electricity in Rural Districts," NELA, *Proceedings* 40 (1911): 520.

Acknowledging that the subject of farm electrification had "to some extent, always been considered a joke" among utility men, according to one NELA member, the committee in 1913 demonstrated more nuance and understanding of problems and opportunities.[12] Broken into discussions of rural electrification in the western, central, and eastern states, the group's report highlighted marked regional differences. The representative of the eastern states committee commented on

Fig. 7.2. Cartoon suggesting that electricity will help retain kids on the farm, 1911. *Source:* John G. Learned, "Report of Committee on Electricity in Rural Districts," NELA, *Proceedings* 40 (1911): 521.

the obstacles to getting the large number of small farms in his area to use enough electricity to make service profitable.[13] By contrast, western states' utilities that sold power to irrigators had little difficulty in arguing for the value of electricity to farmers and to power companies.

Reflecting a different way of thinking about rural electrification, the authors of the central states report observed that utility companies should *not* seek farm customers directly. Despite the benefits that electricity could offer farmers, the managers concluded simply, "The 'farm business,' as developed at present, does not furnish the central-station companies with returns commensurate with the necessary capital investment."[14] But all was not lost. The document observed that utilities should remain eager to string lines to operators of electricity-intensive activities in rural settings, such as coal mining, stone quarrying, gold dredging, grain elevating, and wood sawing.[15] Once these wires had been installed and largely paid for, extensions to nearby farms would make more economic sense. The farm business, in short, would be served as a by-product of working with rural industries that clearly warranted lines built to them.[16]

Rural electrification received sporadic attention in subsequent years, and especially during and after World War I, when farmers prospered from selling crops to war-torn Europe. One manager of an Ohio utility reported that an increasing number of farmers had requested service, but the company had not yet determined the best way to provide it. A few options existed for dealing with the large expense of building the lines, such as encouraging farmers to organize themselves into associations and then selling power to them as a single entity; the groups would resell power to individuals using their own equipment. But that option required farmers to have specialized (and not widely held) technical abilities, according to power company men. Alternatively, farmers' associations could build lines at their own expense and make a gift of them to utilities, which would then sell power and maintain the lines. That option also encountered likely problems, because it required unexpected cooperation among the farmers in the organizations. Finally, and seemingly most practically, utilities could deal with farmers on a piecemeal basis, having each finance the cost of the lines' construction and then selling power at rates that compensated for maintenance, taxes, and depreciation. Of course, these rates would exceed those paid by town folk.[17]

As the Ohio executive discovered, utilities still needed to determine the best means of financing high lines. In general, managers preferred to have farmers compensate companies in advance to install wires, transformers, meters, and associated equipment. Moreover, ruralites needed to learn that, because of the extra expenses of providing equipment to them, they must pay (and not complain about) higher rates than urban customers enjoyed. They also had to employ more electricity-using equipment so their overall load (and bills) increased.

At the peak of farm wealth in 1920, utility managers heard (or read) about scattered efforts to deal with rural customers. Many uncertainties existed relating to the cost of service and the ability of farmers to employ more load-consuming appliances and equipment, suggesting that central station men would need to pursue further investigations of them.[18] But with two thousand farmers in Wisconsin already connected to systems and more asking for service, wrote an *Electrical World* editorialist, the demand for rural electrification did not constitute a short-term craze. In several other states, the problem of serving farmers had started to become acute. "Central-station managers cannot afford to ignore the situation," he concluded.[19]

Simultaneously, demand for electricity in rural areas came from a different source—from people taking part in the "back to the farm" movement. As early as 1913, a manager observed that an increasing number of urbanites, who had become accustomed to "city conveniences," had applied to utility companies for lines to their rural abodes, though the "investment cost is obviously the thing that stands in the way of supplying this service."[20] Another utility official in 1921 reported that the same movement stimulated "an insistent demand for the conveniences provided by modern electric service which were formerly enjoyed only by the residents of city sections."[21] Though not likely a major motivator for pushing rural electrification—since the population grew much more slowly in farming districts than in urban centers—the trend nevertheless attracted some attention and thinking about supplying power to these previously neglected customers.[22]

To provide for rural demand, some power companies strung wires to points where individual farmers (or groups of them) constructed extensions to their homes and workplaces. But too often, these farmers built inadequate lines that did not measure up to utility standards.[23] While NELA's Overhead Systems Committee realized that it needed to establish better standards, Grover Neff, the Wisconsin utility executive (and committee member) admonished his colleagues in 1921 to do more. He observed that the industry generally "did not look with favor on the building of farm lines or the supplying of electric services" to ruralites, regarding this service as "a troublesome and unprofitable business."[24] But he argued that "utilities have been somewhat asleep in not studying" rural electrification more effectively, and he urged them to cooperate with farmers to build better lines and gain their business.[25]

To pursue this work, NELA created a "Rural Lines Committee" in August 1921 with Neff as chairman.[26] The *Electrical World* complimented the organization for taking this step, noting that it would perform a "truly national service of immense social and economic value." Beyond that reason, rural electrification offers "a means of pulling him [the farmer] out of some of the economic difficulties he is facing and enabling him to employ labor saving devices to a degree comparable with other industries."[27] (By 1921, the farm recession had already hit hard, with many crop prices falling by more than half from their 1920 peak.) Committee members reported on at least one problem facing utilities as they considered serving farmers, namely, the lack of understanding of the actual costs of stringing wires to them. Apparently, everyone recognized that service in rural districts proved more costly than in cities, but little reliable data existed about maintenance and operation expenses. The committee would try to determine these costs, a *NELA Bulletin* article noted, though the author also cautioned, "Unless rural electric service is worth more than it costs[,] it should not be supplied."[28]

Neff campaigned to help his colleagues realize that farm service could have greater value than commonly believed. He argued in early 1922, for example, that growing rural demand for power should not be viewed as simply a "fad due to the war prosperity of the farmer," partly because electrical use could reduce labor and animal expenses, thus making the cost of electric power justifiable to agriculturalists.[29] More significantly, he warned that rural electrification should be undertaken by utility companies before farmers developed their own approaches to obtain service. He specifically pointed to the means by which ruralites haphazardly strung up telephone lines, without adhering to the communications industry's best practices. A similar effort to bring electricity to the farm "would be a menace that no electric light and power company can afford to incur."[30] He therefore urged his colleagues to address farm electrification in the same manner they had introduced electricity into factories fifteen or twenty years earlier, when they helped manufacturers transition from the use of steam and water power to electricity. By taking heed of the lessons learned in that experience, central station companies could avoid mistakes that might occur when electrifying farms.[31]

In another address to his colleagues, Neff integrated arguments of social equity and good business. At the end of his presentation of the Rural Lines Committee report at the 1922 NELA conference, he observed that from an ethical point of view, the industry "cannot electrify the cities and small communities and leave nine tenths of the area of this country and one-fourth of its population without electric service." After all, he continued, "farmers constitute a large and important part of this nation and must be given proper and deserving attention."[32] But this notion of fairness also had a practical side, since neglecting farmers early in the 1920s would drive rural citizens to find other means to obtain electricity, such as by building their own distribution systems that would ultimately connect to the utilities' grids. The result of such a program would be disastrous, because farmers would control (within a decade) "ten times as many miles of distribution system as would be owned by the regular utilities."[33]

In an eerily prescient manner, Neff warned his colleagues of what ultimately occurred through the later intervention of the TVA and REA. "If the utility companies neglect to develop a satisfactory and reasonably standard fundamental plan for supplying rural service[,] some other agency will be forced to work out a plan for such service[,] and this work will be undertaken by those who are not as close to or as vitally interested in the subject as ourselves."[34] J. C. Martin, *Electrical World*'s western editor, reiterated Neff's concern, observing that unless utilities supply power to farmers, they will obtain service another way. "The question now is whether we will be able to come to a definite plan, to give the service on a satisfactory basis," he continued, "or whether they will be left to get it as best they can."[35] Further supporting this argument with more nuance, another manager argued that as farmers saw more of their urban cousins get power and as transmission lines passed over their land (without providing service to them), they would ultimately compel the regulatory commissions or legislatures to force utilities to sell electricity to them. But the government bodies would do so in a way that would disadvantage the power companies. Utility firms, he admonished, needed to take the lead in rural electrification.[36]

Public Power Concerns

Little doubt exists that Russell, Neff, and like-minded managers firmly believed in the value—to farmers and power companies—of increasing the size of the rural market. But some utility men also sensed that rural electrification fit into a larger framework of challenges faced by the commercial power industry. In particular, they realized that the neglect of farmers might spur political and social pressure on utilities that would harm the industry's standing.

Apprehension about government control of the central station power business had a substantial history by the 1920s. Emerging during the Progressive Era of American politics, the rapidly growing electric utility business invited comparison to the railroad industry. Apparent abuses by some transportation firms, which had become de facto monopolies that charged high freight fees in several markets, spurred establishment of federal and state regulation.[37] To prevent similar corporate exploitation of electricity customers, policymakers pursued two approaches. In one, states established regulatory commissions (or expanded the scope of already existing railroad oversight bodies) to supervise companies' operations and rates. But even before regulation of electric companies became commonplace after 1907, many city governments adopted an alternative approach, by creating utility departments to provide electricity in the same way as they supplied water and other necessities.[38] Reformers and muckraking journalists often touted the benefits of these "public power" (government-run) entities, which paid no dividends to shareholders and which remained immune (in theory) to corruption, suggesting that they more closely aligned their interests with those of the people they served. By 1900, several metropolises, such as Chicago, Detroit, and Columbus, Ohio, had established electricity providers.[39] Many smaller cities joined the public power movement; in the decade after 1896, the number of municipal electric systems tripled.[40] Seattle and Los Angeles joined the ranks of public system operators in 1910 and 1917, respectively. From 815 in 1902, the count of city-run power providers jumped to a peak of 2,581 in 1922.[41]

While growing in numbers, these public power systems declined in overall influence as private electric companies consolidated with others and extended their reach beyond individual cities. Exploitation of technological change made some of this growth possible: since the 1890s, engineering firms developed means to transmit electricity at higher voltages, as noted earlier in discussions of transmission from Niagara Falls and from the Sierra Nevada mountains. During World War I, electric utilities began interconnecting with those in other cities as a way to provide emergency backup power and for improving overall efficiencies in converting fuel to electricity.[42] The ability to link companies in this manner also made management of formerly autonomous power systems feasible, encouraging efforts to integrate several small firms into large entities. Entrepreneurial managers often used the holding company structure, promoted by S. Z. Mitchell and his Electric Bond and Share firm, as an effective tool to make the power producers more financially and operationally efficient.[43]

Through the use of new transmission technologies and holding companies, commercial utilities flourished, especially as regulatory bodies in most states provided the appearance of keeping them in check.[44] Municipal utilities, by contrast, remained small and geographically constrained, offering power to a small fraction of the nation's customers: even during the heyday of their existence in 1922, municipal utilities served only about 13 percent of all electricity users and earned just 8 percent of total revenue from electricity sales.[45]

Despite their firms' apparent dominance, leaders of the commercial central station companies viewed the threat of public power as real and significant. Advocacy groups (such as the Public Ownership League of America), academics, and policymakers kept the threat alive by asserting the superiority of government-owned utility networks.[46] Meanwhile, new challenges emerged occasionally. Most important, perhaps, the federal government made inroads into controlling the nation's waterways—the source of energy for hydroelectric plants—during the first decades of the twentieth century.[47] In some cases, public power advocates cleverly introduced the idea that government-run electricity-producing entities could serve rural citizens in a way that profit-seeking power companies would not. By doing so, promoters joined the ideas of rural electrification with public power and put utilities on the defensive.

George Norris of Nebraska, a Progressive Republican US senator and backer of President Theodore Roosevelt's efforts to regulate big businesses, was an ardent advocate for public ownership of hydroelectric resources. Since before World War I, when Congress debated a proposal to dam the Hetch Hetchy valley of Yosemite National Park to supply water to the growing San Francisco population, Norris had argued that river systems constituted national treasures that should not be relinquished to private industry. Though he endorsed construction of the dam in 1913 legislation (something that environmental preservationists opposed), he fought against the interests of commercial power companies, especially Pacific Gas and Electric Company, which he felt would obtain favorable treatment for making and transporting power.[48] Norris continued supporting public ownership of government-protected resources by effectively blocking Henry Ford's attempts (begun in 1921) to purchase the Muscle Shoals dam on the Tennessee River.[49] Built during the war to produce nitrates for manufacturing explosives, the hydroelectric facility became the source of congressional controversy; the senator led Congress to pass legislation for establishing a federal power system in the Tennessee Valley, with the dam as the centerpiece, but Presidents Coolidge in 1928 and Hoover in 1931 vetoed the initiatives.[50]

In his opposition to private takeovers of water resources, Norris expressed misgivings about the "power trust," made up of large corporations that, according to critics in the 1910s and 1920s, held excessive control over society.[51] In particular, Norris referred to large manufacturing and financial firms (such as General Electric) and central station power companies, which allegedly charged high prices that hindered the socially beneficial potential of widespread electrification.[52] The same distrust of power firms spurred him to suggest, in 1925, that the Federal Trade Commission (FTC) scrutinize utility company practices. His idea gained support in 1927 from Thomas Walsh, a Democratic senator from Montana, and in the following year, the FTC started examining utilities' lobbying and public relations efforts to influence attitudes about government-owned and for-profit power entities. The investigation ultimately yielded useful information for utility industry critics.[53]

Norris harmonized in his public power views with Gifford Pinchot, who had served as the first Chief of the US Forest Service under President Theodore Roosevelt and as a member of the Country Life Commission. Because of Norris's work to gain public control of natural resources, Pinchot in 1917 asked the senator to serve as a director of the National Conservation Commission, an organization created in 1909 largely to support federal initiatives in water power and mineral leasing.[54] Unable to take the position, Norris nevertheless worked with Pinchot and supported the latter's unsuccessful campaign to become Pennsylvania's US senator in 1914.[55]

Winning election as the state's governor in 1922, however, Pinchot demonstrated his reformist inclination by proposing a public power plan known as "Giant Power," which incorporated the advancement of rural electrification. Wedding Progressive Era philosophy about government's ability to enhance the lives of everyday people, his distaste for the power trust, and new opportunities made possible by evolving transmission technology, Pinchot sought to establish a novel system: generating plants located near coal mines would produce power for transmission by high-voltage lines to population centers.[56] Among other efficiencies, the approach would reduce the need to haul coal via rail lines to power plants in (or near) cities, a practice that caused logistical problems during World War I.[57] Morris Cooke, an engineering consultant, served as the plan's chief architect, having already earned his activist credentials by forcing the Philadelphia Electric Company to reduce rates in 1916.[58] Both he and the governor felt that the state regulatory body had not done enough to limit the expansion of profits by what appeared to be a "growing electric monopoly" (as Pinchot called it a few years after the public debate).[59] "Either we must control electric power," Pinchot wrote in the Giant Power report in 1925, "or its masters and owners will control us."[60]

Significantly for rural citizens, the proposed system would make electric service more widely available.[61] Given the apparent disinterest by private companies to supply power to farming districts, Giant Power would depend on creation of citizen-owned distribution companies.[62] These organizations would use their members' manpower and other local resources (along with borrowed money) to construct distribution lines and install auxiliary equipment, such as transformers, and then partner with state government to purchase electricity from Giant Power transmission companies. Rural customers would therefore get electricity, but outside the framework of the established utility industry. As one Pennsylvania rural service advocate, Harold Evans, observed, "If the private companies do not do it [rural electrification], public ownership will be tried."[63]

Some of these membership groups already existed in the United States. Farmers in Petersburg, Illinois, enjoyed success in 1911 when building thirty miles of lines that served fifty farms. Paying a minimum charge of $2 per month (in addition to the cost of the poles, transformers, lightning arrestors, and associated equipment), the owners bought power from a nearby central station company. When the utility planned to build a transmission line between two communities, other farmers offered free access across their properties if they could install step-down transformers, bought by the landowners, and obtain power for their own use. The utility agreed, and it sold electricity at "town rates." According to a census report, the arrangement worked well for the firm because all the equipment was "owned by the customers, so that the company is without investment on their account." But farmers benefited too, because they "fitted up their places with electrical equipment, installing pumps, motors, fans, irons, bathroom heaters, and other appliances."[64]

Such arrangements succeeded when farmers had enough capital to invest in distribution lines and when they could manage the system effectively. To rationalize this process, some farmers in the early twentieth century established cooperative consumer associations, also known as co-ops.[65] Based on models in American agriculture (pursued by the Granger movement starting in the late 1860s, for instance) and by practicable examples in Europe and Canada, the co-ops raised capital with membership fees and often purchased wholesale power from nearby central stations. They then sold electricity at attractive prices, but high enough to produce a surplus for maintenance. One co-op in Idaho, formed in 1919, required its original eighty-two members to pay membership fees of $125 each, which enabled it to build nineteen miles of distribution lines. It purchased power at an average rate of 2.5 cents per kWh and sold it at a rate comparable to that paid by city customers. Beyond charging reasonable prices, the organizations returned any excess above costs to the farmer-owners.[66] Before 1935, forty-five similar co-ops existed in the United States, aided by establishment of laws in a few states that enabled the associations to operate legally as nonprofit entities. The preponderance of them distributed power in Idaho, Washington, Iowa, Minnesota, and Wisconsin. In Washington State, the co-ops benefited from purchasing power from nearby municipal utilities at relatively low prices.[67]

But not all co-ops fared as well. In fact, most of these organizations failed, including several established in the 1920s in Ohio, Pennsylvania, Iowa, and Nebraska. As noted by the Federal Emergency Relief Administration in May 1935, many co-ops lapsed because of poor financial and operational management.[68] In one dramatic case, an Ohio co-op lacked sufficient capital to maintain its lines, such that a wire-bearing pole collapsed during a storm and killed a child.[69] Additionally, laws in several states did not permit the creation of the nonprofit organizations.

And, of course, the co-ops suffered because private power companies discouraged their existence. The government report noted some examples in which utilities sold power to the co-ops at such high rates that members could not afford to care for the lines, giving companies the opportunity to purchase them at a discount. Some of the firms then raised rates or discontinued service on unprofitable lines.[70]

Pennsylvania did not get the chance to employ the co-op concept because of the ultimate defeat of the Giant Power proposal in the state legislature in 1926.[71] Critics successfully attacked the initiative, arguing, for example, that it constituted an effort to convert a private industry into a state-run enterprise.[72] The utility trade journal, *Electrical World*, branded the plan as "radical" and paraphrased the governor as saying, "If the present trend toward monopoly by electric utility companies is not soon checked, . . . the people in self-defense may be forced to public ownership of these utilities."[73] One utility official attacked the governor on grounds that the plan simply constituted a means for him to gain support for his ambitions to become the US president. He attempted to discredit the governor by associating Pinchot with public-ownership advocate Carl Thompson, described as a "socialist" and "communist."[74] The governor's promise to bring electricity to farms drew censure too, since (according to opponents of Pinchot's plan) the utility industry already provided power to farmers who consumed enough electricity to make the use practical for customers and companies.[75]

Supporters of Giant Power and other public power plans, such as intellectually influential members of the Regional Planning Association of America, often pointed to the apparent success of the government power network in Ontario, Canada.[76] Created in 1906, the provincial government's Hydro-Electric Power Commission (HEPC) initially derived power from a plant at Niagara Falls. Most notable for its impact among American utilities, the Canadian system, since 1911, explicitly took measures to bring power to farmers.[77] This work gained the attention of rural electrification proponents in the United States, especially after 1921, when Ontario offered subsidies of up to 50 percent for construction of transmission and distribution lines while also lending money to citizens to purchase electrical equipment.[78] Meanwhile, customers appeared to pay less for electricity than their counterparts on the American side of the border. In the eyes of many observers, the commission's work symbolized the great value of government control of electric power resources.

To combat the possibility of similar publicly owned systems emerging in the United States, the utility industry decried the HEPC as socialist and repugnant to American values. Funded by NELA, consulting engineer William Murray argued in a widely distributed report that private companies produced cheaper and more reliable power without the burden of government ownership—a type of management that "eliminates all incentive for gain and . . . initiative."[79] Though Ontario utility managers refuted the report, NELA officials continued releasing studies critical of the Canadian system. One publication asserted that the HEPC may have offered cheaper electricity but only because of unfair government advantages and by pricing power at below cost.[80]

In another effort to disparage the Canadian system, NELA paid University of Minnesota professor Earl A. Stewart (an agricultural physicist who directed the work done by his state's Committee on the Relation of Electricity to Agriculture) to travel to Ontario in 1925 and 1926.[81] As a supposedly disinterested academic, Stewart wrote a report that NELA officials described as "authoritative."[82] It noted that the provincial government subsidized the cost of electricity to rural customers, which explained why rates appeared lower than those in the United States. Moreover, the province-run utility only extended power to the "richest agricultural districts," which included some suburban areas that had a relatively large number of customers per mile, ensuring copious consumption and making the average distribution cost look smaller.[83] At least one editorial writer used Stewart's arguments as evidence against Senator Norris's claims that the Canadian power system constituted a "brilliant success." (The legislator drove to Ontario in the summer of 1925 to gain firsthand experience of the provincial system.)[84] "Personal investigation by Professor Stewart, a recognized authority upon application of electricity to farm uses," a 1927 *Boston Herald* news piece recounted, "shows that Ontario farmers have no advantage over American farmers."[85] Despite further rebuttals from Canadian authorities to Stewart's report (and others that followed), NELA continued to denigrate HEPC's rural electrification work, as did prominent utility officials.[86] Wigginton Creed, president of California's Pacific Gas and Electric Company, observed that Ontario's farmers paid twice as much as farmers in his territory while "obtaining precisely the same service."[87] (Of course, Golden State farmers consumed huge amounts of power for irrigation purposes and therefore paid exceedingly low rates.)

With similar intentions to discredit the idea of public power, the private utility industry attacked other ventures. In 1924, for example, it focused its ire on a proposal to extend service from the Seattle and Tacoma municipal electric systems to rural areas beyond city lines. Arguing that such extensions would have served as "the entering wedge for a much wider program of putting the State into the light and power business to supersede private enterprise," public relations officials of a regional industry association worked diligently to sway voters.[88] The successful effort to defeat the plan may have benefited from the influence of newspaper editorialists, some of whom argued that ruralites would end up paying for unprofitable investments made by city utilities. "The farmer is again the goat," observed the Shelton (WA) *Mason County Journal*; worse, the initiative constituted "a bit of socialism which has for its real object the advancement of state ownership."[89] An editorial in a rural California newspaper commented on the trend of ballot measures to give increasing control over the electric power system to municipalities, and it advised readers to vote against them.[90]

Despite defeating several public power propositions, utility men realized that the threat of government intervention remained real, especially in the realm of rural electrification. NELA's director of publicity, George F. Oxley, said as much when he communicated with a Nebraska colleague in February 1924. Commenting on efforts within the state that sought, like those in Washington, to erect power lines from municipal organizations to rural customers, Oxley observed that "power companies sooner or later must extend their lines into rural communities" though they should not do so at a loss. But if companies do not supply the power, then farmers will eventually demand some form of government entity to supply it to them—an undesirable outcome for the private industry, obviously.[91] A utility official in 1925 noted candidly that his colleagues supported rural electrification efforts as a public relations move, in part because of his feeling "that the next political and economic cyclone will probably originate on the farm, and . . . we want to be pretty good friends of the farmers when that starts."[92]

Happily, observed Oxley, the industry had already begun to address this problem by creation of state affiliates to the Committee on the Relation of Electricity to Agriculture (see next two chapters for details.) The organizations had begun partnering with several stakeholders, and their establishment constituted "a good piece of public relations work" to examine farm applications of power that would make the business profitable.[93] Earl White, director of the national CREA, harmonized with the value of such work. Attending an NELA committee meeting in 1927, he observed "that rural electrification is one of the most important public-relations activities and public-relations problems of to-day."[94] The twenty-three state CREA organizations then in existence, he noted, had been carrying out investigations that should build electricity consumption on farms in such a way that the rates charged to them will become more reasonable. To help maintain the research and public relations elements of the CREAs' work, utility companies should appoint at least one person in each firm to work with farmers, White urged.

He also encouraged NELA to continue advertising in agricultural publications. Other industry men concurred, noting that rural electrification should "be considered a very important part of the public-relations-work."[95]

In sum, the demand for electrical service to farmers appeared to cause a public relations quandary for utility officials who also confronted attacks by public power advocates. Utility men responded—at least in part—by studying problems associated with farm electrification as early as 1910 and by establishing the CREA in 1923. In thanking a bulletin editor for sending him an article describing the organization's formation, NELA's Nebraska director of public relations in 1926 candidly confessed, "As strange as it may seem[,] the electric industry is not anxious to have the farmers calling for electric service" because of the great expense to establish it. However, he also noted that "when public opinion becomes centered on any subject[,] it demands careful and thoughtful attention"—attention that would be provided by further research on rural electrification.[96]

The Global Perspective on Rural Electrification

Focusing on American political pressures helps explain some of the utility industry's motivation for seeking, albeit less-than-enthusiastically, greater rural electrification in the 1920s. But the movement to extend power lines to farmers did not only occur in the United States. In several other technologically advanced countries, rural advocates promoted efforts that can be viewed as an international phenomenon. In other words, political and industry leaders throughout industrial nations had become attuned to the benefits of providing central station power beyond the cities. The pursuit of rural electrification occurred within a global dialogue, one in which American utility managers participated.

Opponents of farm electrification in the United States could have gained encouragement from events occurring in England in the 1920s. Suffering a period of "rural stagnation" after World War I,[97] England operated a fragmented power network, which improved when the government established an integrated and more efficient national grid in 1926.[98] But progress in rural districts did not come easily, in large part because of farmers' supposed reluctance to use much electricity. The country's most prominent farm electrification champion, R. Borlase Matthews, acknowledged as much. An electrical engineer who tested various applications of electricity on his six-hundred-acre estate, Matthews observed, "The fact is often entirely overlooked that agriculture is a very speculative business, and further, the farmer has lost much money in trying out new inventions." Moreover, he commented, since the farmer "does not properly understand mechanical contrivances—especially if they are electrical—it is but little wonder that he is a bit chary when electrical schemes are propounded to him."[99]

Matthews nevertheless thought electrification of farms would ultimately prove beneficial, and he spurred the Royal Agricultural Society of England to examine the subject.[100] A research report published in 1924 concluded that not enough evidence existed to verify claims that electricity stimulated the growth of seeds nor that electricity could sterilize milk economically, for example. Neither did high-voltage electric currents enhance crop production more than other methods (such as plant breeding or improvements in loosening the subsoil).[101] And while electricity could power small motors, which clearly had advantages over gasoline engines in terms of simplicity, cost, and robustness, they would only be used for a few hours each week, making them perhaps less valuable than horses or tractors. Meanwhile, Matthews noted that the consumption of electricity on farms—for lighting and related amenities (such as for irons and some household appliances)—would remain "absurdly small," making central station power service to farms unattractive.[102]

Most English authorities seemed to agree with this assessment. Harmonizing with the view of many American utility managers, an English critic of rural electrification via distribution lines believed that a more promising avenue consisted of using isolated power plants to provide electricity for lighting and occasional motor use.[103] And while electricity eventually found its way through the national network to small rural towns, the lines rarely extended to farms. By 1942, only about 25 percent of farms in England and Wales had access to central station power, and more than half of them used electricity for just lighting and a few household appliances.[104]

Outside of England, however, the practice and prospects of rural electrification work appeared more encouraging, as utility officials related at the first World Power Conference, convened in London in June and July 1924.[105] In one conference session, speakers from Denmark explained that their country had a long history of financing agricultural pursuits through cooperative societies, with the first electricity co-op established around 1900. These groups of ruralites pledged their farmlands and building assets as collateral for obtaining loans from banks and private investors, enabling large numbers of farmers to obtain power for lighting and motors. In an unanticipated fashion, the organizational work helped establish a class of consulting engineers who further advanced rural electrification work in the country.[106] By 1924, about one-third of farms and rural areas obtained electricity, even though supplying power in the small country evidenced similar problems that power distribution entities discovered elsewhere—namely, that it required a considerable investment to serve sparsely populated customers.[107]

Other European nations also demonstrated increasing commitments to rural electrification. In Norway, the abundance of waterfalls and rapidly flowing water provided the resources for several hydroelectric plants. As early as 1906, the government took steps to retain control of them and to distribute power to a dispersed population.[108] With more than half the country's population living in rural areas until after World War II, the nation's farmers held significant political power, and many gained access to electricity.[109] A similar environment existed in Switzerland, also rich in water power, where residents (rural and urban) had largely displaced all other lighting systems with those operated by electricity in the early 1920s.[110] And even in countries that depended on fossil fuels, many seemed to offer electricity to a greater proportion of farmers than in the United States. A 1934 survey of rural electrification sponsored by the US Department of Agriculture observed that about 60 percent of German farmsteads obtained central station power in 1927. Meanwhile, about 71 percent of French communes (the equivalent of civil townships and municipalities, half of which remained rural) received electric service in 1930.[111]

In many of these and other countries (such as Canada [in Ontario], Australia, New Zealand, Denmark, Finland, Sweden, France, and Czechoslovakia), governments took a leading role in owning, subsidizing, or regulating electric power systems. Several provided electrical services as a supplement to those offered by private industry. Frequently, planners in these nations viewed the creation of interconnected electrical systems (serving urban residents, industries, and rural customers) as elements of national security and industrial policies.[112]

The political importance of providing central station service for farms grew in Russia too, with the Soviet government extending power lines beyond cities. Historian Jonathan Coopersmith noted that rural generating stations "served as visible signs of progress, showing the peasant that the interests of the people were the interests of the party." The national government provided technical and administrative assistance to regional electrification companies, funded by large industrial firms and cooperatives, such that 651 rural stations existed by the end of 1926. Along with power that came from seventy-seven factory plants, the generation capacity for farm use totaled 17.9 MW in that year.[113]

Local and unique situations spurred rural electrification in some countries. In Sweden, the onset of hostilities during World War I (despite the country's neutrality) caused it to lose access to traditional supplies of oil and coal, which it used to

produce most of its power. The nation turned to exploiting its hydroelectric resources, located in rural areas, which provided power to nearby farms, such that
40 percent of the country's tilled land (not necessarily 40 percent of farms)
obtained access to electricity by 1925.[114] And in Germany, where about half of
farms already enjoyed electrical lighting in 1921, the enthusiasm for purchasing
electrical wires and equipment grew during the country's hyperinflationary period
in the early 1920s. According to one observer, the rapid loss in the purchasing
power of the nation's currency "encouraged farmers to spend their money [for electrical wares] instead of hoarding it."[115]

At the same time, national governments often provided services for rural residents to promote establishment of electric cooperatives (as was done in Denmark).
The organizations benefited by avoiding the capital costs incurred when building
their own power plants, since they purchased power generated by government-
owned and private companies, as was the case in Sweden. In Germany, co-ops
also took advantage of the ability to erect distribution lines from one farmhouse
to another over private property, in a way that would likely not be permitted if for-
profit companies owned the wires.[116]

Farm electrification advocates in the United States employed examples of foreign accomplishments as rhetorical and political weapons. Ontario's HEPC served
as the most potent example during the 1920s because it provided power from the
same hydroelectric source at Niagara Falls as an American utility and because of
obvious similarities between the United States and Canada. In an influential 1934
report on revitalizing the Mississippi Valley, Morris Cooke spotlighted Ontario's
public utility system and the value of government's involvement in stringing lines
to farms.[117]

Soon after its creation, the Rural Electrification Administration promoted itself
as a way to help the United States gain parity with the more enlightened countries
in the world, often by shaming the private power industry. Perhaps expressing
widespread racist attitudes, Morris Cooke in a 1935 radio address noted that
Americans commonly viewed "Orientals as backward." Nevertheless, "in Japan,
over nine homes out of every ten are benefited by electric service." He also
observed that (by selectively choosing facts) "in Canada, we find the publicly
owned Ontario Hydro-Electric System . . . providing current free of charge in rural areas to operate electric washing machines, radios, and electric household
pumps."[118] And in a 1936 pamphlet that provided answers to those seeking to start
electric cooperatives, the REA indicated that only about 12 percent of farms in the
United States enjoyed central station service. That number compared unfavorably

to Germany, where 90 percent of ruralites obtained power. Farmers in Belgium, Holland, and Switzerland received "almost universal service," while more than half did in Sweden, Norway, and Denmark. (The REA publicized foreign superiority in rural electrification with the graphic shown in figure 7.3.) Even New Zealand served more than half its farmers with electricity. Providing a little nuance to these numbers by observing that the "rural population of most of these countries is more congested than that of the United States," the publication nevertheless asserted that "the American farmer should share in the relative prosperity and high standards of living" afforded by widespread electricity use.[119]

That refrain appeared as the preface of an article written in the 1940 *Yearbook of Agriculture* (published by the US Department of Agriculture, into which the REA had been absorbed in 1939). REA economist Robert Beall observed that 95 percent of French farm families received central station power in 1935, while the number reached 85 percent in Denmark and 100 percent in Holland. Ninety percent of German and Japanese farms obtained electricity as well, he noted. In the article's

RURAL ELECTRIFICATION
IN VARIOUS COUNTRIES

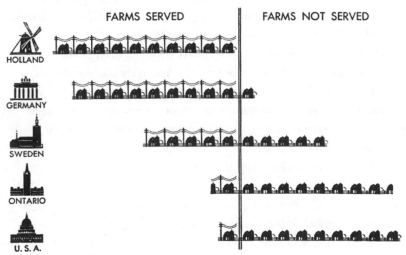

Each farm represents 10% of all farms

Fig. 7.3. REA graphic suggesting that electrification of American farms lagged behind that in other countries. *Source: Rural Electrification News* 1, no. 8 (April 1936): 7.

first sentence, the author made the significant point that "[t]he most advanced country in the world in the use of modern methods in industry and agriculture . . . has lagged astonishingly in making electricity available to farm communities." Beall qualified this assertion (on the second page) by observing that conditions in these countries differed from those in the United States, including "density of population, type of farming, per capita income, and form of government."[120] Even with these caveats, the publication made it appear that rural electrification in America lagged woefully behind that in other countries.

Rural electrification did not top the list of priorities for power company managers in the United States. Utility executives realized that serving farms would be expensive and not immediately rewarding in most cases. Nevertheless, an undercurrent movement of support for farm power emerged, spearheaded by people like Grover Neff. Moreover, leaders of the movement tapped into, consciously or otherwise, other forces and circumstances surrounding them. When important political figures (such as Senator Norris and Governor Pinchot) advocated government intrusions into the power industry—with explicit ramifications for rural electrification—industry officials felt obliged to respond publicly and effectively. They did so in part by establishing formal committees, starting in 1910 within NELA, to examine the farm electrification problem more systematically. The people involved in these committees took their work seriously, as the detailed reports produced for NELA and other publications suggest.

These informational and research efforts assumed even greater meaning as American utility engineers and managers looked at their counterparts' work in other countries. While all nations faced similar economic challenges in electrifying sparsely distributed farms, many developed responses that enabled larger percentages of rural populations to obtain power. Of course, government support proved critical in some countries where rural electrification constituted elements of political, social, and industrial policies in ways that did not comport with American notions of free enterprise.[121] Nevertheless, the publicizing of such experiences bolstered utility critics' arguments that private firms disadvantaged rural citizens in the United States by controlling too much of the electric system.

Not surrendering to private power detractors, company officials took issue with rural electrification advocates. Aside from their rebuttals to claims made about the Ontario power system in the 1920s, utility leaders responded emphatically to assertions made by Morris Cooke, for example. In a 1935 article, trade association manager Howard Bennion countered the REA administrator's claims, such as the assertion of Japanese supremacy in electrifying farms, which he argued compared dissimilar data. (Cooke noted that 90 percent of *all* [rural and urban] Japanese homes had been electrified, while only 10 percent of American farms received central-station power.)[122] He also reiterated the notion, made by Canadian researchers in 1924, that the densities of rural populations differed in other countries, making comparisons between the United States and elsewhere unrealistic.[123] Overall, Bennion argued that America had not become "a backward nation in rural electrification," despite Cooke's claims that drew on incomplete or even nonexistent statistics from other countries. Instead, American power companies had pursued rural electrification efforts vigorously and effectively.[124]

But such subtle arguments did not carry much weight in the public discourse, and utility managers likely felt considerable pressure to electrify more farms, or at least to explore visible ways to do so. Perhaps the greatest impetus to accelerate rural electrification initiatives came from industry critics who pointed to the work done in the Canadian province of Ontario. Indeed, historian Ronald Kline has argued plausibly that creation of NELA's Rural Lines Committee in August 1921 "hardly seems coincidental" with Ontario Hydro's measures (the result of a law passed in April 1921) taken to construct additional transmission and distribution lines to rural areas and to subsidize farm electrification.[125]

In short, concerns about the public power movement and claims about the "success" of foreign countries' work on rural electrification understandably put utility managers on the defensive. These pressures suggested that even modest attempts to advance rural electrification work, such as creation of committees within NELA, would have significant public relations value.

Chapter Eight

The Industry Organizes the CREA

There are no data available now to show the exact part that electric service can play in increasing farm efficiency.

—*ELECTRICAL WORLD* EDITORIAL, 1922

While it may seem obvious today, the practical value of electricity had not yet become evident to farmers in the 1920s. To be sure, most of them acknowledged that the energy form would illuminate rooms safely and power pumps to bring water into their homes. In addition, electricity would enhance the lives of farm women, allowing them to enjoy the conveniences of urban life, such as machines that cleaned clothes and cooked meals. Electric power would make life more enjoyable for children too, reducing their incentive to flee to the city when they grew up.

But farmers, especially the millions who suffered economic hardships in the 1920s, did not necessarily view these advantages as worthy of investments in a distribution line and new electrical equipment. Even when some ruralites paid for wires to their properties, they did not necessarily use electricity productively to do much more than light their houses and barns. In fact, their patterns of consumption vexed and disturbed the experts, as noted by Wisconsin agricultural engineers in 1923: "Too many farmers are literally 'killing the goose that laid the golden egg,'" they wrote. "That is, after they have gone to the expense of from $300 to $500 a piece to pay the cost of building a line through their territory, they hang a 50 watt bulb at the far end when it is completed." At the same time, these researchers highlighted an enigma faced by farmers; they observed that "there is not enough electrical machinery developed at the present time which the farmer can use profitably to make it possible for him to become a large consumer."[1] In other words, while farmers may have known of potentially labor-saving electrified equipment, they lacked confidence in its economic viability.

Rural electrification proponents within the power industry sought to resolve the quandary by creating the national Committee on the Relation of Electricity to Agriculture (CREA) and its state affiliates. The establishment of these groups marks the industry's commitment to pursue electrification of farms through research and demonstration. The effort may have been stimulated by increasingly vocal criticism from outside the industry. Even so, some managers, such as Grover Neff, appeared to harbor pure motives, and they applied scientific approaches to overcome hindrances to widespread use of electricity on farms. Cleverly, the rural electrification proponents established relationships with other stakeholders, yielding more substantive outcomes than could have been achieved by the industry alone.

Formal Organizations Created

The undercurrent advocates of farm electrification won institutional victories with the creation of formal research groups. As noted earlier, the National Electric Light Association established the Committee on Electricity in Rural Districts, which metamorphosed into the Rural Lines Committee in 1921, chaired by Grover Neff. Perhaps motivated by fears of public power expansion, rural electrification promoters within the industry took further action: on 11 September 1922, committee members met with leaders of the American Farm Bureau Federation (AFBF), a national advocacy organization founded in 1920.[2] The AFBF had strong connections with the extension services of land-grant colleges and appropriated many issues of modernization that had been pursued by state and local farm associations since the early 1910s. According to historian James Shideler, the AFBF emerged as part of a movement that aimed to acquire "for agriculture a place in the business world."[3] At the end of 1922, the federation represented forty-six state bureaus and claimed 772,634 individual members.[4]

At the meeting, NELA committee members admitted that the "demand for rural electric service has produced an acute and embarrassing situation for the electrical industry due largely to the new and undeveloped state of the use of electrical energy on the farm."[5] (The word "embarrassing" in the statement suggests that utility leaders recognized the significance of growing criticism.) Perceiving a need to involve other stakeholders, NELA and AFBF agents met on 8 March 1923 with representatives of the US Department of Agriculture and the American Society of Agricultural Engineers (ASAE), who agreed to create a national organization that would coordinate various groups' work on farm electrification.[6] That effort gelled with the establishment on 11 September 1923 of the national Committee on the Relation of Electricity to Agriculture, consisting of twelve members. Three acted on behalf of the AFBF, four came from NELA, and one each represented the American Society of Agricultural Engineers, a trade group of isolated electric plant manufacturers, and the Departments of Agriculture, Commerce, and Interior.[7] (In 1924, NELA added two members, and the National Grange won initial representation.) The Farm Bureau Federation's original membership on the committee included its second president, Oscar E. Bradfute, and its secretary-treasurer, J. W. Coverdale. Grover Neff, one of NELA's representatives, also served as CREA's secretary-treasurer. Earl A. White was appointed CREA's director, receiving a salary paid by the supporting groups and office space in Chicago provided by the AFBF.[8]

The choice of White (figure 8.1) to direct the national CREA appears judicious. After receiving the nation's first PhD in agricultural engineering from Cornell University in 1917, White taught at the University of Illinois for seven years and became an expert in the mathematics of tractor plows. A published biographical sketch commented that "Dr. White probably has done more than any other man . . . to bring plow design from an art to a science."[9] In 1921, he won election as president of the American Society of Agricultural Engineers, an organization founded in 1907, whose members labored at land-grant institutions and a few Canadian schools that had similar educational traditions.[10] With such an academic and professional background, White held prestige as a formidable and presumably unbiased researcher. NELA's leaders chose a professional who could rapidly establish credibility for the newly formed CREA.

As a leader of the ASAE, White belonged to a community of agricultural engineering academics who sought to modernize farm life. Like other professionals early in the twentieth century, agricultural engineers pursued specialization as a way to distinguish themselves from others in the technical realm and to accrue social status.[11] Besides creating a professional society (the ASAE) that published research and convened meetings, the engineers performed studies in several areas—from civil engineering (focusing on rural roads, irrigation, and drainage, for example) to mechanical engineering (designing machinery powered by animals and engines) and electrical engineering (even though few farms had electricity yet).[12] In another form of specialization, the agricultural engineers established academic departments in universities. Iowa State University inaugurated the first agricultural engineering program in 1905 (as a "subdepartment" of the Agronomy Department).[13] Cornell University organized its department in 1909, originally called the Department of Farm Mechanics,[14] and Virginia Polytechnic Institute

Fig. 8.1. Earl A. White. *Source: Edison Electric Institute Bulletin* 4, no. 2 (February 1936): 49. Used with permission of the Edison Electric Institute.

followed in 1913. VPI's catalog observed that "the connection between agriculture and engineering has grown more intimate, until to-day many agricultural problems prove, upon analysis, to be essentially problems of engineering."[15]

Agricultural engineers shared the optimistic view that they could apply novel technical and management principles to enhance rural life. Earl White expressed such confidence in his 1921 ASAE presidential address, in which he observed that resolving problems on the farm constituted a "situation made to order for an engineer."[16] As noted by historian Deborah Fitzgerald, these new professionals believed they could bring agriculture into the twentieth century through the use of rational studies and controlled experiments.[17] The engineers saw farm electrification as one element of the modernization process. As early as November 1923, the ASAE held a day-long "rural electrification program" at its annual meeting, with reports delivered by Grover Neff, manufacturing representatives, a professor, NELA's managing director, and the AFBF secretary.[18] In a self-serving manner, the engineers also saw work in the young field as a means to broaden their scope of practice and impact within society. Not afraid to have a little punning fun, the editors of ASAE's journal observed that rural electrification was "a field 'charged' with great 'potential' possibilities" that would not only raise farmers' standard of living but also help "in opening up a larger field of service for the agricultural engineer."[19]

The motivation and training of land-grant engineers made them perfect partners for the research and demonstration activities envisioned by CREA managers. While the organization provided the overall superstructure for rural electrification investigations, the schools offered the expertise and critical resources for doing the work. The land-grant colleges, after all, housed experiment stations and employed extension agents to bring new knowledge directly to farmers.[20] Perhaps because of the perceived importance of these schools for the success of the investigatory programs, CREA's state affiliates usually chose an agricultural engineering professor to serve as director and secretary, as was the case in Virginia, Washington, Idaho, North Carolina, Wisconsin, and elsewhere. And because of their work with farmers on other matters, the land-grant college agricultural engineers had supposedly established trusting relationships with them.[21] At the same time, the college-based professionals commanded the respect of manufacturers' representatives.[22]

CREA director White adopted many of the commonly held cultural attitudes of academics working in land-grant colleges. In particular, he valued good research and deliberation of results before undertaking new projects, as can be inferred from an article written about him in 1925. Though describing great potential benefits of electrified farms, especially for performing work previously done by numerous farmhands, White cautioned against bringing power too quickly to rural areas. The use of electricity in agriculture still remained in an experimental stage, such that "[o]ne of the dangers of the moment is that rural electrification may proceed faster than our ability to acquire the proper knowledge and technique regarding its efficient use." Before farmers would become willing to invest in electrified farm appliances, he argued, they needed to understand the costs of the equipment and of electricity so they could calculate whether the savings and returns justified expenses. Too often, he observed, farmers had been exploited and had "thrown away far too much money on half baked schemes frequently concocted by . . . so-called friends. In this undertaking," he cautioned, "we should not make the mistake of rushing in blindly and repent at our leisure."[23]

Stated Goals of the CREA and Formation of State Affiliates

When founding the national CREA, the organizers explicitly described several initial goals. Working with partners at the Department of Agriculture and NELA in particular, the committee sought to collect data on how farmers currently employed power—from steam engines, gasoline-powered motors, windmills, animals, and humans—and where opportunities lay for electrical technologies. The committee would identify the number and location of farms served by central station power (and by isolated plants) as well as determine the length and cost of distribution line extensions to farms. Another survey would obtain information on agricultural uses of electricity in other countries.[24]

But the chief activity of the committee quickly emerged as support for experimental research on farm uses of electricity that promised economic advantages to all parties.[25] As noted in a 1923 AFBF report recounting the creation of the CREA, the "one big problem . . . is to find the maximum economic uses of electricity in agriculture." Farmers sought electrification, but not "at the expense of having millions of dollars [sic] worth of partially developed equipment placed in [their] hands that will not stand the test and will not be of economical use."[26] The way to avoid such problems, it seemed, consisted of having the most intelligent people—namely, agricultural engineers—conduct practical research. That work, according to Director White, would establish the foundation needed to make rural electrification a practical reality. In the inaugural issue of the *CREA Bulletin* (published in September 1924), he noted that knowledge (or "know how") about farm uses of electricity remained scarce and that he and his colleagues needed to perform "systematic, intelligent and hard work. Merely running lines into the country will not solve the problem," he continued. "We have set out to find this 'know how' in an orderly, scientific manner."[27]

CREA's leaders framed the rural electrification problem as two questions: "(a) How service can be supplied to the farmer and what is involved in its establishment?" and "(b) How can service be utilized by the farmer so it will be profitable to him?"[28] To answer the queries, the national CREA depended on the state organizations to perform research. As the first one established (in September 1923), the Minnesota committee consisted of representatives of the state's Farm Bureau Federation, three power companies, three farmers (including one state senator), and the dean of the University of Minnesota's College of Agriculture. Earl A. Stewart, working in the Agricultural Engineering Department, served as the group's secretary and project director.[29] The Iowa committee, established in February 1924, operated administratively in a similar manner. Chaired by the president of the state's Farm Bureau Federation, it included members from the utility industry and had Jay B. Davidson, head of Iowa State College's Agricultural Engineering Department (and the ASAE representative to the national CREA), sitting as secretary. In a slight divergence from other states' practice, Frank Paine, an electrical engineer, supervised the first research project in Iowa, though Franklin Zink, an agricultural engineer, assisted him. Notably, because she appears to be one of the few women having significant standing in electrical research at the time, Eloise Davison, a professor of home economics at the state college, served as another project participant.[30] Similar forms of organization existed at the other state committees, ten of which had been organized by the middle of 1924. (Ultimately, groups in twenty-six states participated.)[31] While the national committee gave "purpose and general direction" to investigations, the state entities, with the land-grant college professors in charge, assumed responsibility for performing actual research.[32] (In a 1924 advertisement shown in figure 8.2, NELA highlighted some of CREA's objectives.)

The goals of the state committees appeared eminently legitimate and scientific, fitting the modern view of engineers and educators of the 1920s, who identified problems and sought rational solutions to them. Members of the Alabama CREA affiliate in 1925 described their state project's objectives eloquently: "To find facts, develop new uses, promote mutual understanding, disseminate reliable information, [and] substitute 'directed development' for 'cut and try' methods." Most important, "[u]nless ways are developed so that power companies can sell electric service on a sound economic basis and the farmer can use it at a profit, this work will not be a success."[33]

Resources for the cooperative efforts came from various stakeholders. The CREA received in-kind support from the American Farm Bureau Federation, such as Director White's office space in Chicago. NELA also provided funds to the national organization, which it then distributed to state committees. For example, the CREA sent annual payments in the range of $5,000 to $6,000 each to schools such as the University of Minnesota (starting in 1923) and Washington State College (beginning in 1926).[34] By the late 1920s and early 1930s, more than $153,000 went to the state bodies, reaching almost $39,000 in fiscal years 1930 and 1931 alone.[35] Additionally, state universities contributed through their colleges of agriculture and departments of agricultural engineering; power companies built test lines at no or low cost to the committees; manufacturing firms donated equipment and appliances as ways to build load and to develop favorable reputations among customers in a new market; and farmers often paid for wiring their farmsteads. (Table 8.1 provides a sense of the diverse funding sources for the Minnesota experiment, which received support in the amount of about $75,000 from the end of 1923 to the end of May 1928.)

A useful approximation of the overall funds spent for the organized efforts to promote research on rural electrification comes from the Rural Electrification Administration's Harry Slattery. In his 1940 book, *Rural America Lights Up*, the agency's third administrator cited E. A. White as having estimated "conservatively" that more than $1 million came from local power companies. States and colleges provided other resources, suggesting that the "real cost" of the CREA's work "may safely be set at $2,000,000."[36] NELA (and its successor organization, the Edison Electric Institute) spent $400,000 to support the committee's work.[37] In sum, Slattery calculated that CREA research costs amounted to about $2.4 million, which included the time and facilities offered by colleges and their staff. Gifts from manufacturers may have added another $1 million to this amount.[38]

Early CREA-Supported Efforts in Minnesota

The Minnesota committee provides a good example for understanding the initial research activities performed by state CREA organizations. Taking advantage of resources offered by the University of Minnesota's Agricultural Engineering Department and extension service, the state committee began, in September 1923, work on what it described as "the first experimental rural electric line in the world."[39] Managed by Earl Stewart (figure 8.3), an associate professor of agricultural physics who had previously taught at Kansas State Agricultural College,[40] the project sought to "determine the optimum economic uses of electricity in agriculture, and to study the value of electricity in improving living conditions on the farm."[41] Though rural electrification promised great opportunities to increase productivity, Stewart cautiously observed, "Investigation of the economic use of electricity on the farm . . . should precede the establishment and use of electrical equipment in order to insure maximum results at minimum cost."[42] As an integral part of the experiment, the local utility company, Northern States Power, extended a distribution line about six miles to eight farm customers near the town of Red Wing, Minnesota, with power flowing initially on Christmas Eve 1923.[43] Importantly, the committee gave some farmers electric ranges, water pumps, incubators, and brooders; others received laundry equipment, cream separators, and milking machines.[44]

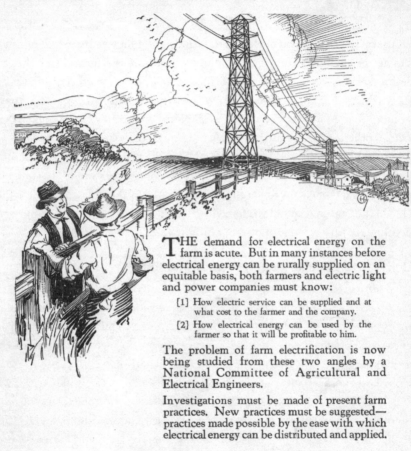

No Stone Will

THE demand for electrical energy on the farm is acute. But in many instances before electrical energy can be rurally supplied on an equitable basis, both farmers and electric light and power companies must know:

[1] How electric service can be supplied and at what cost to the farmer and the company.

[2] How electrical energy can be used by the farmer so that it will be profitable to him.

The problem of farm electrification is now being studied from these two angles by a National Committee of Agricultural and Electrical Engineers.

Investigations must be made of present farm practices. New practices must be suggested—practices made possible by the ease with which electrical energy can be distributed and applied.

NATIONAL ELECTRIC

Fig. 8.2. (*above and opposite*) NELA announces establishment of the national CREA. *Source: Agricultural Engineering* 5, no. 4 (April 1924): 92–93. Used with permission of the Edison Electric Institute and the American Society of Agricultural and Biological Engineers (ASABE, formerly ASAE).

Be Left Unturned

In accordance with the plan drawn up by the National Committee, these investigations are to be conducted by independent state groups of farmers who will have the cooperation of the country's leading agricultural engineers and of the state agricultural colleges.

In Minnesota fourteen farmers on the Red Wing transmission line are now applying electrical energy and keeping records of energy consumption, time, and labor. They are supervised by the engineers of the Minnesota Agricultural College.

Under the direction of the Alabama Polytechnic Institute, a similar experiment has been started in Alabama. A fact-gathering electrified farm community will be established in each of the leading agricultural sections of the State.

An experimental line has been built in South Dakota. The State College of Agriculture has drawn up an investigation program which the farmers will follow.

Kansas has recently joined the movement.

In other states similar groups are being organized.

Even a foreign survey is being conducted by the United States Department of Commerce to find out what is being done with electricity on European farms.

No stone will be left unturned.

Thus the investigation will include actual farm electrification in ultimately as many as ten or a dozen states and a survey of farm electrical practice all over the civilized world.

This is the most comprehensive, practical task ever undertaken by agricultural and electrical engineers and economists. When it is completed, farming will receive an impetus perhaps even greater than that imparted by the introduction of mechanical tilling and harvesting implements. For the result will not be simply more efficient, more economical farming, with a greater return for time and labor expended, but home and working conditions comparable with those of the city, and the richer life that comes with contentment.

The National Committee in charge of the work is composed of economists and engineers representing the American Farm Bureau Federation, the Department of Agriculture, the Department of the Interior, the Department of Commerce, the American Society of Agricultural Engineers, the Power Farming Association of America and the National Electric Light Association.

A booklet on the work of rural electrification has been published by the National Committee. Read it and pass it along to your neighbor. It will be sent free of charge. Write for it either to Dr. E. A. White, care of the American Farm Bureau Federation, 58 East Washington Street, Chicago, Ill., or the National Electric Light Association, at 29 West 39th Street, New York City.

LIGHT ASSOCIATION

Table 8.1.   Contributors to the Fund for Expenses of the Minnesota
CREA's Red Wing Project, September 1923–1 June 1928

University of Minnesota	$9,374
State Committee on the Relation of Electricity to Agriculture	$17,500
Northern States Power Company	
Line costs	$12,308
Meters loaned	$1,000
79 manufacturing companies	$21,632
Cooperating farmers at Red Wing	$13,344
Total	$75,158

Source: E. A. Stewart, J. M. Larson, and J. Romness, *The Red Wing Project on Utilization of Electricity in Agriculture* (N.p. [St. Paul?]: University of Minnesota Agricultural Experiment Station, no date, but likely 1927 or 1928), 2, University of Minnesota Libraries Digital Conservancy, http://conservancy.umn.edu/handle/11299/48685, accessed 15 September 2014.

Collecting extensive data—not only on electrical use but also on farm finances—the researchers learned that power consumption increased dramatically over four years, as did electric bills. But happily, expenses for farm tasks fell more than electricity costs rose, yielding productivity improvements and net income.[45] Studies of individual tasks also demonstrated great labor economies; electrically operated water pumps, for example, saved the equivalent of a month of eight-hour days, since people did not need to draw water and carry it by hand from outside the home or barn. Farmers conserved a similar amount of time by not needing to fill and clean kerosene lamps and lanterns because turning on electric lights required almost no labor.[46] One farmer displayed the amazing benefit of using electric heaters to reduce the mortality rate of baby pigs: before their use to warm and dry newborns, 40 percent died; after installation of the electrical devices, the number dropped to 3 percent, with the cost of electricity for treating ten to twelve litters amounting to less than 50 cents.[47]

Even as only preliminary data appeared, the Red Wing experiment received national attention. As early as July 1924, when the test had run just over a half-year, Charles F. Stuart of the Northern States Power Company (and member of the state CREA) described the Minnesota work in *Forbes* business magazine as "a sincere, determined effort to solve the problem of rural electrification."[48] He noted that surveys of already-connected farmers "in a neighboring state" (not Minnesota) indicated that average homestead consumption totaled a measly 28.7 kWh per month, suggesting that "farmers were bewildered to know just what to do with electricity after it reached them."[49] But with equipment lent to the participants,

Fig. 8.3. Earl A. Stewart. *Source: Agricultural Engineering* 5, no. 6 (June 1924): 126. Used with permission of the American Society of Agricultural and Biological Engineers (ASABE, formerly ASAE).

farmers began discovering new possibilities. One of the article's photographs showed the use of an electric drive on a cream separator, which "takes more cream out of the milk and leaves less milk in the cream, which is economy all around" (figure 8.4). As important, the job is now cleaner, such that the farmer "can wear his Sunday clothes for performing the operation."[50] Despite this progress, Stuart ended the article by warning farmers not to jump on the electric bandwagon until the experiments have been completed. "The farmer has all too frequently been the loser," he commented, "because of buying costly equipment that he was not in a position to use, or that was not suited to his uses." One needed to avoid "invest[ing] his money in rural electrification and purchas[ing] . . . equipment until necessary facts have definitely been ascertained."[51]

Fig. 8.4. Users of an electric stove and cream separator show how life became easier with electricity in Red Wing, Minnesota. *Source: Agricultural Engineering* 5, no. 7 (July 1924): 153. Used with permission of the American Society of Agricultural and Biological Engineers (ASABE, formerly ASAE).

While still exuding caution, the state CREA members nevertheless felt confident enough to promote their work in a public manner—by erecting a display at the Minnesota State Fair in September 1924. The exhibit did not have enough space to demonstrate all the applications used in the test, such as feed grinding or corn shelling, but it included a model farm with miniature electric lines that illuminated the house, yard, and other buildings. Not simulated, real chickens ran into and out of an operating electric brooder. *Public Service Magazine* observed that the exhibit "visualized the proposition [of farm electrification] very impressively and was the center of intense interest throughout the fair" (figure 8.5).[52] A notice in *Electrical World* told a similar story, while also remarking that nearly a quarter-million people stopped by the display (a huge number, if true), many of whom obtained a booklet describing the benefits of using electricity on the farm.[53]

Good publicity about the Red Wing project continued arriving. An article in the August 1925 issue of *Popular Mechanics* magazine noted that the Minnesota investigations had already demonstrated how farmers could employ electricity to both

Problem of Farm Service Being Solved

Fig. 8.5. "Miniature Farm Equipped with Electric Light and Power," first public exhibit of the Red Wing experiment's results at the Minnesota State Fair, 1924. *Source*: *Public Service Magazine* 37, no. 4 (October 1924): 107.

improve convenience (through the use of electric drives on washing machines, for example) and increase income by employing electric motors for cream separation and ensilage cutting. Electric pumps not only brought water into homes, reducing huge amounts of labor; they also "added revenue by having drinking basins in the barns, for it is well known that dairy cattle with adequate water facilities will give more milk than animals that are led to the trough on a cold winter's morning, when the farmer has to break the ice."[54] With such evidence, even so early in the experiment, the article carried the enticing title "Electricity to End Farm Drudgery."[55]

The significance of the Red Wing experiment may also be measured by a *New York Times* article published at the end of the four-year study. Drawing on the words of utility executive Charles Stuart, the report observed that the farmers employed electricity "liberally and intelligently" in ways that benefited both themselves and the power company. The article noted that ruralites who earlier had access to electricity complained about $5 monthly bills when they only used 25 kWh for a few lights and small appliances. By the end of the experiment, farmers who paid about $15 per month for much greater usage expressed few objections, because electricity enhanced productivity and convenience.[56] And for the power company, the higher revenue made farmers "a feasible class" for service.

In a discussion with utility colleagues, Stuart observed that during the experiment's four years, the Red Wing participants saw their revenue increase about

43 percent while operating expenses grew by only 1 percent. Undoubtedly, some gains arose because of useful advice offered by project members and because farmers had become better businessmen during the experiment. In other words, electricity consumption alone likely did not produce all the gains. Moreover, Stuart rightly observed that much of the experimental equipment had come as donations or loans and that farmers might not have bought all of them without aid. But the financial picture still looked good if the farmers had paid for the equipment themselves, he argued. As noteworthy, the power company also profited because farmers used increasing amounts of power. "[F]rom our point of view," observed a utility executive, "those [farm] customers are very acceptable to the company. We are not losing any money on them," he happily observed, while electricity consumption had raised the farmer's standard of living "up to, and I should even say higher than that of his urban cousin" and left the agrarian family with "a nice little profit besides."[57] In short, farm productivity surged along with electricity usage, and the utility profited as well—an ideal outcome for all involved.[58]

Economic analyses did not capture all the benefits accruing from the Red Wing experiment, however. As Professor Earl Stewart observed in 1924, the "arrival of electricity at the farm is accompanied by many changes which cannot be measured by labor records or cost accounts. It frees the women as well as the men from rural bondage." He even noted that electricity convinced one family to remain on the farm. "The wife is now getting along without the ever-troublesome hired girl question staring her in the face, and her husband now smiles." Another farmer told Stewart that he had considered moving to town but later planned to remain on the farm. "Electricity now provides hot water for his bath on tap at the tub, water at the barn for his stock, cooks his meals, and lights his journey through life. Watch him live," wrote Stewart. Everyone seems happier with electricity, he reported. "Even unborn chicks in the shell are touched to greater endeavors by the magic wand of electricity," as electric incubators increased the yield of hatched chickens.[59]

To publicize the value of the experiment themselves, the Red Wing managers produced a silent film observing (as text) that "while the magic power [of electricity] has brought cheer and convenience to urban home and industry, progress in country homes has not kept pace with that in the city."[60] Of course, the University of Minnesota and its partners sought to change that situation, with the movie showing earnest-looking businessmen talking about the potential value of the Red Wing experiment. The film explained that farm advocates often made lofty claims about the benefits of various types of electrical equipment, but without providing proof. By contrast, the Minnesota investigators would test those assertions. "We propose to determine facts," the movie insisted.

Midway through the film, the tenor changed from descriptions of administrative actions to views of ruralites employing electricity to improve their lives. In one scene, a young country man brought home his new bride—a city woman—to an electrified farmstead. After driving through snow to the farmhouse, the groom gave his wife a light kiss in her modern kitchen and showed, with his arms wrapped lovingly around her, some of its electric appliances. Later, she rested on his lap in the sitting room after he turned on the overhead light. The scene ended as she rose from this position to turn off the light. But she returned to plant a more meaningful kiss on her adoring husband, as seen through the open door. Aside from intimating the romantic significance of electricity, the movie illustrated the uses of the new energy form for cooking, cleaning, sewing, elevating grain, watering animals, milking cows, separating cream, bathing, and other activities. Among other bits of wisdom, the film noted: "If electricity can be made *To Pay Its Own Way*[,] it will mean the electrification of the American farm and the building of a better civilization."[61]

Throughout the first decades of electrification, utility managers generally viewed the farm market as uncertain and experimental. The uncertainty arose not because managers thought farmers would shun electricity use. Clearly, ruralites would employ electricity to provide illumination in homesteads and barns. But lighting did not consume much electricity (or yield much revenue), such that central station power companies would not profit from serving rural districts. Hence, the industry would need to identify heavy-load-creating equipment and demonstrate its economic viability to farmers.

To achieve these dual goals, leaders of the utility business acquiesced to the undercurrent rural electrification advocates and created institutions to promote their work. Managers such as Grover Neff helped build alliances with other stakeholders, and in 1923, they established a national organization to pursue research and demonstration activities that benefited everyone. In the process, the industry and its partners spent more than $3.4 million over several years.

To place this spending in perspective, the utility industry received gross revenue (not profits) in 1929 of $1.955 billion from more than 24 million consumers.[62] The industry disbursed $853 million on construction alone, split almost evenly for generation, transmission, and distribution facilities.[63] As the Federal Trade Commission investigations revealed, NELA spent about $1 million each year in the

1920s on propaganda efforts to sway newspaper editors, college professors, and the public to discourage government takeovers of power companies and to maintain positive views toward private utilities.[64] That amount dwarfs the expenditures for rural electrification efforts through the CREA and the state affiliates.

In other words, while rural electrification took on great meaning to a number of people within the utility industry at the time, it clearly did not constitute a top priority—at least as measured by the amount of money devoted to it. But I have not argued in this book that stringing high lines to farmers constituted the greatest concern of NELA or utility managers. Rather, rural electrification remained an undercurrent movement that served the companies' interests during the contentious 1920s. Nevertheless, the industry effort resulted in more significant results than historians and promoters of the REA gave it credit for. The Red Wing experiment in Minnesota constituted only the first of many CREA-sponsored activities in more than half the states of the Union, providing useful information that even REA administrators would later acknowledge.

Chapter Nine

State Committees Work
to Resolve Uncertainties

The use of electricity on the farms of America is only in its infancy.
Fortunately[,] the great power companies realize the opportunity
for development in this field and are studying the problem of farm
electrification very thoroughly.
— ROBERT STEWART, DEAN OF THE COLLEGE OF AGRICULTURE,
UNIVERSITY OF NEVADA, 1928

The Red Wing experiment served as an initial, broad-brush effort to determine whether farmers and power companies might benefit from widespread electrification of tasks. It showed that ruralites could improve productivity when using electrical equipment while also raising their material standards of living. But because the Committee on the Relation of Electricity to Agriculture partners paid for so much of the hardware and wiring costs, the study did not determine whether farmers, without such subsidies, would pursue electrification on their own. Moreover, interested observers noted that the Minnesotans may not have represented farmers nationwide, having good incomes and more potential for profit (largely from dairy activities) than most. The Red Wing participants also enjoyed expert supervision to help them make sound decisions about equipment use. In other words, even with the apparent success of the Minnesota demonstration, utility managers and ruralites still harbored doubt about the value of employing electricity on the farm.

Aware that one set of experiences would not remove all the uncertainties about farming with electricity, the national CREA and state committees elsewhere pursued other investigations. If nothing else, a diverse group of tests would establish the value of electricity under different local conditions, where contrasting climates and farming activities could have substantial impacts on power consumption.[1] The affiliates therefore performed geographically relevant research that would give farmers confidence to purchase the most productive electric equipment available.

Work of Exemplary State Committees

Electrical and agricultural experts in Wisconsin eagerly joined with the national CREA to establish a state committee in early 1924. It is easy to understand why: several of the state's utilities had already shown an interest in rural electrification and had begun extending lines to farmers. At the same time, the companies worked with the state regulatory commission to design standardized rates for rural customers (described in the next chapter). Furthermore, faculty and staff at the state's land-grant institution, the University of Wisconsin, had been performing research on the use of electricity on farms for several years, as epitomized by the extension service's 1923 publication of *Turn On the Light*. A primer on employing power coming from isolated electric units and central stations, the booklet's authors acknowledged support from Grover Neff (in his capacity as chairman of NELA's Rural Lines Committee), several utilities, equipment manufacturers, staff members of the state regulatory commission, and the secretary of the Wisconsin Utilities Association.[2] The latter organization had been created in a 1922 merger between the Wisconsin Gas Association and the Wisconsin Electrical Association—both energy company trade groups—and Neff served as its first treasurer.[3] In early 1924, the organization provided funds for a fellowship to support a researcher working on rural electrification under the direction of the newly formed state CREA body, directed by University of Wisconsin's agricultural engineers.[4]

The Wisconsin CREA adopted an approach similar to that of its Minnesota counterpart, with the cooperating power company (Wisconsin Power and Light) stringing a distribution line to several farms near the town of Ripon.[5] Like those performed in other states, this experiment sought to determine which uses of electricity proved most economically viable. As stated by E. R. Meacham, a Utilities Association fellow, the "problem of fitting electricity into the farmer's scheme of life is such that the effect is not immediately apparent." After being convinced that a new application of electricity to farm equipment actually works—from a technical standpoint—the farmer next asks, "Will it pay?" The question had relevance, since farmers needed to advance $300 each toward construction of the distribution lines—an amount not required in the Minnesota trial and one that made this test more realistic.[6]

Research initially focused on the use of electricity to run motors for various tasks, to incubate chicks, to pump water, and to perform household activities.[7] Pea-processing machines employing electric motors saw energy costs decline by

17 percent, for example, compared with costs experienced when using steam engines. Even better, because the electrically run devices processed the vegetables more quickly, they produced total savings of more than 36 percent. Electric milking machines replaced hand milking, yielding reductions of between twenty-one and sixty-one minutes of work to obtain one hundred pounds of milk at an energy cost of only 2 to 7 cents. Electric ranges drew large amounts of current, as did electric water heaters, and they gave "entire satisfaction."[8] This last comment suggests that the Wisconsin investigators explored more than just financial considerations, realizing that electrical applications may "pay in a variety of ways that are entirely beyond statistics." Better lighting in barns reduced the amount of dirt getting into milk pails, providing an intangible benefit. Likewise, "[r]efrigeration may affect the eye and the stomach but not the pocketbook." Overall, electricity offered psychological benefits, such as "pride," "self-respect," and "[a]dded cheerfulness" along with improved living conditions, reduced waste, and increased spare time—all benefits that proved difficult to quantify.[9]

Agricultural engineers at Alabama Polytechnic Institute (later known as Auburn University) worked to demonstrate cost-effective uses of electricity in an environment where temperatures and farm operations differed dramatically from those in northern states. Formally established in January 1924, the state committee partnered with the Alabama Power Company to install high lines to test farms.[10] As in Wisconsin, the committee and utility did not offer subsidies for wiring or for equipment. Extension agents and the university's agricultural engineers furnished advice and supervision, based in part on research performed at the experiment station, dealing with electrical devices employed for various home and production applications.[11]

Alabama farmers first used electricity mostly for lighting, and the typical farm under study at the end of 1925 had installed ten lamps. On average, illumination required just 18 kWh per month. Such small consumption clearly did not justify service from the utility company, since it brought in less revenue than needed to compensate for the cost of extending lines. However, after about six months or a year, farmers had overcome the financial stress of wiring their homes and buying light fixtures, and they started becoming more amenable to other uses of electricity.[12] A survey of rural customers who had gained experience with the new energy form showed that 74 percent bought electric irons, and from 12 to 14 percent purchased electric ranges, water pumps, curling irons, and fans. Some farmers had become more ambitious, using electricity outside the home for "milking, ensilage cutting, feed grinding, threshing, pumping water for livestock and for irrigation, milk cooling, separating cream," and several other purposes—thirty-seven in all besides lighting.[13] By 1926, Alabama Power noted that publicity about the first test lines created a demand among unserved farmers to obtain electricity, such that the company erected 183 miles of wires, up from just 39 miles two years earlier. (The number of rural consumers jumped from 240 to 1,796 in the same period.)[14] Dairy farms proved to be valued power company customers, employing about 2,200 kWh annually, while poultry farms drew an average of 2,640 kWh.[15] On the other hand, Alabama's farmers of cotton (a crop grown by more than 80 percent of the state's agriculturists, according to the 1930 census) still wanted electricity largely for lighting.[16]

In Virginia, which had a more diversified rural economy than Alabama, researchers also sought to determine how farmers could profitably increase their power consumption. Meeting in February 1924, representatives of electric utility companies, the state's Department of Agriculture, farmers' organizations, the state chamber of commerce, and Virginia Polytechnic Institute established a CREA partner. VPI agricultural engineering professor Charles Seitz served as chairman and secretary.[17] In one dramatic (though not original) experiment, Virginia researchers collaborated with a farmer who separated his flock of 320 hens into a group that obtained natural daylight for about ten hours per day and another that received electric light for thirteen hours per day. The electrically illuminated animals produced about 45 percent more eggs. If both groups of hens would have had the benefit of electric light, Seitz projected, they would have produced additional eggs, netting the farmer about $270—more than enough to cover the farmer's expenses (of $227) to equip the barn with wires and bulbs. In another experiment, a tomato farmer employed electric lights to trap tomato worm moths, reducing destruction of more than 50 percent of his crop and saving between $2,000 and $4,000.[18] Such results clearly offered excellent selling points for investing in farmstead wiring and some types of electrical equipment. (The cover of VPI's Agricultural Engineering Department's 1929 annual report featured a photograph of an electrified dairy farm, shown in figure 9.1.)

The Idaho affiliate of the CREA performed research on a farm operated by the state university's agricultural experiment station at Caldwell using more than $5,000 worth of electrical equipment.[19] Beyond the several-year-long experiments, the land-grant institution used the farm for demonstration of devices such as feed grinders, hay choppers, hoists, milking machines, dairy refrigerators, and water heaters.[20] By the time of these experiments, Idaho farmers already had a longer history of electrification than elsewhere, with "progressive electric utilities" extending power lines to farms that used electricity for irrigation purposes. Approxi-

Fig. 9.1. An electrified dairy farm in Virginia. *Source:* From the cover of the 1929 VPI Agricultural Engineering Department's annual report. Used with permission of Virginia Tech University Libraries, Special Collections and University Archives. Photograph courtesy of Dr. W. Cully Hession, Virginia Tech Department of Biological Systems Engineering.

mately 14 percent of Idaho farms enjoyed electric service by 1925, about three times the national average.[21] Despite the relatively widespread use of power, the state CREA sought to help farmers employ other forms of economically rewarding equipment that would raise electrical load further and augment farmers' and utilities' income. (The cover of the Idaho CREA's 1927 report is shown in figure 9.2.)

In its research, the Iowa CREA initially examined sixty-three rural customers served by distribution lines emanating from the town of Garner. Already more advanced than their peers, these farmers proved "ready to work with those in charge of the project and assist in whatever ways they can."[22] (Plate 7 shows the cover of an Iowa farm electrification research report.) But unlike other CREA state-run operations, the Iowa project involved formal studies in home economics. Eloise Davison, a leader in this burgeoning field in the early twentieth century, cooperated with Vivian Brashear and Harriet Brigham, instructors in household equipment at Iowa State College (later known as Iowa State University), who

Fig. 9.2. Prosperous Idaho farms, thanks to electricity. *Source*: Cover of Idaho CREA report written by Hobart Beresford, *Progress Report of the Idaho Committee on the Relation of Electricity to Agriculture*, August 1927. Courtesy of University of Idaho Library, Special Collections and Archives.

conducted experiments in Garner's farm homes.[23] The researchers explained in 1928 that they sought to employ electrical devices to lessen "labor and drudgery of house work on the farm and [to conserve] the strength of farm women."[24] From surveys, they learned that women first chose to use electricity for lighting once power lines reached their homes. Electrically pumped water followed quickly because women viewed it as "perhaps the most important factor in making a farm home modern" by eliminating the work of manually drawing water from wells and by enabling indoor plumbing and bathrooms.[25] Other popular appliances included hand irons, ranges (figure 9.3), refrigerators (figure 9.4), radios, and electric fans. Though not widely used, reflector heaters still merited attention in a research report that included a heartwarming photograph of a pair infants comforted by the device, with a caption that read, "Two reasons for using the glow heater on the farm" (figure 9.5).[26]

In several homes, women previously used laundry washing machines, sometimes powered by gasoline engines attached to shafts (by belts) that also turned cream separators and other appliances. Replacing the engines with electric motors made it easier to start the equipment and resulted in less noise. But bigger gains arrived when women used new washers with electric motors directly incorporated in them (without belts attached to external power sources and thus taking up less space). The more compact machines allowed the cleaning process to move from a detached wash room (or dairy house) into the main residence, near a supply of heated water.[27]

Small electric appliances offered the means to reduce unpleasant work, but as specialty items, they received a mixed reception. Electric ironing machines, for example, could press twenty-five towels, tablecloths, and sheets in just one hour. But with costs ranging from $160 to $175, these devices remained expensive; worse, shirts, dresses, and other clothes still required hand ironing.[28] Meanwhile, women saw coffee percolators as attractive enough to use at the dining table, but their high cost (between $13.75 and $18.00 each) served as "an apparent drawback to their more widespread use." Automatic electric toasters used at the table saved time by reducing the number of trips to the kitchen, though some homemakers complained that the devices handled only small slices of bread.[29]

The Washington State CREA diverged from other committees by eschewing the approach of establishing test farms on which new equipment could be demonstrated and studied. Rather, working on a budget of about $6,000 annually, agricultural engineers at Washington State College in 1925 surveyed owners of the thirty thousand already-connected farms to learn which existing electrical applications proved most productive.[30] They found, for example, that electric motors

Fig. 9.3. Cooking is cleaner and easier with electricity. *Source:* Harriet C. Brigham, Frank J. Zink, and Frank D. Paine, "Electric Service for the Iowa Farm," Report no. 6, *Iowa State College of Agriculture and Mechanic Arts Official Publication* 27, no. 12 (18 July 1928): 18. Courtesy of Iowa State University Library, Special Collections and University Archives.

provided considerable labor savings when moving hay. Compared with using horses and hiring a boy to drive them, an electric hoist reduced costs by $2.31 per day when lifting thirty to forty tons of unbaled hay.[31] Committee leaders hoped that publicity of such research results would help expand awareness of novel electrical applications and encourage unserved farmers to seek high-line power.[32]

Fig. 9.4. "Electrical refrigeration is considered one of the greatest helps to the farm home." *Source*: Caption for photograph published in Harriet C. Brigham, Frank J. Zink, and Frank D. Paine, "Electric Service for the Iowa Farm," Report no. 6, *Iowa State College of Agriculture and Mechanic Arts Official Publication* 27, no. 12 (18 July 1928): 27. Courtesy of Iowa State University Library, Special Collections and University Archives.

California held a wide lead over other states' electrification efforts in the 1920s, as noted earlier, but its agricultural engineers still participated in CREA's work. Establishing a local committee in May 1924 after national director Earl White visited the state, the University of California, Davis chose agricultural engineer Ben D. Moses as the project director.[33] Through 1940, the committee pursued about

Fig. 9.5. "Two reasons for using the glow heater on the farm." *Source*: Caption for picture published in Harriet C. Brigham, Frank J. Zink, and Frank D. Paine, "Electric Service for the Iowa Farm," Report No. 6, *Iowa State College of Agriculture and Mechanic Arts Official Publication* 27, no. 12 (18 July 1928): 32. Courtesy of Iowa State University Library, Special Collections and University Archives.

thirty-five experiments with funding of approximately $110,000; supplemental monies came through the agricultural experiment station, yielding a total of about $250,000 invested in the projects.[34] Studies focused on using electricity for warming piglets after birth, increasing the efficiency of irrigation pumps, catching insects, refrigerating fruits and vegetables, air conditioning greenhouses, and handling dairy cows.[35] In this last activity, the various uses of electricity—for lighting barns, sterilizing equipment, milking, and refrigeration, for example— contributed to high loads. California's CREA chairman H. B. Walker calculated in 1928 that, because of the many dairy applications of electricity, each cow required the consumption of about 200 kWh per year—a useful rule of thumb for agricultural engineers and farmers.[36]

The success of several state experiments may have motivated national CREA leaders to demonstrate their work to a larger audience. They did so by establishing a "National Rural Electric Project" in College Park, Maryland, in 1928. Working with annual budgets of $25,000 and support from Baltimore's Consolidated Gas, Electric Light and Power Company, the project's managers performed experiments on topics having a national—not just state-specific—interest at five farms equipped with household and farm machinery. While scientific research would remain a top priority, so would publicity to politicians and others in nearby Washington, DC. "It is planned to make the electrified Maryland community a show spot," noted a 1928 report; it would serve as "a district of infinite educational value to everyone interested in agriculture." To ensure that no one passing the experimental site missed its significance, the operators planned to erect a "large electric sign . . . to call attention to the nature of the project."[37]

From inauspicious beginnings, formal research efforts pursued by CREA state committees expanded dramatically in the 1920s and into the first years of the next decade. To highlight the progress and output of the state organizations, the national CREA published a compendious *CREA Bulletin* issue in January 1928, self-described as an "elementary treatise of a development which promises to grow with astounding rapidity."[38] (The CREA had been publishing issues of the *Bulletin* since September 1924 as one way to disseminate results of investigations.) The 136-page volume, titled *Electricity on the Farm and in Rural Communities*, contained tables, photographs, and text explaining the various electrical experiments performed in farm homes, barns, and fields around the country. It also included summaries of scores of research topics, and the index listed more than two hundred entries dealing with practical uses of electricity.[39] In June 1928, CREA director White reported that twelve thousand copies of this issue had been distributed at no cost while another twenty-three thousand copies had been purchased, largely by utility companies.[40] (Some farm applications of electricity, as illustrated in the issue, are included in figures 9.6, 9.7, 9.8, and 9.9.)

As more knowledge accrued about valuable uses of electricity on farms, the national CREA produced two more comprehensive documents highlighting activities pursued by the state organizations and partners. The June 1931 *Report on Farm Electrification Research* counted 211 investigations either "connected with or stimulated by" the national CREA, 493 projects going on in educational institutions and the Department of Agriculture, 118 studies pursued by utilities and commercial companies, and 39 projects undertaken in various laboratories.[41] Describing the state committees' research as "pioneer work," the publication noted that investigators identified problems with electrical service to ruralites and "shaped programs to overcome them."[42] It provided data on the outcomes of experiments and

Stifling kitchens need
refreshing breezes~

Fig. 9.6. An electric fan in the kitchen offers extra comfort at little cost. *Source:* NELA, *Electricity on the Farm and in Rural Communities, CREA Bulletin* 4, no. 1 (30 January 1928): 24. Used with permission of the Edison Electric Institute.

suggested some unusual, but valuable, uses of electricity, such as to operate burglar alarms and lights for deterring petty thievery.[43]

More palatable for general reading, the revised edition of the national CREA's *Electricity on the Farm and in Rural Communities* appeared in November 1931 as an issue of the *CREA Bulletin*. Containing 322 pages of research summaries, it also included numerous photographs illustrating how women and children enjoyed electrically operated farm appliances. One chart's caption stated that "[a]n electric range[,] in addition to being a great convenience[,] is a great labor saver" because it required only one-seventh the time needed to operate a coal stove.[44] Aside from devoting about 20 percent of its pages to the use of electricity in homes, the volume offered information on employing electric water pumps for fire protection, using electric milking machines to raise dairy product output, retarding bacterial growth in milk with dairy refrigerators, and operating electric motors to cut and store ensilage.[45] It also reported more definitive results on the value of electric lighting for poultry: not only do chickens lay more eggs when illuminated,

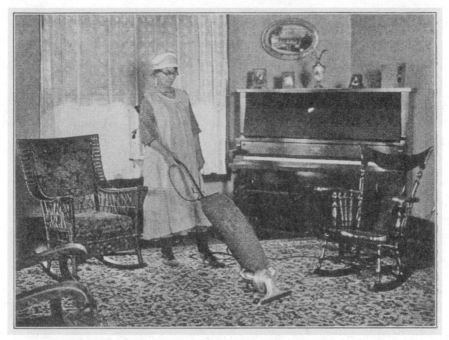

Fig. 9.7. Saving labor and effort with an electric vacuum cleaner. *Source*: NELA, *Electricity on the Farm and in Rural Communities, CREA Bulletin* 4, no. 1 (30 January 1928): 25. Used with permission of the Edison Electric Institute.

especially in winter months; lighting also encourages old hens to gain weight at times when seasonal meat prices peaked.[46]

Scores of other topics received attention: employing electric motors for irrigation, which increases production of vegetables and fruits; using electric heaters to prevent chilling of newly born animals; sawing and planing wood with electric motor–operated devices; and electrification of country stores, mills, greenhouses, churches, and schools. The trade journal *Electrical World* described the volume as invaluable, such that farmers should read it "from cover to cover" and "keep it as a reference book."[47] Containing information on the headway made in recent years, the issue included a preface observing that "this promises to be only a substantial beginning" toward more electrical progress in the future.[48]

CREA's Work to Eliminate Myths and False Hopes

Besides performing research to determine productive ways to employ electricity, industry and CREA-supported stakeholders also contributed to dispelling

Fig. 9.8. An electric motor attached by belt to an ensilage cutter. *Source:* NELA, *Electricity on the Farm and in Rural Communities, CREA Bulletin* 4, no. 1 (30 January 1928): 55. Used with permission of the Edison Electric Institute.

Fig. 9.9. The radio brings the world to rural Indiana. *Source:* NELA, *Electricity on the Farm and in Rural Communities, CREA Bulletin* 4, no. 1 (30 January 1928): 22. Used with permission of the Edison Electric Institute.

exaggerated claims made by some promoters of rural electrification. In doing so, they fulfilled their mission to help provide definitive knowledge that would allow farmers to make wise choices about electrical devices. If ruralites became disheartened too often by purchasing expensive electrical equipment that did not live up to expectations, they would not adopt practices to make high-line extensions valuable enough for themselves or for power companies.

Plowing fields with electricity received much attention early in the century. In theory, electric motors offered great advantages: unlike horses or other traction animals, they never tired. Moreover, they turned on and off quickly, and they required little maintenance. Such benefits encouraged experimenters in Europe to design electric motor–powered tilling devices. A 1911 NELA report suggested that in Germany "[t]illing the soil is now accomplished with electric power," with a plow and disk cultivators pulled across a field by a wire cable attached to an electric motor."[49] Frank Koester, a Swiss-American consulting engineer, wrote enthusiastically (as noted in chapter 4) about the technology, arguing in 1912 that it had reached a state "far beyond the experimental stage."[50] Though expensive in first costs, these plows appeared cheaper to operate and more productive than using horse-drawn or steam tractor–pulled plows, especially on soft ground, where the lightweight electric plow could operate tirelessly.[51] A steam-powered plow, he claimed, cost about $14,000 to $15,000, whereas electrical plows started at $8,000.[52]

The popular and technical literature in the early 1920s also suggested that electrified equipment for traction would soon populate farms. Accompanied by an illustration of a field spanned by overhead wires, a *Science and Invention* article in 1922 predicted widespread use of electric plows, cultivators, and planting machines. The demand for such electrified machines would multiply as power companies extended their interconnected transmission lines throughout the land, making the new energy form so easily accessible that it will become the "king of all powers on the farms, as well as in the cities."[53]

Electric plows never became popular, however. As US Department of Agriculture researcher Robert Trullinger explained in 1924 (with published remarks appearing in an early issue of the *CREA Bulletin*), the success of electric tractors or trucks remained stymied by the difficulty in obtaining power from either a cable coming from a transmission line or from an on-board battery. "The problems in either case," he observed, have "never been solved with any marked degree of satisfaction."[54] Perhaps as significant, farmers rapidly adopted a more effective form of motive power, namely, the gasoline engine–powered tractor. Emerging during the first years of the twentieth century, these lightweight machines replaced animal- and steam engine–powered tractors while also becoming relatively cheap; the Henry Ford and Sons Company introduced a mass-produced tractor in 1917 for $750. As the country came out of recession, the firm dropped the price to $395 in 1922, and by 1923, the Fordson tractor (branded as such in 1918) had won 77 percent of the nation's market.[55] Farmers using gasoline-fueled tractors still needed additional equipment for plowing, but the total cost still dwarfed (by an order of magnitude) the price of the electric plow described by Koester a decade earlier.

While the cable-connected electric plow may have lost credibility in the United States, another type of electricity-using device gained greater interest from state committees and the national CREA. Drawing on supposed "knowledge" of the effects of electric sparks in the atmosphere, Pittsburgh inventor Hamilton Roe designed a machine that sent static electricity into the soil with the expectation that the energy would convert atmospheric nitrogen into fertilizer. Pulled by a tractor that transmitted power to a generator for production of the high-voltage spark, the device also killed weeds and deleterious insects while flipping the soil.[56] Promotional literature observed in 1927 that the Roe plow "is accomplishing More for Agriculture than any Other Discovery Heretofore Invented."[57] Hyperbole notwithstanding, the apparatus appeared promising as one of many proposed approaches to exploit electroculture (figure 9.10).

Several farm-region newspapers published glowing accounts of the Roe plow, often including a picture of the device and descriptions of its claimed virtues. "This little machine with its box of wires and generators for supplying a current into the earth may be destined to change completely the whole system and trend of farming as it has been practiced since time immemorial," noted the picture caption in the *Caledonia (NY) Advertiser* on 14 July 1927.[58] Better yet, the "cost of the treatment is negligible," observed the inventor according to the *Reading (PA) Eagle* on 12 August 1927, since the energy for the generator comes, via a belt, from the tractor to create electricity that enters the soil.[59]

National publications also celebrated the apparently marvelous machine. In "Electricity Speeds Up Crops," *Science and Invention* magazine described the remarkable results derived from its use.[60] *Popular Science Monthly* further gushed, "If the new method proves practical[,] it would save millions of dollars now being spent to fight insect pests."[61] *Time* magazine also reported on Roe's device, paraphrasing the inventor, who claimed it would "electrocute weeds, grubs, soil bacteria" and cause crops to spring "from the volt-purged ground in record time and abundance."[62] Perhaps offering the most credibility, the *New York Times* in August 1927 published a substantial account of the device, which sent "103,000

"The Law was our school-master"

ROE WIRELESS
ELECTRIC PLOW

The Plow That Tills Ground Nature's Way.
It Kills Quack-grass, and other foul growth.
It Eradicates All Insect Life in the Soil.
It Makes Seed Beds as fine as Powder.
It Eliminates the use of Sprays.
It Germinates Plants In Less Time.
It Matures All Growing Crops Quicker.

It Segregates Air and "Fixes" Nitrogen, Oxygen and Carbon on and Enriches the Soil with these Essential Elements. It offers a System of Drainage that Conserves Soil Moisture. It Makes the Only Plant Food that Nature Uses to Feed the World. Electricity is the Only "Farm Relief."

The Roe Wireless Electric Plow is accomplishing More for Agriculture than any Other Discovery Heretofore Invented.

Fig. 9.10. Promotional brochure for the Roe electric plow, undated, but likely 1927. *Source:* Courtesy of Lynne Belluscio and Terry Guilford of the LeRoy (NY) Historical Society. Earlier published by Lynne Belluscio in the *LeRoy Pennysaver & News,* 26 August 2018.

volts of electricity into the soil as it moves along." Somewhat uncritically, the news-paper reported on early experiments showing that plants germinated substantially sooner than those in unelectrified soil, and they grew in fields that had previously been unproductive. Meanwhile, weeds died off without other treatments. It quoted Roe as claiming that his device restored worn-out soils with electricity, the essential element of plant productivity. "As the two plow blades penetrate the soil and pass along," he observed, "there is an intense electrical field created between them which produces an effect like lightning," eliminating the need for fertilizers since the current liberated nutritious chemicals from the soil. To gain further benefits, Roe plowed in the north and south directions, as the device supposedly then worked "in harmony with nature and the magnetic poles of the earth" (figure 9.11).[63]

Because the machine held out hope for such unusual and wonderful results, "a good many farmers are asking questions about it," noted the managing director of NELA in October 1928, suggesting that systematic research be conducted to test the claims.[64] But from the start, some CREA state organizations remained skeptical. M. L. Nichols, the head of agricultural engineering at Alabama Polytechnic

Fig. 9.11. Promotional photograph of Hamilton Roe (*left*) with his electric plow. *Source*: Photograph distributed August 1927 by Press Studios, Buffalo, NY. Courtesy of Lynne Belluscio and Terry Guilford of the LeRoy (NY) Historical Society. Previously published by Lynne Belluscio in the *LeRoy Pennysaver & News*, 26 August 2018.

Institute, observed in a letter to a USDA engineer that he had "no hopes . . . for this method" and felt that "the energy required [to operate the plow] seems to be out of proportion to the effect."[65] As reported in the *CREA Bulletin* in 1931, Michigan State College agricultural engineers described the device as "an oversized spark gap radio transmitter" that "made uncertain results almost inevitable." In field tests, researchers found that the electrically plowed fields offered inconsistent outcomes.[66] As master's thesis work, University of Maryland student Sam Winterberg (working in conjunction with the National Rural Electric Project test facility) performed controlled experiments comparing plants grown in soils plowed with and without the Roe device; the study failed to "indicate any consistent advantage from the use of the electric plow."[67] After a description of similar experiments on the high-voltage stimulation of soil, the editor of the *CREA Bulletin* admitted that electrical currents sent through the soil may play a role in boosting plant growth, but that the mechanisms remain poorly understood. The approach remained full of possibilities rather than concrete results.[68]

Ultimately, agricultural experts agreed that the Roe electric plow did not deliver on its highly trumpeted promises. Writing for an audience of bankers in 1928, an academic commented on the existence of "many absurd and extravagant claims" by advocates of equipment such as "a specially equipped electric device for sending powerful currents of electricity into the soil." He observed, "Such a claim may be an excellent basis as a stock selling scheme in a promotion company, but it has no basis in scientific fact."[69] A few years later, the *CREA Bulletin* editor concluded, "There seems to be little to gain by further field tests of the electric plow as now built."[70] In the same volume, he noted, "Occasionally some zealous inventor-promoter has focused public attention on impractical devices with claims for weird accomplishments. These have had publicity value if not practical merit."[71] Perhaps the author was referring to the Roe plow as a prime example.

Unresolved Problems

Electrified threshing also seemed—at least in theory—to have great merit. Before the use of machines, starting in the late eighteenth century, the process of removing grain kernels from the husks of plants such as wheat and barley remained a time-consuming and laborious process. With the advent of steam engine–powered threshing devices used in fields, farmhands loaded dried grain stalks onto conveyor belts for delivery to a device that separated the components. But advocates of rural electrification saw electric motors as superior replacements for steam engines, owing to their light weight and ease of maintenance. The electric motor offered enormous advantages compared with steam- or gasoline-powered engines, according to Iowa researchers in 1914: lower first cost, reduced operating costs, and no fire hazard. In this case, farmers took power from a 2,200-volt distribution line and transformed the voltage to 220 and 110 volts for use with the motor at a cost of about $800, compared with about $2,000 for a similar steam- or gasoline-powered machine.[72]

The potential value of using electric motors for threshing received careful study in the CREA-sponsored Red Wing experiments. In June 1924, Minnesota researchers installed a 15-horsepower electrical motor on a threshing machine, with a transformer and cable reel carried on another vehicle (and with wires attached to a 2,300-volt distribution line).[73] The thresher and the transformer truck needed to be transported by another prime mover, such as a horse or a tractor.[74] Several problems ensued, with the motor overloading at times and the circuit breaker opening frequently, thus interrupting operations.[75] A series of improvements in the equipment followed, including the elevation of voltage in the high line to 6,900 volts, yielding a machine that supposedly proved "feasible and economical."[76] Despite the large effort spent on this element of electrifying the farm and some positive experimental results, it appeared that the electrical thresher's use would not become widespread. To make the machine operate well, farmers needed to ensure that all its components "be made fool proof"—a difficult requirement to meet—and that they processed only the best quality grain.[77] Moreover, the machine suffered from too much complexity, by requiring a separate transformer truck, for example.[78]

At the same time, gasoline-fueled tractors in the mid-to-late 1920s had been improving rapidly in ways that allowed them to serve as better power sources for threshers. Fordson tractors, along with those made by competitors such as International Harvester, contained exterior belt pulleys and take-off shafts that allowed the engines to operate a variety of farm equipment. Advertisements and articles suggested that their use with threshers required less labor than alternative approaches while achieving excellent results.[79] And the tractors, of course, did not need external sources of power and associated equipment, such as transformers, cable reels, and nearby high lines. Finally, data collected from three years of experiments did not show great advantages in cost between threshing with electric motors and tractors. In fact, for several tasks, such as oat threshing, tractor power appeared cheaper than using electric motors.[80] In short, the case had not been made for electrically threshing in the field. (The use of small motors for threshing in a barn, near an easily accessible source of power, however, continued to be discussed and employed.)

Equally notable, CREA researchers made it clear that some highly publicized electrical appliances had great value in certain circumstances, but deficiencies in others. In Minnesota, Professor Julius Romness reported in 1934 on trends in farm electrification in the decade after the pathbreaking Red Wing experiment began. He noted that seven of the eight original farms retained the electric cooking ranges they received, yet few found broad use because farmers had easy access to wood fuel, which proved cheaper than electricity for food preparation. More important, the wood-burning stoves provided heat for the home during long, frigid winters. Farmers used the electric ranges largely during warm weather and for preparation of light meals.[81] And because of the low cost of other available fuels, Romness observed, farmers discontinued their use of loaned electric water heaters even before the experiment ended in 1928.[82] Electric space heating had likewise not become popular. Presenting data showing that one pound of coal contained about four times the heating energy produced by a kilowatt-hour of electricity, he concluded that "the use of electric heating appliances must largely be justified from a convenience and labor saving standpoint rather than from an economic one."[83] In Alabama, by contrast, CREA researchers found that families enjoyed using electric ranges for general cooking, but they proved extremely expensive for heating large volumes of water.[84] If nothing else, the rural electrification groups realized that one size did not fit all—that successful electrical applications in one part of the country did not always suit farmers in others.

In a similar form of critical introspection, some CREA researchers also acknowledged the illusory nature of some of their own hoped-for uses of electricity. Put differently, they did not simply discredit the fanciful proposals of overenthusiastic advocates of electricity on farms; they also showed that some their favorite notions had not yet proven viable. The 1931 CREA Bulletin, for example, speculated on the health benefits of employing "artificial sunshine" (especially in the winter) produced by ultraviolet light–emitting electric lamps. While publishing a picture of happy children playing under these devices, the editor noted that their effectiveness and safety still had not been demonstrated.[85] Likewise, researchers found that ultraviolet light did not boost egg production in poultry, even though some people thought the radiation would spur better use of calcium and phosphorus consumed by chickens.[86] Nor did the transmission of radio waves raise flies' internal temperatures enough to kill them; despite having great theoretical merit, experiments failed to yield anticipated results.[87] Growing tomatoes in electrically heated water and nutrients also did not show an economically attractive outcome.[88] Overall, California's CREA chairman H. B. Walker observed retrospectively in 1940 that "in these investigations, our answers have not always resulted in positive endorsements of electrical energy for certain farm uses, but our batting average has been well over fifty per cent."[89]

At the same time, seemingly obvious uses of electricity still needed experimentation and confirmation of their value. The chick brooder, a heater to keep newborn chicks warm and healthy, for example, appeared easily electrified and gained attention among farmers, but the Wisconsin tests did not initially prove their superiority over oil-fired devices.[90] The California group pursued improvements of the technology as well, using an electrically operated incandescent light, which produced a substantial amount of heat. Agricultural engineer Ben Moses observed that farmers using heat lamps, even those controlled by thermostats, sometimes lost many chicks because of poor air circulation. As a result, some farmers expressed "prejudice" against electric heating systems until other farmers devised approaches, tested by Moses and his colleagues, that eliminated the problems in the early 1930s.[91]

Electric refrigeration on farms also received much attention, but not much usage, especially in households. In the 1920s, the technology still remained in its infancy, and manufacturers often produced components (compressors, condensers, and evaporators) that needed installation into custom-crafted cabinets originally designed for cooling by ice.[92] Though several companies entered the business, the machines remained unreliable and required frequent adjustment and service.[93] Companies started manufacturing self-contained refrigerators in the early 1920s, but they remained unpopular, in part because they employed toxic sulphur dioxide, ammonia, and methyl chloride, which sometimes killed customers after leaking.[94] The devices seemed ideal for building electrical loads—using double the amount of energy previously consumed in the average home. Unfortunately, as a NELA committee reported in 1924, they still proved "unsatisfactory," such that central station companies "have been much slower to interest themselves in" them, using their resources to promote lower-consumption appliances.[95] Companies such as Frigidaire and General Electric produced improved devices, with the Iowa CREA researchers reporting in 1928, for example, that women greatly enjoyed the machines, which decreased food spoilage and enabled increased diversity in diets because more fruits, vegetables, and meats could be kept on hand. But the appliances remained expensive, costing between $210 and $535 for those used in the tests.[96] Despite a rapid rate of growth in the 1920s, mechanical refrigerator sales totaled about 468,000 nationwide (to urban and rural customers) in 1928.[97]

Only in the 1930s did the price of the appliances decline substantially, aided by New Deal–era efforts to spur manufacturers to build inexpensive units (see chapter 12).[98] They also benefited from the use of safe chlorofluorocarbons as refrigerants, the most popular of which General Motors patented as Freon-12 in 1931.[99] Not unexpectedly, the comprehensive 1931 CREA publication, *Electricity on the Farm*, devoted several pages to describing the virtues of domestic refrigerators.[100] Even with refrigerators becoming less dangerous and cheaper, however, resistance to the machines remained pervasive for cultural reasons. Farmers had already used natural sources of cooling (such as wells and cellars) for food storage, and they canned fresh vegetables instead of buying them in cities. "Outside of the South," historian Ronald Kline observed, "year-round refrigeration was seen as something farm people could do without when it came time to decide which electrical appliances to buy in the 1930s."[101]

The electric refrigerator, in other words, may have been more of an aspiration than reality in the 1920s. From today's perspective, when almost every urban and rural home contains the appliance, it appears obvious that farmers would rapidly welcome its arrival, such that their absence existed only because the greedy, urban central station companies deprived ruralites of electricity. In fact, the refrigerator's acceptance on the farmstead occurred gradually and not just because of difficulties in obtaining electric power.

To make rural electrification economically feasible for utility companies, farmers needed to increase their power consumption. Realizing that most ruralites only thought of electricity for a few low-demand applications—especially lighting—the utility industry collaborated with agricultural engineers at land-grant institutions to discover and demonstrate ways to make electricity use financially attractive to all parties. These efforts clearly bore fruit, such that the executive secretary of the American Farm Bureau Federation, a member of the CREA, observed in 1930: "A few years ago most farm people thought of electricity on farms only in terms of the convenience that would come from being provided with electric lights. Today," he added, "farmers everywhere are clearly recognizing the fact that electrical power for lighting purposes, while important, represents only a very small use to which this energy can be put." In fact, the committees had "tested and found practical" about 250 economically productive applications of electricity.[102] This number dwarfed the thirty supposedly cost-effective uses of electricity outside the farm home that a 1913 NELA study had identified and the thirty-five applications that the Washington State CREA counted in 1925.[103] And for the enhancement of life in general (in the farm home, where many tasks, particularly those done by women, did not receive economic analyses), the use of electricity for lighting, water pumping, ventilation, and other tasks seemed indisputably appreciated.

As important, the CREA groups did important work to discount the value of hyped—but ultimately useless—technologies. The organizations showed that certain electric appliances, such as water heaters and cooking ranges, might not appeal to farmers living in cold climates. Likewise, electric refrigerators remained unattractive because of safety and price concerns along with preferences for other forms of food preservation. On a larger scale, the committees saved farmers huge investments in equipment that seemed too good to be true. The Roe electric plow serves as the epitome of such a technology, which I describe not to belittle a machine that appears implausible today but to depict the occasionally sketchy understanding of electricity for farm applications at the time.

Acknowledgment of the industry's contributions to boost productivity-enhancing uses of electricity came from a variety of sources—some unexpected. A 1934 issue of *Editorial Research Reports* noted that efforts to spur rural electrification, despite impediments, had been largely successful. With the "impetus" of the Red Wing, Minnesota, experiment, begun in 1923, "there has been a relatively large and steady rise in the number of electrified farms during the last decade," the document observed.[104] And while the first REA administrator, Morris Cooke, generally castigated private utilities for their insufficient farm electrification work, he conceded in 1935 that the CREA's state organizations have "done much in educating farmers in the possibilities and need for rural electrification."[105] The next REA administrator, John M. Carmody, additionally acknowledged the role played by land-grant colleges (if not their collaborator, the CREA) for assistance in rural electrification. While taking a swipe at utility companies for their lackluster endeavors to electrify farms, Carmody nevertheless observed in 1938 that the "agricultural experiment stations and universities have generously made available for our use . . . results of their research into specific uses of electricity on the farm. Much of this technical work," he added, "antedated the inception of our agency."[106] He also recognized the CREA for "its stimulation of research work" that involved the agricultural experiment stations in more than half the states, including the "ambitious" Red Wing project.[107] Carmody's successor, Harry Slattery, gave further credit in 1940 to the national and state committees for the pioneering efforts that offered "very great educational value."[108] He noted that their "activities provided an effective stimulant to the agricultural colleges, experiment stations and electrical industry."[109]

Even if overly charitable, such statements suggest that the national CREA and its state associates won recognition for doing useful work. If nothing else, they (and the documentation of the committees' activities) discredit the assertions that the industry's efforts to pursue rural electrification research constituted window dressing. Historian D. Clayton Brown contended, for example, that private companies proved "slow or unwilling" to serve farmers and that its primary, but feeble, effort to do so—the establishment of the CREA—made it "clear that the farmer could not rely on the power industry for service."[110] As noted earlier, many other historians and commentators have accepted Brown's conclusion as authoritative, and they parroted the same storyline, explicitly noting (as did Laurence Malone) that power companies "ignored the rural market." This account of the activities pursued by the CREA, its state affiliates, and industry partners suggests otherwise.

Chapter Ten

Regulation and the Extension of Lines to Rural Areas

> Long before the recently renewed interest in rural electrification, the Commission was, where possible, ordering into effect more liberal rates and extension policies, and in other cases informally urging the various utilities to liberalize their rural rates and extension rules.
> —WISCONSIN PUBLIC SERVICE COMMISSION, 1936

The expansion of rural electrification required more than just research on value-enhancing technologies that would elevate farmers' power consumption. It depended on the creation of a social and technical infrastructure—a web of stakeholders that sought to expand efforts to serve farmers. Using the terminology of Thomas Hughes and building on it, the *overall* utility system had, by the 1920s, established a sophisticated and well-integrated cohort of actors and institutions in the business, educational, legal, political, and technical realms. That system achieved maturity, with participants working toward increasing its growth, profitability, influence, and momentum.

A similar level of development did not yet exist in the rural electrification *subsystem*, with many elements still being constructed by its undercurrent advocates and partners. It remained open and subject to variables outside the control of utility managers. Among these variables stood state utility regulation, which in the early years of the century failed to provide the legal machinery to facilitate service to rural customers. The absence of standard rules and procedures impeded utilities and farmers from working with each other, contributing to managers' complaints about the difficult class of customers. The environment improved considerably in the 1920s and early 1930s, however, with regulatory bodies facilitating the extension of high lines to an increasing number of farmsteads.

Franchises and Regulatory Functions in Cities and Rural Areas

The entrepreneurs who established power companies in cities during the 1880s and later followed the same legal path as others who earlier built urban railway and telephone networks: they obtained franchises—formal concessions from municipal governments to string wires over and under streets in return for the distribution of beneficial services to citizens.[1] In New York City, electricity providers obtained franchise grants, made by the Board of Aldermen, to operate for limited amounts of time.[2] As partial compensation, the city received free public lights and assurances of reasonable customer rates.[3]

At first, policymakers viewed dispensation of franchises—often to many suppliers within a city—as a way to promote competition and better, cheaper service. But several years of practice disabused them of that notion.[4] Company officials found they could easily bribe officials for favorable franchise conditions,[5] while technological change in the industry often overtook the limits of agreements. The use of alternating current networks and large generating plants, which enabled long-distance power transmission and scale economies, for example, made consolidation of firms (along with their franchise areas) attractive. Many cities consequently ended up with de facto monopolies immune from the forces of competition or public control.[6]

While some city governments gained oversight by buying private companies and establishing municipal utilities, another model of public control materialized. In 1907, Wisconsin and New York established state commissions that approved electric companies' rates, requirements for service, and issuance of corporate securities. Importantly, Wisconsin's regulatory law gave its commission the authority to issue "indeterminate" franchises to companies.[7] No longer holding the right to sell power for a limited amount of time (such as twenty-five years), utilities could operate in assigned territories for an indefinite period that depended only on how well they performed. The threat of municipal takeovers remained a stick to help companies maintain their eagerness to serve customers properly. By 1920, thirty-five states had established similar institutions that exercised jurisdiction over electric utilities;[8] despite imposition of government supervision, power companies and commissions developed a stable relationship that appeared to benefit all parties.[9]

Significant for this account, the establishment of a utility as a legal monopoly within a franchise service area came with the company's "obligation to serve" all entities within it. Drawing on English common law and statutes, this commitment enabled city residents and businesses to request (and receive) power from a local electricity provider at published rates and without discrimination within various classes (residential users, manufacturers, etc.). Nor could potential customers usually be denied service if they initially appeared to be unprofitable. Moreover, new and old customers obtained equal treatment and quality of service (in theory), all of which came under the purview of regulatory commissions.[10]

The obligation to serve had little meaning to most rural citizens, however. Unlike city folk who lived in incorporated entities holding the right to issue franchises, ruralites generally inhabited unincorporated areas managed by larger administrative entities such as counties. In many of these areas, local governments did not retain the authority to offer franchises. As noted by historian Forrest McDonald in his discussion of Wisconsin utilities, the legal ability to serve residents in unincorporated areas remained "vague" and "almost non-existent" in the years during and after World War I.[11] Today, we can read maps that illustrate the neat allocation of geographical territories to specific power companies, rural co-ops, and municipal utilities. But early in the twentieth century, many nonurban areas remained electrically barren and unclaimed by utilities.[12] Nothing compelled firms to accept new customers in these undeveloped regions in the early 1920s, and firms could reject applications for distribution line extensions at their discretion.[13]

If companies wanted to sell electricity to individuals in unassigned territories, usually contiguous to those already served, they generally could do so.[14] But the rules for doing so varied greatly throughout the country. In some states, noted a government survey published in 1936, "utilities may extend their lines into contiguous unincorporated territory without obtaining the permission of either the State commission or the local authorities."[15] In other jurisdictions, companies needed to receive regulatory approval. And in some cases, cities or towns held the limited right to serve ruralites. Municipal utilities in Indiana, for example, could string lines to farms, but only within six miles of corporate limits.[16] This inconsistent fabric of authority persisted even as farm electrification grew more common after 1935. It remained until 1950, for example, before Virginia's legislature granted the state commission the ability to designate service areas among power suppliers.[17] Florida's regulatory body, meanwhile, only acquired that power in 1965 through a court ruling.[18]

While regulatory commissions could not force companies to serve rural customers outside of franchise areas, they acquired (in many states) oversight jurisdiction once the firms chose to provide electricity. As an early example, the California

Railroad Commission in 1916 established its authority to review and alter agreements between agricultural customers and companies.[19] The Illinois commission likewise exerted its right to approve conditions imposed by utilities on rural customers. In issuing its General Order 59 in 1920, state regulators sought to clarify procedures for handling the increasingly "extensive demands for electric service in rural communities."[20]

When power companies agreed to extend high lines to rural areas, they typically expected farmers to pay, in advance or within a short amount of time, a large portion of the initial costs. The Southern California Edison Company in 1911, for instance, charged high prices during the first year of service to recoup up to half of its total costs.[21] Other California companies also wrote contracts requiring farmers to defray most of the distribution line expenses within the first year, with the balance repaid in subsequent years. But as these utilities became convinced of the profitability of irrigation loads, they assumed the cost of the lines and transformers themselves, knowing they would recover expenditures quickly. The move encouraged more rapid electrification among farmers, who previously found it onerous to compensate firms for high-line expenses and to pay for irrigation pumps and home wiring. Happily, as noted by a Pacific Gas and Electric Company official in 1913, the new "liberal policy" contributed to making "electricity extremely popular, and in many sections[,] it is the rule rather than the exception to see the farmer doing a great part of his work by means of electricity."[22]

Outside the West, the cost of extensions constituted a greater hardship for companies, especially those obliged to negotiate individual agreements with every new rural customer. Throughout the next decade, however, regulatory commissions advanced rural electrification efforts by working with utilities to provide more uniform and simpler-to-administer terms for farmers. In May 1920, for example, the Wisconsin regulatory body heard a case in which farmers sought extension lines from a newly constructed hydroelectric facility. Instead of entering into unique contracts with each farmer, the Wisconsin Power, Light and Heat Company won approval of a trial plan that included uniform terms for all users. Following the utility's lead, other firms in the state estimated the cost of line extensions; customers would then pay the full cost before the company offered service, with the understanding that if the estimated amount exceeded the actual charges, the companies would refund the excess.[23]

Once the lines became energized, new customers paid "regular urban rates" plus surcharges for depreciation of transformers and other fixed expenses. In making this arrangement, the Wisconsin commission explicitly acknowledged the high cost of serving rural customers. It also appreciated the fact that utilities needed to make their extensions self-supporting.[24] In other words, regulators generally concurred with company officials who appeared unwilling to assume all the perceived risks of serving low-consumption customers.[25] In reviewing various arrangements for financing extension lines, a 1922 NELA report concluded, "Neither the utility nor the farmer knows very much regarding the varied uses to which farmers can put electricity, nor does either fully appreciate its actual value."[26] (Of course, NELA created research committees and the CREA to help overcome these gaps of knowledge.)

Easing Inhibitions to Rural Electrification

As utilities and commissions saw greater rural demand for electricity in the following years, new arrangements emerged that provided more certainty and less burdensome terms for farmers. In Wisconsin, the Milwaukee Electric Railway and Light Company contributed to the cost of extension lines equal to the estimated revenue from customers over a three-year period; customers paid the balance.[27] Further liberalization of the rules came in 1926, when the Wisconsin regulatory commission approved the plans of eight major companies to spend up to $400 per customer for high-line construction. Regulators also accepted in 1926 the utilities' application to standardize the rules and rates it offered to rural customers outside of 194 cities in 33 counties.[28] Not remarkably, perhaps, Grover Neff signed the agreement as vice president of the Wisconsin Power and Light Company and as the representative for the other firms. By doing so, he helped the state rationalize a previously haphazard process of providing service to farmers.

Novel approaches for bringing power to rural citizens emerged in Pennsylvania within the Giant Power plan introduced in 1923. A report produced by its advocates in 1925 observed that existing formulas, in which rural customers paid higher rates than city customers, deterred rural electrification.[29] The proposal, if enacted, would have helped mitigate such constraints by requiring utilities to extend distribution lines to unserved customers, some of whom might have established cooperative distribution companies to purchase and distribute power to their owners (namely, to the consumers themselves).[30] Though Governor Gifford Pinchot and Morris Cooke lost political battles within the state to make the Giant Power proposal a reality, suffering from intense utility industry opposition, some of the ideas persisted. The notion of using farmers' cooperatives became a key element of the REA after 1935—no surprise, since Cooke helped convince President Roosevelt to create the organization and then served as its first administrator. And while the proposal's provisions giving the state regulatory commission extra powers to advance rural electrification never became law, the oversight body still pursued creative means to make it easier for farmers to obtain high-line service.[31]

One such innovation consisted of General Order 28, the result of negotiations managed by the commission between the State Council of Farm Organizations, a group of agricultural and government organizations, and the Pennsylvania Electric Association (which represented utilities).[32] Under the 1927 ruling, power companies would construct, operate, and maintain distribution lines at their own expense, but rural customers would guarantee revenues over a three-year period to make the service viable to the firms.[33] Policymakers hoped that when farmers gained more familiarity with electricity, their usage would increase amply, so that "the question of the monthly minimum will cause no difficulty."[34] To reduce cash outlays, consumers could supply labor and materials toward the lines' construction.[35]

The approach brought positive results. Within seven months of the new policy's implementation, Pennsylvania's utilities built 1,441 miles of rural lines and initiated service to almost 18,000 previously unelectrified rural customers, of which 4,317 were farms. The chairman of the public service commission observed in 1928 that companies connected nonurban power consumers at a rate of fifty-eight per working day, with the expectation of building another four thousand to five thousand miles of rural distribution lines in the next four years.[36]

Virginia adopted a comparable plan after its General Assembly created a commission in March 1928 to accelerate the state's rural electrification efforts. Using the Pennsylvania plan explicitly as a model, the body (chaired by Julian Burruss, president of Virginia Polytechnic Institute) recommended various measures, adopted by the state's regulatory authority as Rule 18, to extend power lines to rural customers.[37] As in Pennsylvania, the plan offered farmers the ability to contribute labor to reduce cash costs, but it established four years (instead of three) as the contracted period for minimum payments. For a five-mile-long extension costing $5,000 and serving ten customers, the required outlay on a four-year contract would be about $100 annually, or $10 monthly per customer.[38]

This type of arrangement became increasingly popular. Farmers could acquire electricity without a huge initial expense, such that they had money available to invest in wiring and equipment, which would lead to increased electricity demand.[39] (Utilities, on the other hand, won assurances they would earn enough revenue to avoid financial losses.) As noted in 1929 by a manager of a northern Illinois firm that employed a similar rate design, a farmer would make "liberal use of electricity if he can obtain it without being obliged to finance extension of lines."[40] Concurring, the executive secretary of the American Farm Bureau Federation observed in 1930 that by developing means for utility financing of the lines, "the farmers have been able to invest available funds in electrical devices and power consuming equipment." Consequently, the newly installed lines "have rapidly developed very satisfactory current loads throughout the whole year, and as another consequence, rates for this current have been reduced, thus further benefiting the farmer."[41]

Some state regulators exerted more power—by requiring companies to completely finance the construction of rural distribution lines and without asking for any financial contributions from customers. Understandably, California's regulators appear to have led efforts to make its line construction policy more flexible, seeing that farmers in the state used disproportionately large amounts of power. In 1929 alone, the state regulatory commission helped broker an arrangement whereby utility companies paid the entire expense of rural distribution lines.[42] Beyond that, a few midwestern firms in 1929 had also "gone so far," according to a New York State agricultural professor, "as to underwrite the wiring and equipment charges simply as a means of building load," seeing that electrification increasingly proved beneficial to the farmer and would yield reasonable income for the utility.[43]

A 1933 academic review reported that the new, more permissive method of company financing of distribution lines had become increasingly prevalent, contrasting with earlier methods of forcing farmers to pay all the construction costs themselves. To be sure, companies still sought minimum returns, but with prodding and approval from regulatory commissions, they also implemented rate structures encouraging increased power consumption and greater utility revenue. In states such as Virginia, Wisconsin, Pennsylvania, Vermont, Alabama, and New York, utilities and oversight bodies established "far sighted policies" for stimulating wider use of electricity on the farm.[44]

By the late 1920s and early 1930s, commissions had approved various formulas for rural customers to obtain line extensions. Aside from authorizing contracts for irrigation customers, regulators ratified rate structures for farmers who used electricity for lighting and appliances.[45] The burgeoning interest in rural electrification and the involvement of state commissions led the American Society of Agricultural Engineers to issue a report in 1928 summarizing recently developed arrangements for farm customers. Though large variations existed in them, reflecting different local conditions, the plans retained common features: in most cases, farmers paid part of the cost of line extensions in advance as well as per kilowatt-hour rates that exceeded those of urban consumers.[46] Another study, performed in 1932, continued to highlight differences. Nevertheless, the companies, often with state commission endorsement, sought to share the construction cost of line extensions with new rural customers who would be expected to use increasing amounts of power.[47]

That expectation already had gained some empirical backing. The trade literature frequently published encouraging stories about companies and commissions that stimulated demand by eliminating up-front remittances for line extensions and by reducing rural electric rates. Grover Neff reported that per-customer electricity use on his company's connected farms grew 44 percent between 1925 and 1927, for example.[48] Likewise, evidence from New York State showed that farmers with access to electricity boosted consumption by 59 percent between 1926 and 1928.[49] Some New England utilities, meanwhile, observed in 1929 that more generous construction terms and cheaper energy prices had helped initially unpromising farm customers increase their demand, pushing them into a "profitable class."[50] In further good news for utility companies, ruralites quickly demonstrated greater use of electricity than "domestic" (that is, urban residential) customers did. A study of twenty-three companies serving more than ninety thousand farm customers in 1932 noted that average use exceeded that of city residents nationwide by a factor of about three.[51] Rural use continued to grow, *Electrical World* concluded, and in rural areas, "the potential market is enormous."[52]

Well before creation of the REA, then, the regulatory framework for dealing with power companies' extension of lines outside of franchised, urban areas had become more standardized and generous to rural customers. Compared with the piecemeal manner in which utilities and commissions dealt with rural electrification in the 1910s, circumstances had improved significantly by the late 1920s and early 1930s. Instead of holding expectations for full payment of rural lines before or soon after construction, many utilities contributed to building costs and gave farmers a few years to increase their consumption and provide companies full restitution of their original outlays. Of course, the legal and regulatory status of electrification efforts in rural districts had not yet achieved the consistency that existed in cities. Nevertheless, state commissions had begun playing increasingly supportive roles as stakeholders in the rural electrification subsystem.

Chapter Eleven

Momentum in the Rural Electrification Subsystem

> We can say encouraging progress has been made in *all directions*, and that remarkable progress has been made in some features of the [rural electrification] movement. The encouragement comes from the increased interest in the matter by all parties, and the remarkable material progress in the extension and use of the service, the increased activities of our educational institutions, the increased publicity, the better understanding, and changed mental attitude [among utility managers and farmers].
>
> —EUGENE HOLCOMB, CHAIRMAN,
> NELA RURAL SERVICE COMMITTEE, 1929

As regulatory elements of the rural electrification subsystem began to co-alesce in the 1920s and 1930s, so did other components. In fact, government-approved accommodations for farmers seeking electric service paralleled—and became co-constructed with—the activities of proponents in educational and corporate institutions. At land-grant colleges, for example, professors offered short courses on and demonstrations of rural applications of electricity for traditional students, businesspeople, and farmers. And as interest in nonurban electrification expanded, utility companies established a growing number of rural service departments within their firms, often staffed by agricultural engineers. The industry's trade organization continued research on critical problems—especially means to reduce the cost of stringing distribution lines—in efforts to make rural electrification more economically feasible. Manufacturers also expanded efforts to perfect and promote equipment that farmers would desire once they received high-line service. Even publishers saw ways to reach a market of country readers interested in electricity. In short, a host of actors contributed to efforts resulting in a fourfold increase in the number of farms that obtained central station–generated electricity between 1923 and 1933.

Developing Manpower for Rural Service Departments at Power Companies

Unsurprisingly, land-grant institutions played leading roles in creating the educated manpower that advanced rural electrification. The University of Idaho was among the first colleges providing formal training on electrified agricultural machinery, which constituted part of the Agricultural Engineering Department's "Farm Motors" course in 1911.[1] A decade later, the department offered a dedicated course on "Electricity on the Farm," which focused on the "general operation of electric generators and motors, . . . construction and operation of storage batteries, farm lighting units, and house wiring."[2] At North Carolina State College (now North Carolina State University), students who enrolled in "Agricultural Engineering 404" in 1921 learned to operate isolated lighting plants, such as those made by the Delco-Light Company.[3] Virginia Polytechnic Institute also offered courses dealing specifically with farm applications of electricity within the Agricultural Engineering Department. In a 1923 article, VPI's Charles Seitz described his program's farm electric plant laboratory and included a photograph illustrating its up-to-date electric motors, isolated generating equipment, and storage batteries (figure 11.1).[4] At about the same time, professors at the Washington State College taught an "Electricity on the Farm" course that introduced students to "farm lighting outfits and installation of same; care of storage batteries; the farm telephone; and . . . different types of gas engine ignition."[5] Other universities offered classes on agricultural power that dealt with (along with other sources of energy) wind and electricity.[6] By the late 1920s and early 1930s, a few colleges listed specific courses on "Rural Electrification."[7]

Upon graduation, many of the agricultural engineering students worked for power companies, managing the firms' rural electrification efforts. In 1929, VPI's Professor Seitz proudly informed his institution's president that he sent six graduates to pursue farm electrification activities at four of the state's leading utilities.[8] Some of the graduates obtained jobs at the Appalachian Electric Power Company (based in Roanoke, about forty-five miles from the school), which increased its efforts to attract rural customers. Former student R. R. Choate became a rural service engineer with the firm in 1928, a time when the company served only 194 nonurban customers. Meeting with farmers, he learned of their hard-headed determination to know whether their investment in electrical equipment would earn more money than they spent. The farmer "wants to know what the equipment will cost in dollars and cents and what it will save for him in dollars

Fig. 11.1. Farm electric plant section of Virginia Polytechnic Institute's agricultural engineering laboratory. *Source*: Charles E. Seitz, "Agricultural Engineering Development in Virginia," *Agricultural Engineering* 4, no. 4 (April 1923): 60. Used with permission of the American Society of Agricultural and Biological Engineers (ASABE, formerly ASAE).

and cents or hours of labor." He further observed that dairy farmers seemed most interested in electrification and that electric water pumping systems had become among the first electrical devices installed outside the home. Beyond his visits to existing and potential customers, Choate remarked that his company had set aside a lecture room in its Roanoke office where customers could evaluate appliances exhibited by local dealers and manufacturers. The ultimate goal of all these activities, of course, was "to build up the rural load which will be profitable to the consumer as well as to the Power Company."[9]

Just as the Roanoke firm had done, the Alabama Power Company took advantage of trained specialists from its state's land-grant college, Alabama Polytechnic Institute, to sell farmers on electricity. By 1932, the utility had established five regional offices for assisting rural customers, each staffed by agricultural engineers who advised clients about using electrical appliances and equipment in ways that would increase their net income and improve their families' standard of living. Making 5,429 visits to individual rural customers and interviewing another 902 prospective customers during 1932, the engineers worked with company salesmen to help farmers buy appropriate equipment to suit their particular situations. "The farmer soon learns that the agricultural engineer is his consulting engineer," the company boasted in a paper that won an industry award for rural electrification.[10] Partly because of the company's activities, and despite the depressed agricultural conditions in the state since the early 1920s, annual electricity sales to rural customers rose significantly—from 294,000 kWh in 1924 to 5,957,000 kWh in 1932— an increase of 1,926 percent. That impressive gain occurred as the firm erected increasing lengths of nonurban high lines, jumping from just 78 miles in 1924 to 1,914 miles in 1932.[11]

As early as 1927, fifty-seven companies had established rural service departments, while forty-three others had trained people to address the farm market without creating dedicated organizations.[12] Like the new "special farm service department" established in 1927 by Nebraska Gas & Electric Company, these corporate bureaus employed men who "will frankly tell the farmers what electric service is practical and in what way service can be obtained."[13] A year later, 160 companies had established such programs, populated by 403 agricultural experts.[14] In June 1928, Grover Neff asserted that creation of these rural service divisions "proves that the utility companies of this country recognize that farm electrification is a big job," requiring large financial and managerial commitments. He added, "Those utility companies which first organized separate rural service departments are taking the lead in farm electrification."[15] In presenting one company's program as an example, NELA's Rural Electric Service Committee in 1927 observed that employees provided farmers with various forms of assistance and cooperated closely with manufacturing firms and contractors, agricultural schools, and the state CREA affiliates.[16] Another NELA committee reported that the people staffing these rural service departments "know both the electric business and the farming business" and can make "intelligent recommendations about the application of electric energy on any given farm" while, simultaneously, inspiring "a feeling of confidence."[17]

Complementing these efforts, the General Electric Company began in 1927 a rural electrification course, which the firm claimed was "the first of its kind fostered by a large electrical manufacturing company." Its inaugural class contained five men who had grown up on farms and who already had taken university classes in agriculture departments, such that they appreciated "farm problems at first hand." The four-month-long program included instruction in electrical engineering, manufacturing, and sales of small electric motors, lighting equipment, electrical household devices, and other useful implements. Though some graduates would likely work for General Electric, the firm "expected that this course will make available to public utilities a group of men who have become specialists in the commercial end of farm electrification."[18] In other words, the company provided another source of professionals devoted to farm electrification.

By 1929, about two hundred power companies had hired employees specifically to pursue rural electrification activities, with many coming from programs at agriculture schools.[19] Two years later, the number of "especially qualified rural service men" totaled approximately one thousand nationwide.[20] All of Wisconsin's major electric utilities had, by 1931, established rural service departments, in which agricultural managers gained assistance from field workers, manufacturers' representatives, and equipment distributors.[21] To encourage creation of these organizations within utilities, NELA's Rural Electric Service Committee published case studies of firms that had already done so. In its first report, the group highlighted the work of the Wisconsin Power and Light Company and the Alabama Power Company, providing examples of rate structures for farm customers and newspaper advertisements.[22] Using "the facts as developed by the experimental lines and put[ting] them to work on thousands of other farms," members of these rural service departments "are in a pioneer field and are doing great work both for the industry and for agriculture," according to a 1929 report.[23] L. E. May of the Cleveland Electric Illuminating Company harmonized with this view in 1932, observing that by working with farmers and manufacturers, a company's rural service department enabled "the advancement of the farming community it serves, and [the firm] derives a share of the benefits which accrue to the community."[24]

R&D on Distribution Line Technology

The extension of lines to farms benefited from incremental improvements in distribution technologies. A NELA committee on rural service observed that manufacturers pursued such efforts beginning in 1913. Several manufacturing companies, for example, worked to design efficient outdoor, weatherproof substations in which transformers lowered voltages from 33,000 volts or higher to about 2,300 volts for use on distribution lines. (Other transformers near farms stepped down the voltage to 440, 220, or 110 volts for use by various machines and appliances.)[25] Further advances enabled General Electric to claim in 1929 that its new substation technology effectively resolved the "problem of supplying scattered rural central-station loads economically."[26] The company also crowed about its new factory-prefabricated switching equipment that reduced costs without sacrificing engineering excellence.[27]

As individual companies continued to develop lower-cost hardware, NELA's Overhead Systems Committee in 1921 announced efforts to establish guidelines for distribution line construction.[28] Based on reported experiences, the committee concluded that companies could build lines to farms more simply and cheaply than those employed in urban areas. They assuredly needed to remain safe, but as committee member Grover Neff pointed out, "farm lines do not have to be built as heavy and strong as lines in cities."[29] Two years later, as chair of a NELA subcommittee on rural lines, Neff reported on successful attempts to increase the distances between poles suspending distribution lines. Some newly installed wires carried electricity at 2,300 to 17,000 volts, traversing distances of greater than 250 feet between poles, in contrast to the more typical rural spans of about 150 to 200 feet.[30] As distances increased, the companies needed fewer poles and associated equipment (such as porcelain insulators used to ensure that electricity does not flow into adjacent wires or physical supports, especially when wet), resulting in lower costs. Neff explicitly noted that extending the distance between poles produced large savings and could "aid in the solution of the economic problems which are so serious a feature of rural line construction."[31]

Taking a novel approach, USDA agricultural engineer George Kable reported in 1928 on the design of lightweight frames (made of two 2-by-4 hinged timbers to form an "X" shape) that could replace poles and be installed quickly; spans cost as little as $333 per mile in Oregon.[32] Summaries of other research projects appeared in the comprehensive June 1931 *CREA Bulletin* issue. One experimental rural line in Florida saw poles spaced at a notably long distance of four hundred feet; impressively, it survived a 1929 hurricane that delivered wind gusts of 117 miles per hour. More generally, NELA had begun surveying the best construction practices that included longer spans and higher voltages. While some approaches may not have yielded positive results, the *CREA Bulletin* editor concluded, "Considerable progress has been made in reducing costs, and improvements are still under way."[33]

Similar advances in design of rural distribution lines received attention in a 1932 *Electrical World* article detailing work done in Michigan. Using strong wires made of aluminum reinforced with steel cores, a utility built economical, no-frills lines with spans of three hundred feet. The author concluded "that long-span construction with high strength conductor is very definitely cheaper than short-span construction." But no one should be surprised, he remarked, since "it has been demonstrated so repeatedly in the last few years that long spans are in use by the majority of power companies building rural lines."[34] More striking, an engineer for a Tennessee company explained that his firm had begun erecting lines that spanned six hundred feet between poles, made possible partly through the use of "copperweld" wire. Relatively new and strong, the copper and steel conductor also cost less than traditional all-copper wires.[35] Expenses ranged from under $500 to about $900 per mile.[36]

In other words, utility managers had become increasingly familiar—in the days before the TVA and REA—with new ways to decrease the cost of stringing lines in rural areas. As an *Electrical World* editorialist contended in 1933, while referring specifically to an article describing the use of copperweld lines, "the industry knows how to build rural lines at a minimum cost to give maximum service."[37] Alex Dow, president of the Detroit Edison Company, concurred, observing that companies had learned much in previous years, such that they employed lightweight poles separated by four hundred feet (or, as in the case of the Tennessee utility, up to six hundred feet) to reduce costs. Additionally, they found that they could transmit power along existing roadways at higher voltages than previously imagined, allowing for less current loss from electrical resistance. (Higher voltages generally meant diminished resistance and line losses.) The lower costs gave managers hope that they could serve more rural customers and at rates comparable to those paid in cities. "Even in districts where farms run four to the mile," Dow observed, "the differences in costs are tending to disappear as between rural service and metropolitan service, and the differentiation of rates is yearly less warranted."[38]

In a similar acknowledgment that companies had found ways to reduce costs, Philip Sporn, vice president of the American Gas and Electric Company, claimed in 1934 that his firm had simplified the design and construction of distribution lines for rural areas, such that the company pared the cost per mile of line to about $500.[39] The assertion comported with the generally accepted notion that the power industry had greatly lowered line expenses. In an article highlighting the passage of the Rural Electrification Act of 1936, the *New York Times* explained that companies paid between $830 and $1,000 per mile in 1936 to construct a mile of rural high line in that year. For comparison, the average cost of the first 4,500 miles of REA-financed lines came to slightly more than $1,000 per mile.[40]

Some of these improvements received attention when Arthur E. Silver, a retired engineer who worked for the Electric Bond and Share holding company, accepted an award in 1951 for contributions to rural electrification. Like others, he noted that utility managers had realized by the 1920s that they needed to reduce costs if rural electrification were to expand significantly. He pointed to pioneering efforts pursued by Idaho Power Company, then a subsidiary of the holding company, which performed comprehensive surveys of its rural territories and coordinated engineering, sales, rate, and construction efforts for entire communities instead of just dealing with individual farmers. (In part owing to its successful work to serve a largely rural territory, the firm won the utility industry's Charles A. Coffin Award in 1936.)[41] He further described the use of high-strength, low-cost wires to extend span lengths to six hundred feet, reducing the number of poles by half compared with previously standard practice. Cooperating with manufacturers, he also contributed to the design of a simplified and cheaper farm transformer. The efforts, Silver proudly stated, helped force down the cost of service to farms by more than 35 percent in the decade ending around 1935. Overall, this work proved so successful, he claimed, that the Rural Electrification Administration "adopted almost unchanged the designs and construction practices developed and established by the private power companies." Perhaps taking more credit than the REA would have offered, he observed that the government agency and private utility companies "together made rapid progress toward completion of the task" of bringing electricity to farmers.[42] (Some examples of Idaho Power's promotional materials are illustrated in figures 11.2, 11.3, 11.4, and 11.5.)

Short Courses and Other Forms of Spreading the Rural Electrification Message

Central station managers had complained for years that farmers needed to learn how to employ electrical equipment more productively and increase power consumption. In addition to having its rural service agents consult with rural customers, the utility industry leveraged the value of its land-grant partners to disseminate knowledge to various stakeholders. Most notably, the NELA, CREA, and experiment stations at land-grant schools often collaborated with power firms to sponsor various educational events dealing with rural electrification.

Purdue University offered the first short course in October 1927, where professors and experiment station staff members demonstrated the newest electrified farm equipment to utility representatives. The dean of the state's agricultural experiment station, who also served as the chairman of Indiana's rural service committee, and Eloise Davison, a member of the Iowa CREA affiliate, welcomed the attendees.[43] Grover Neff, chair of NELA's Rural Service Committee, addressed an evening banquet, while electrical experts spoke on various topics. The University of Wisconsin's Floyd W. Duffee, for example, discussed the "general purpose farm motor," while F. G. Riley of Purdue's Department of Poultry Husbandry described the "use of electric lights to increase winter egg production."[44]

At least three other land-grant colleges—in Wisconsin, Oregon, and Washington—offered similar learning experiences in early 1928.[45] California's earliest short course

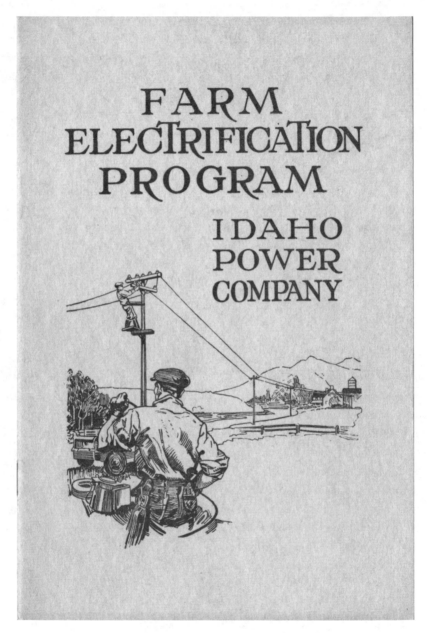

Fig. 11.2. Cover of Idaho Power Company brochure on farm electrification, noting (inside) that as of 31 July 1928, the company furnished power to 7,208 rural customers. *Source:* Used with permission of Idaho Power Company. Photograph courtesy of Washington State University Libraries, Manuscripts, Archives, and Special Collections.

Fig. 11.3. Frontispiece of Idaho Power Company booklet, originally captioned "Electricity does the farmer's work." *Source: Farm Electrification Program* (Boise: Idaho Power Co., 1928). Used with permission of Idaho Power Company. Photograph courtesy of Washington State University Libraries, Manuscripts, Archives, and Special Collections.

Fig. 11.4. Electricity for barns, water supply, and irrigation systems on Idaho farms. *Source*: Idaho Power Company, *Farm Electrification Program* (Boise: Idaho Power Co., 1928), 6. Used with permission of Idaho Power Company. Photograph courtesy of Washington State University Libraries, Manuscripts, Archives, and Special Collections.

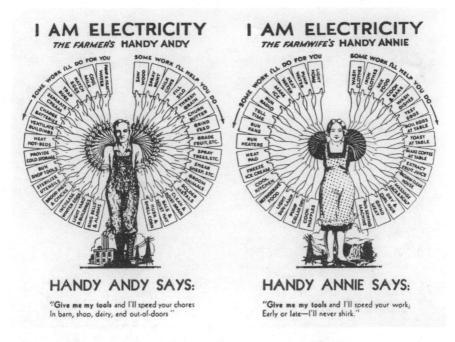

Fig. 11.5. Graphic used as part of a 1933 sales program by Idaho Power Company showing how electricity makes life better for men and women on the farm. *Source*: Used with permission of Idaho Power Company. Photograph courtesy of Tyrone Corn of Idaho Power Company.

took place in December 1928, with utility employees playing the role of students who "spent three days sitting at the feet of university engineers so they might learn how to make electricity serve agriculture better," according to a newspaper editorial.[46] Increasingly typical, many of these courses devoted some time for addresses on women's issues, such as the talk "What Electric Service Means to the Farm Housewife," given at the 1928 Purdue event and another on electric cookery at Washington State College in 1929.[47] In such specially designated "Women's Sessions," experts demonstrated, for example, the benefits gained by baking vegetables in electric ovens and by using electric washing and ironing equipment.[48]

Virginia Polytechnic Institute sponsored its first short course in June 1929 (the program is shown in figure 11.6). The event, according to the Richmond, Virginia, *News Leader*, "is said to be the first short course of its kind in the Southeastern states,"[49] and it reflected the view among members of the recently formed Joint

Rural Electrification Short Course

OBJECT

The course is designed mainly for men and women engaged in the advancement of rural electrification. It is open to all interested in the subjects under discussion.

REGISTRATION

Those desiring to attend the course are requested to register by mail at once. This is desirable as it is necessary to know the approximate number of persons to provide for. There will be no registration fees connected with the course.

ACCOMMODATIONS

Lodging can be secured at Green's Hotel, Old Brick House, University Club, or private homes. Meals can be had at the hotel, restaurants, or college dining hall.

Write the Agricultural Engineering Department, V.P.I., Blacksburg, Virginia, indicating preference for accommodations.

To reach Blacksburg, take the Norfolk and Western to Christiansburg, where automobile transportation can be procured to Blacksburg, a distance of eight miles.

First Annual Rural Electrification Short Course

TO BE HELD AT

Virginia Polytechnic Institute

Wednesday, Thursday, Friday
June 12, 13, 14, 1929

Agricultural Engineering Department
Virginia Polytechnic Institute

The Public Utilities Association of Virginia
The N. E. L. A. and Manufacturing Companies
Cooperating

SESSIONS IN AUDITORIUM
AGRICULTURAL EXTENSION BUILDING
V. P. I.
BLACKSBURG, VIRGINIA

Fig. 11.6. (above and opposite) Virginia Polytechnic Institute's short course program on rural electrification, 1929. *Source:* Used with permission of Virginia Tech University Libraries, Special Collections and University Archives. Photograph courtesy of Dr. W. Cully Hession, Virginia Tech Department of Biological Systems Engineering.

Committee on Rural Electrification, convened by Governor Harry Flood Byrd, that education on electricity constituted an important element in solving the "farm problem."[50] College president Julian A. Burruss greeted the guests upon their arrival; he had demonstrated his commitment to rural electrification by chairing the governor's committee.[51] Luminaries included Earl White, director of the national CREA, George Kable, director of the National Rural Electric Project, and represen-

Program

Wednesday, June 12th

MORNING SESSION

8:30—9:30—Registration.
9:30—Address of Welcome, Dr. J. A. Burruss, President, V.P.I.
9:50—"Rural Electrification in Virginia," Chas. E. Seitz, Head Agricultural Engineering Department, V.P.I.
10:10—"A Program for Rural Extensions"
W. E. Wood, President, Virginia Electric Power Company.
J. W. Hancock, Division Manager, Roanoke and Lynchburg, Division, Appalachian Electric Power Company.
Lewis Payne, General Manager, Virginia Public Service Company.
L. E. Long, General Manager, Shenandoah River Power Company.
10:50—"How Can the County Agent Assist in the Rural Electrification Movement?" John R. Hutcheson, Director, Extension Division, V.P.I.
11:15—Discussion.
11:30—"Electric Power a Logical Development in American Agriculture," George W. Kable, Director, National Rural Electric Project.
12:00—Discussion.
12:30—Intermission.

AFTERNOON SESSION

1:30—"Requirements of Electrical Equipment for the Farm Home," Miss Eloise Davidson, Research Department, National Electric Light Association.
2:00—Discussion.
2:15—"Rural Electrification in the United States," Dr. E. A. White, Director, National Committee on Relation of Electricity to Agriculture.
2:45—Discussion.
3:00—"Rural Line Construction," W. I. Whitefield, Manager, Roanoke Division, Appalachian Electric Power Company.
3:30—Discussion.
4:00—"Contracts and Rates," C. N. Schoonmarker, Virginia Public Service Company.
4:30—Discussion.
7:00—"Good and Bad Practice in Farm Lighting," W. C. Brown, National Lamp Works, General Electric Company.
7:30—Discussion.
8:00—Rural Electric Movies, "Romance of Sleepy Valley," American Farm Bureau Federation; "Yoke of the Past," General Electric Company.

Thursday, June 13th

MORNING SESSSION

8:30—"Irrigation by Electricity," W. H. Coles, President, Skinner Irrigation Company.
9:00—Discussion.
9:30—"Electricity and the Poultry Industry," H. L. Moore, Extension Division, V.P.I.
10:00—Discussion.
10:15—"Electric Brooding, Incubation and Poultry House Lighting," Geo. W. Kable, Director, National Rural Electric Project.
11:00—Discussion.

11:30—Inspection of V.P.I. Poultry Plant and Electrical Equipment.
12:30—Intermission.

AFTERNOON SESSION

1:30—"Electric Water Systems for the Farm," Professor P. B. Potter, Agricultural Engineering Department, V.P.I.
2:00—Discussion.
2:30—"The General Purpose Portable Farm Motor," F. T. Smith, Industrial Department, General Electric Company.
3:00—Discussion.
3:30—"Peak Loads on the Farm" (Methods of Building a Profitable Rural Load), Geo. W. Kable, Director, National Rural Electric Project.
4:00—Discussion.
4:30—"Demonstration of the Rural Electric Truck," L. T. Wood, Agricultural Engineer, Virginia Electric Power Company.
5:00—"Demonstrations of Farm Electric Equipment," Agricultural Engineering Laboratory.
7:00—Dinner. Address, F. W. King, Vice-President, Virginia Public Service Company.
"Cooperation of Individual Light Plant Dealer and Electric Utility Companies," J. E. Waggoner, Public Relations Department, Delco Light Company.

Friday, June 14th

MORNING SESSION

8:30—"Electricity and the Dairy Industry," Professor C. W. Holdaway, Head Dairy Husbandry Department, V.P.I.
9:00—Discussion.
9:30—"Electric Milk Cooling and Storage on the Farm," C. W. Pegram, Dairy Manufacturing Specialist, Extension Division, V.P.I.
10:00—Discussion.
10:30—"Electric Milking Machines and Separators," P. M. Reaves, Dairy Husbandry Department, V.P.I.
11:00—Discussion.
11:30—"Electric Feed Grinding Equipment," Professor V. R. Hillman, Agricultural Engineering Department, V.P.I.
12:00—Discussion.
12:30—Intermission.

AFTERNOON SESSION

1:30—"Methods of Merchandising Electrical Equipment to the Farmer"
As Viewed by Sales Manager of Electric Company," L. F. Riegel, Virginia Electric Power Company.
2:00—Discussion led by N. F. Lawler, Virginia Public Service Company.
2:30—"As Viewed by Manufacturers of Farm Electrical Equipment," J. W. Savage, Merchandising Department, General Electric Company.
3:00—Discussion led by C. G. Hillier, Mansfield Works, Westinghouse Electric and Manufacturing Company.
3:30—"As Viewed by Rural Service Field Man," R. R. Choate, Agricultural Engineer, Appalachian Electric Power Company.
Remainder of afternoon to round table discussions and demonstrations of equipment.

tatives from General Electric, Westinghouse, and Delco-Light.[52] A report written after the course conclusion noted that attendance reached more than seventy-five people.[53]

The courses and demonstrations quickly evolved to reach beyond an audience of experts and to the general public. Private companies sponsored many of the events, sometimes with the cooperation of the land-grant colleges and their extension services. At a series of rural exhibitions (eighteen during March 1929), Wisconsin Public Service Corporation representatives displayed milking machines, a feed grinding machine (employed to process more than one thousand bushels of grain at no cost to visitors), a lavatory with hot and cold running water, an electric range, and other household appliances. Eschewing any explicit effort to sell equipment, the meetings served largely educational intents, often including refreshment and various amusements. The company's director of the Home Service Department, Zella Patterson, demonstrated electric baking, followed by free distribution of electrically baked cookies, which participants may have enjoyed during the screening of the film *The Yoke of the Past*. Vividly contrasting the "ancient methods of farming with the up-to-date modern methods," the movie caused "one white-bearded old patriarch . . . bubbling over with mirth" to remark " 'that's right, that's the way we used to do it.' "[54] Overall, the meetings seemed successful, with attendance at one event (of 1,011 people) constituting about 15 percent of Kewaunee County's rural population. At a demonstration in Menominee County, the company reported an audience of 925 people, or one-quarter of the county's rural residents. Beyond pleasing farmers, the meetings also accomplished "the double purpose of enlisting the valuable services of your county agents in spreading the gospel of rural electrification."[55] (Figure 11.7 shows activities outside the Northern States Power Company's exhibit, possibly at a fair held in Marathon County, Wisconsin.)

Even as the Depression deepened, such educational activities continued, especially at state and county fairs, where electric company demonstrations competed with exhibits featuring livestock and several forms of unelectrified farm equipment. At a January 1930 event in Sheboygan, the Wisconsin Power and Light Company showed a comedic movie, *Ride 'Em Cowboy*, before screening *The Yoke of the Past*. The program continued with discussions by power company officials, a University of Wisconsin agricultural engineer (Joseph Schaenzer), and utility employee Charlotte Clarke, who provided "Practical Home Making Hints." It also included testimonials of farmers who already employed electrical applications.[56] A month later, the firm hosted "Electric Farming" conferences in the towns of Belleville and De Forest. Each included various demonstrations (with Schaenzer displaying and discussing electrically operated water systems) and with separate programs for women to learn about electric cooking and refrigeration. Moreover, the power firm distributed merchandise certificates (with slogans "Electrify for Better Living" and "Electricity is a Cheap Hired Hand") to encourage purchases of load-building appliances. Holders could use the coupons for discounts of up to $20 toward electric ranges, radios, clothes washers, and vacuum cleaners (figure 11.8).[57]

In 1931, University of Wisconsin agriculturalists sponsored a "Farm and Home Week," which offered special lectures concerning poultry, marketing, and livestock.

Fig. 11.7. Northern States Power's "Electricity on the Farm" exhibit, undated, but possibly 1928 in Marathon County, Wisconsin. *Source*: Used with permission of Xcel Energy, the holding company parent of NSP. Photograph courtesy of University of Wisconsin–Madison Libraries, Archives and Records Management.

Organizers noted that "five forces are at present working to enhance the value of the farm home," namely, "a Steady Job, Power, Transportation, Leisure, and Beauty." "Power" included electricity, which pumped water into the home and provided energy for other appliances that "certainly help to make the farm attractive." The main program included a talk, "Electric Power Transforming the Farm Home," while the women's event demonstrated small appliance use. Eloise Davison, NELA's home economics adviser, discussed "Laws to Follow in Lighting the Home" and "What the Homemaker Needs to Know about Choosing Electrical Apparatus." CREA director Earl White lectured on "Electricity for Farm and Home," while the University of Wisconsin's Joseph Schaenzer elaborated on mechanical ventilation and "What Can Electricity Do for You."[58] In a similar fashion, the annual instructional events held at Texas A&M University (starting in November 1931) displayed new labor-saving electrical equipment, effecting "a vast cultural and societal change," according to

Fig. 11.8. Promotional certificate for electric appliances, 1930. *Source:* Used with permission of Alliant Energy, the holding company parent of Wisconsin Power and Light. Photographs courtesy of University of Wisconsin–Madison Libraries, Archives and Records Management.

historian Henry Dethloff and agricultural engineer Stephen Searcy, as rural people learned about the advantages of electrification.[59]

Power companies that participated in demonstrations and fairs clearly exhibited their increasing commitment to rural electrification. Of course, their activities reflected self-interest rather than a desire to provide social equity or welfare. In a survey of forty electric firms that sponsored state and county exhibits, the magazine *Electricity on the Farm* (discussed later in this chapter) observed in 1930 that companies achieved at least three objectives. First, they educated prospective customers about the benefits of electrifying their farmsteads. "The old adage that

Plate 1. Cover of Minnesota CREA report, *Presentation of Purpose of the Red Wing Experimental Rural Electric Line, 1924. Source:* Courtesy of University of Minnesota Libraries, Archives and Special Collections.

Plate 2. REA poster designed by Lester Beall. *Source*: Courtesy of Library of Congress, Prints & Photographs Division.

Plate 3. Federal Theatre Project poster for *Power*, Arthur Arent's play that premiered in 1937. *Source*: Courtesy of Library of Congress, Prints & Photographs Division.

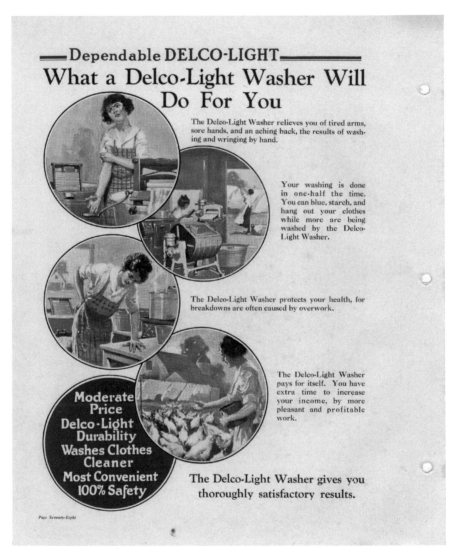

Plate 4. Delco-Light washer. *Source*: *The Delco-Light Story* (Dayton, OH: Delco-Light Co., 1922), 78. Used with permission of General Motors Media Archive. Photograph courtesy of the Richard P. Scharchburg Archives at Kettering University.

Plate 5. Delco-Light advertisement for farm home lighting. *Source: System on the Farm* 4, no. 3 (March 1919): 165.

The road to health and happiness is found in making the home modern and up-to-date.

Bright, safe, convenient Delco-Light conserves the health of every member of the family and makes better, happier homes.

It is the way of progress.

Plate 6. The Delco-Light Way provides health, happiness, and progress; undated image, likely 1922. *Source*: Used with permission of General Motors Media Archive. Photograph courtesy of Richard Backus, editor in chief, *Gas Engine Magazine*.

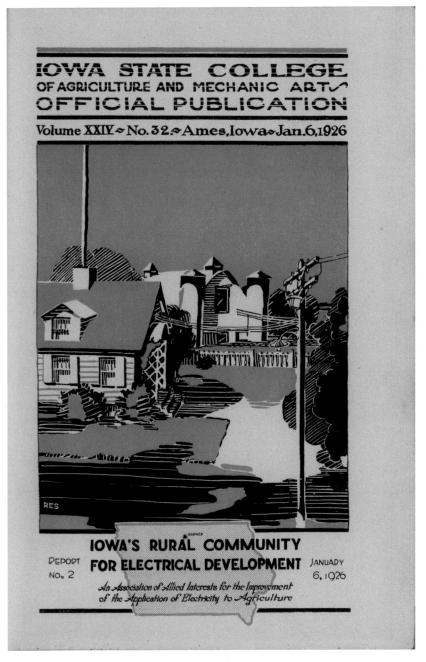

Plate 7. Electric lines help Iowa's farmers. *Source:* Cover of the Iowa's Rural Community for Electrical Development, Report no. 2, *Iowa State College of Agriculture and Mechanic Arts Official Publication* 24, no. 32 (6 January 1926). Courtesy of Iowa State University Library, Special Collections and University Archives.

Plate 8. Cover of the first issue of *Electricity on the Farm* (July 1927).

'seeing is believing' applies particularly well" to rural electrification, the magazine noted. Second, the demonstrations helped existing customers learn of new ways to raise their energy consumption and get more value from a service they already received. Third, the fairs simply provided "good will for the power company," since the events attracted not only farmers but also other rural residents and town folk. Despite the great expense of putting together the short-lived events, utilities obtained considerable benefit. By "taking the public relations angle into account, power company exhibits at fairs may be considered good business moves."[60]

Utility subsystem stakeholders supplemented face-to-face instruction by exploiting the popular new communications technology of radio. At the University of Wisconsin, agricultural engineer Joseph Schaenzer took to the air waves in January 1930 to explain "How 45,000 Wisconsin Farmers Are Using Electricity." Of the 190,000 farms in the state, he observed, 28,000 obtained power from utility companies and another 17,000 produced power themselves with isolated plants. Expressing amazement that these numbers grew from practically zero only five years earlier, he expounded on how electricity raised living standards, saved time in the home and in the barn, reduced labor effort and costs, and enlarged farm incomes.[61] Schaenzer's next radio talk, "More Electricity for Less Money," clarified elements of the electric bill that frequently elicited scorn. The service charge, for example, paid for fixed costs involved in getting electricity to a farmstead, such as the expenses of building the line and for maintenance, depreciation, taxes, interest, and insurance. The demand charge largely covered the transformer, whose cost grew as its output (in kilovolt amperes) increased. By contrast, the energy charge seemed fairly easy to comprehend, as it related directly to the amount of electricity consumed. Utilities commonly used step rates, in which the first 25 kWh (in Schaenzer's case) cost 10 cents each, the next 25 cost 7 cents each, and usage beyond 50 kWh cost 3 cents each. The declining-block rate structure encouraged customers to employ more electricity, with the average price falling as consumption increased. "[T]he more electricity that can be used efficiently and advantageously on the farm," Schaenzer reminded listeners, "the more profitable and economic its use becomes to the consumer."[62]

Radio seems to have been an important element in General Electric's efforts to encourage rural business as well. Starting in 1928, the company used WGY in Schenectady (GE's hometown in upstate New York) to broadcast a series of talks on "Some Practical Solutions of Farm Electrification Problems." Though GE employees gave many of the talks, which numbered 208 by the end of 1932, rural service managers and utility executives also contributed. CREA director Earl White presented "A Horoscope of Farm Electrification" in November 1932, in which he observed that the "signs of the Zodiac clearly say that in the future[,] the farmer will use much more electricity than is the case today—many times more."[63]

Corporate Support

Several manufacturing companies had already set their sights on rural markets, as evidenced by the publications, advertising, demonstration work, and training of farm specialists. In fact, manufacturers such as GE and Westinghouse appeared to show more enthusiasm about rural electrification than did some utilities. These companies likely saw farmers as potential customers for most of the products they already sold to city folk, but they also hoped to sell milking machines, incubators, brooders, pumps, and almost anything that could be hooked up to electric motors. To reach this group of consumers, General Electric (in particular) exploited the growing influence of advertising agencies to pinpoint consumers with specialized messages. In 1925 alone, the firm placed fifty-five advertisements in eighteen farm periodicals, which counted a total circulation of more than 8 million.[64]

Of course, utility managers understood the potential profits arising from power sales to rural customers, but to obtain them, the firms first needed to erect a vast and expensive distribution network. By contrast, manufacturers could sell directly to farmers without making infrastructural investments. Ideal customers received central station power, such that their consumption would not be limited.[65] Even if farmers owned isolated plants, small hydroelectric power stations, or wind-powered generators, which provided modest supplies of electricity, they could ultimately become worthwhile patrons of high-line service after having developed a familiarity with electrical lifestyles. Consequently, the companies' representatives urged utilities to pay more attention to the rural market. GE publicity manager Walter Bowe observed in 1931, for example, that the number of central station farm customers had increased remarkably in the years between 1923 and 1931—from 166,000 to 648,000—with further additions expected. As important, farmers made up one-sixth of all new customers of power companies, a proportion that "indicates the relative and growing importance of this group of customers" and the consequent need to pay attention to them.[66]

Producers of electrical equipment seemed eager to work with the state affiliates of CREA to gain a foothold in a potentially large market. In the Red Wing, Minnesota, experiment, seventy-nine companies contributed equipment worth more than $21,558 from 1923 to the end of 1927. (General Electric and Westinghouse provided hardware valued at $1,078 and $942, respectively.)[67] Manufacturers showed similar interest in other states' CREA partners. In a letter to Professor Charles Seitz at Virginia Polytechnic Institute in March 1924, for example, Westinghouse's "farm plant salesman" offered his assistance to the recently selected state committee chairman. The author urged Seitz to read an article on the Red Wing experiment as well as a piece written by his company's R. C. Cosgrove, who also represented the isolated plant industry on the national CREA. Cosgrove's article, he said, "tends to favor small light and power plants until the farmer has become more educated to the use of electricity," but Seitz should note that the company remained interested in serving customers with high-line service too.[68] Likewise, the Alabama CREA group happily acknowledged in 1925 the cooperation of several companies, such as Westinghouse, General Electric, and Delco-Light, which offered technical advice and discounts for electrical equipment used in experiments.[69]

The large manufacturing firms employed judicious advertising to keep in the good graces of agricultural engineers and others who worked on electrification efforts at land-grant colleges. General Electric regularly bought full-page advertisements in the journal *Agricultural Engineering* (the organ of the American Society of Agricultural Engineers), read by college instructors and researchers. In 1925, the company promoted electricity for spurring increased egg production from hens, for instance (figure 11.9). A later advertisement, "Making a hard job easier," portrayed a farm wife in an electrically illuminated kitchen watching as her husband heads out on a snowy evening with a shovel in hand. The text got to the point quickly: "The time will never come when farming will be listed as an easy job. But groping in the dark is one hardship electricity will abolish. On farms electrically equipped, power lines bring clean, safe lighting to the darkest corners. And the same power drives the motors of many labor-saving machines."[70] Similar advertisements appeared in student-edited publications such as Alabama Polytechnic Institute's *Alabama Farmer*, the University of Arizona's *Arizona Agriculturalist*, and the *Purdue Agriculturist*.[71]

Publishing about Rural Electrification

Beyond the behemoths General Electric and Westinghouse, numerous smaller companies saw potential value in farm electrification. In 1922, the Farm Light and Power Publishing Company (located in New York City) issued a *Year Book* listing manufacturers and wholesalers of electrical equipment for use on farms, along with articles describing features of various appliances, good wiring practices, and sources of isolated power. Consisting of 338 pages, the book named fifty-one

Fig. 11.9. Electricity boosts productivity in henhouses. *Source*: Advertisement in *Agricultural Engineering* 6, no. 10 (October 1925): 253. Used with permission of General Electric and the American Society of Agricultural and Biological Engineers (ASABE, formerly ASAE).

manufacturers of gasoline-fueled generators, including Allis-Chalmers, Delco-Light, and Westinghouse. It also contained information on at least seventeen manufacturers of storage batteries and scores of companies that produced associated devices such as lamp adapters and insulated wires.[72] Its publication suggests a vibrant market of large and small companies seeking to profit from an apparently growing interest in farm electrification.

The same publisher became a more important player in the business of stimulating rural electrification efforts by producing, starting in 1927, *Electricity on the Farm*, "a magazine for the farmer to promote the use of electricity."[73] Featuring colorful covers and illustrated articles, the periodical explained the great value that men and women could obtain by electrifying their farmsteads. The largely apolitical magazine did not seek subscriptions from individuals; rather, the publisher sold issues to utilities for distribution to existing and potential rural customers (plate 8).[74] In an analysis of the magazine's circulation following issuance of the July 1929 number, the editors reported that the publication "reaches the cream of the farm market" with sales of more than 120,000 copies to power companies.[75] Reporting on a survey of advertising to farmers performed by sixty-four utilities in 1930, a General Electric publicity executive noted that companies used newspapers, state or regional farm magazines, and a few other specialized publications, but not with much regularity—except for *Electricity on the Farm*.[76] A manager gloated in early 1931 that the magazine had "been adopted as a vital part of the rural business building program of more than 350 power companies in all sections of the country," with distribution of two hundred thousand copies per month.[77] In a 1934 speech, Grover Neff endorsed the publication as one that "devotes all its editorials and advertising space to the farm uses of electricity." As a supplement to farm newspapers, pamphlets, and CREA newsletters, the magazine got "the farmer thinking about new and better methods of using electric service." After all, he observed, "No matter how old our story may seem to us, there are many of our customers who do not know it."[78] Beyond offering practical articles on "Doubling Garden Crops with Electric Light"[79] and "Electrical Housekeeping,"[80] for example, the magazine promoted the notion that electricity connoted modernity to the farmer. Starting with its July 1928 issue, the publication sported the subtitle *A Monthly Magazine for the Progressive Rural Family*.[81] That tag line changed with the March 1933 issue to *A Monthly Magazine for the Up-to-Date Rural Family*.[82]

Unquestionably, the farm market exhibited a series of challenges for utility companies—challenges that, for largely economic reasons initially made power company managers reluctant to serve it. Nevertheless, a variety of actors saw value in an underserved portion of the nation's population, and they added to the burgeoning momentum of a distinct rural electrification subsystem.

Perhaps the most important subsystem participants consisted of members of the newly created profession of agricultural engineering. Usually working at land-grant colleges, agricultural engineers became the anchor for the subsystem's educational components, and they served as crucial intermediaries between the utility industry and farmers. They did so, in large part, through their leadership of CREA's state affiliates. Beyond performing critical research and administrative activities, the engineers also taught students about farm use of electricity (generated by isolated power plants and central stations), such that many obtained jobs in utility companies' increasingly numerous rural service departments. In their corporate homes, the graduates helped farmers employ electricity in novel and mutually profitable ways. Furthermore, the agricultural engineers in land-grant schools often put together educational events that appealed to utility rural service managers, farmers, and general audiences. These instructional forums and public demonstrations in nonurban settings provided another way to spread the word about rural electrification, complementing the use of radio talks and publications coming from extension services and for-profit companies.

Momentum in the subsystem continued growing because of efforts made by equipment manufacturers. General Electric and Westinghouse understood that they benefited from selling equipment to utility companies and to ultimate customers, and they pursued complementary efforts to support rural electrification through advertising, education of manpower, and donations of equipment for CREA experiments. Private power companies also realized benefits—in terms of income and public relations—as they performed work to bring down the cost of extension lines.

The expanding number of farmers using electricity suggests the establishment of a maturing subsystem. In an optimistic statement of significant progress, the author of the Rural Electric Service Committee's 1927 report (most likely Grover Neff, the committee's chairman) observed that farm electrification had "emerged from the experimental stages" in the previous few years and had moved into a period of rapid growth.[83] In fact, from the end of 1923 through 1926, the number of farms obtaining electricity in a sample of twenty-seven states jumped by almost 87 percent. According to the report, such growth should drive "each power company operating in rural territory to see to it that it is properly organized to foster developments . . . in an efficient and energetic manner." And even as farmers had experienced an "agricultural depression" since 1920, the trend of electrification, if continued unabated, would lead to 1 million electrified farms by the end of 1932 and 3 million by 1938.[84] (Recall that, in 1923, about 6.3 million farms existed in the country, 177,561 of which had electricity supplied by utilities.)[85]

Enthusiasm for farm electrification continued into 1928, with NELA's Rural Electric Service Committee (then chaired by Charles F. Stuart of the Northern States Power Company) reporting more headway.[86] Just a few years earlier, a committee report noted, rural electrification had been viewed by almost everyone as a "gigantic task" that "presented difficulties which seemed insurmountable." But within a half-decade, utilities had started providing rural services that were "bigger, broader and more numerous in character than anyone had any conception of when this work first was undertaken."[87] The CREA had just published a comprehensive 136-page report highlighting the experiments performed in twenty-four states, and its work seemed to have had a positive impact. Stuart noted that in the five years since formation of the Minnesota CREA in 1923, the industry went from a sense of discouragement about rural electrification to great optimism. A major source of the improved conditions resulted from the work pursued "with farm organizations, with governmental departments, with agricultural colleges, manufacturing companies, and others." The combined effort, he continued, sought to "solve this problem [of rural electrification] which seemed to defy the industry, the colleges, and the farmers, and everybody else who tackled it alone."[88]

The first years of the 1930s threatened to disrupt this rapid growth rate. After all, the onset of the Great Depression subdued activity in the entire economy, with agriculturalists hit by lower demand for farm goods and with prices falling to levels not seen in more than twenty years.[89] In most industries, economic contraction translated into the loss of customers. After reaching a peak of 24,555,732 customers in 1930, the central station business gave up billpayers in the next two years in the domestic, commercial lighting and small power, and industrial power sectors.[90]

Contrasting with this decline, the number of rural users never stopped increasing, though the blistering growth rate ebbed. In the peak year of 1928, the industry energized 113,021 new farms. Additions totaled 69,926 in 1929, and 73,751 in 1930.[91] In March 1932, *Electrical World* reported that the number of farm customers grew by almost fifty thousand during the previous year.[92] Subsequent stories in the journal reprised this news, with some highlighting the growing spans of rural lines and greater farm power consumption in Alabama and Virginia, for example.[93] Overall, from the beginning of 1924 to the end of 1932, the number of ruralites who obtained high-line service (709,449) showed an average annual growth rate of 16.6 percent. Put differently, the percentage enjoying high-line service rose from 2.8 percent to 11.3 percent in the period.[94]

With not too much exaggeration, the chair of NELA's Rural Electric Service Committee, Eugene Holcomb, proclaimed in October 1929 (before most people realized the Depression had begun) that utility companies recently made great strides in bringing power to farms in the previous few years: "Farmers are becoming electrically minded. They are gaining confidence in the sincerity of the power companies. And the companies are looking more favorably upon rural business. The electrical industry is ready to deliver power to the farmer's door at reasonable prices, as soon as he can make use of it. But it takes time to adapt himself to the new conditions. He cannot revolutionize his whole practice and habits over-night. The change is coming in an orderly fashion."[95]

The rural electrification advocates, in other words, may have done a good job in creating a subsystem that involved an increasing number of active stakeholders. Impediments to further gains still remained, of course, such as finding ways to finance rural folk so they could wire their farmsteads and purchase electrical equipment, especially during troubled times. If economic conditions improved quickly and if left free to regain its momentum, perhaps the utility industry's farm electrification movement would yield even more impressive gains, in line with Grover Neff's prediction of serving 3 million rural customers by 1938.[96]

PART III / Growth of Rural Electrification Efforts in the 1930s

Unfortunately for industry promoters of rural electrification, the economic malaise starting in late 1929 did not ameliorate quickly, and self-inflicted problems caused the utility industry to incur the public's ire. Consequently, private companies failed to advance their own plans to become the exclusive providers of high-line service to nonurban customers. Rather, by offering financial resources and innovative programs, the New Deal's Tennessee Valley Authority and the Rural Electrification Administration threatened to seize the bulk of the farm market, one that the industry had been cultivating for more than a decade.

Sensing a competitive threat, central station companies moved forcefully to expand service to rural customers in the mid-to-late 1930s. Often using less-than-creditable methods, they tried to preempt government efforts by rapidly electrifying farms themselves. In a perhaps ironic, though rarely acknowledged manner, the private firms wired up more farms until 1950 than did the REA-funded electric cooperatives. Even as late as 1955, when about 94 percent of farms received high-line power, utilities served more than 43 percent of them.[1] Despite making such gains, the industry ceded control of the rural electrification subsystem to government-sponsored entities and never regained it.

Chapter Twelve

Government Innovations in the Rural Electrification Subsystem

> Rural line extension is no longer a piecemeal, haphazard, incidental affair patterned on elaborate urban construction standards which in the past have meant high capital and maintenance costs, high interest rates . . . , small usage, relatively slow returns and an uncertain investment. The new method is [a] planned, long-range, community effort, [with] more customers, [and] lower costs which permit lower rates, higher usage, assured income, amortization[,] and a safe investment, whether private or public.
>
> —*RURAL ELECTRIFICATION NEWS*, 1935

With the inauguration of an activist president during the midst of an economic and social crisis, rural electrification received an immense boost starting in 1933. As a candidate, Franklin Roosevelt touted his credentials as a public power advocate and bashed the leaders of the utility industry who appeared, in the minds of many citizens, responsible for the Depression. A little more than two months after taking office, the chief executive signed legislation establishing the Tennessee Valley Authority. Two years later, Roosevelt authorized creation of the Rural Electrification Administration. Both government organizations (along with associated agencies) spurred farm electrification in ways that dramatically extended the work done by private utilities since the early 1920s.

The accomplishments of the TVA and the REA in the New Deal era have been widely documented (though, as noted in chapter 1, not always accurately). Perhaps unlike other analyses, this chapter emphasizes the advantages held by government agencies that enabled them to achieve their successes. Not meant to constitute an excuse or apology for the utilities' unwillingness to undertake similarly effective activities, this discussion nevertheless explains how government-run projects accomplished feats that eluded investor-owned firms. And while private companies could not offer subsidies or obtain capital at below-market rates, for example, their managers learned important lessons from the government's work, which also served as a powerful competitive incentive for them to pursue rural electrification more zealously starting in the mid-1930s.

Creation of the TVA and REA

The TVA constituted a bold public power experiment that sought to achieve social and economic objectives through the supply of inexpensive electricity. Its origins stem from construction of the Wilson Dam on the Tennessee River at Muscle Shoals, Alabama, during World War I to provide electricity for production of explosives. Completion dragged into the 1920s, and with the war over, debate ensued concerning its use and ownership. Industrialist Henry Ford attempted to purchase the dam in 1921 and create a manufacturing region around it, but progressive Republican George Norris, who had espoused public power programs since the 1910s, resisted that scheme as chair of the Senate's Agriculture and Forestry Committee. As previously described, Norris sponsored proposals for multipurpose development and government control of the dam in bills that garnered congressional support—and two presidential vetoes—in 1928 and 1931.[1]

Norris found an ally for his plan in newly elected President Franklin Roosevelt. A supporter of government control of some natural resources, the Democrat declared during his 1929 inaugural address as New York governor that the state had a duty to produce electricity from the flow of water, "which belongs to all the people" and which should be distributed to residents at low cost.[2] In 1931, he pushed to establish the New York Power Authority and appointed Morris Cooke as a trustee.[3] Though he did not make it a major element of his presidential campaign, FDR nevertheless backed greater government involvement in supplying electricity as a means to increase consumer demand and alleviate rural poverty.[4]

Within the supportive environment created by FDR's New Deal, Senator Norris repackaged his earlier legislative efforts into the Tennessee Valley Authority Act, approved on 18 May 1933. The law created a new administrative agency—the TVA—which had a mandate to build dams for preventing floods, reducing soil erosion, improving navigation, and generating electricity. To aid in distributing cheap power to rural customers, the TVA organized cooperative electric organizations, with the Alcorn County Electric Power Association (in northern Mississippi), about fifty miles from Wilson Dam, constituting the first in 1934. Two others followed in 1935.[5]

President Roosevelt asked Morris Cooke, who had been serving since 1933 as head of a Public Works Administration committee investigating the water and land resources of the Mississippi Valley, to develop a comprehensive approach to extend electric service to rural areas. In a 1934 report, Cooke suggested a national program of rural electrification, observing that the federal government could enhance farm life substantially by allotting $100 million to "build independent, self-liquidating rural projects."[6] He repeated the proposal in his "National Plan for the Advancement of Rural Electrification," which aspired to offer long-term government loans to independent, farmer-owned distribution organizations (similar to the co-ops created by the TVA).[7] The proposal won a positive reception from the president, who (with authority provided by the $5 billion Emergency Relief Appropriation Act of 1935) created the Rural Electrification Administration by executive order in May 1935; Cooke served as its first head.[8] (Figure 12.1 shows Cooke signing documents approving the creation of seven REA projects.) A few months later, the president issued Regulation No. 4, which established the REA as a lending agency instead of a relief organization.[9] Working with Democratic Representative

Fig. 12.1. REA's first administrator, Morris L. Cooke, signing approval of loans to begin seven rural electrification projects, 4 November 1935. *Source*: Courtesy of Library of Congress, Prints & Photographs Division, photograph by Harris & Ewing.

John Rankin of Mississippi, Senator Norris introduced the Rural Electrification Act, signed on 20 May 1936, which put the REA on a firmer, longer-lasting basis as a statutory entity rather than as an emergency measure established by presidential fiat.[10]

Initially, it appeared that the REA would lend its funds to private utility companies. In his first press conference after creation of the agency, Cooke said he expected to work with commercial firms and "did not expect any trouble" since "public sentiment should influence them to extend their lines to the farmers."[11] To pursue this collaborative effort, REA leaders met officially with utility representatives to explore options for transferring money and commencing rural electrification projects.[12] At the end of July 1935, utility company leaders proposed to wire about a quarter-million farms in eighteen months at a cost of more than $238 million. While the firms claimed that much of the "immediate urge for rural electrification" was "a social rather than an economic problem," the utilities agreed to engage in such efforts, with government support, in the hope that farm customers would ultimately become profitable.[13] Cooke stated publicly that the REA would quickly use the initial $100 million to start the work.[14] Undoubtedly comforting utility officials, a New York Times article reported, "As 95 per cent of the electric industry in this country is in the hands of private operating companies, the administration's public funds will be dispensed in that proportion." The article further quoted Cooke declaring that "we are not attempting to change the balance from private to public operation, and could not do so if we wanted to."[15]

For a variety of political and managerial reasons, however, Cooke ultimately rejected the industry's plan.[16] Claiming that utility firms preferred to borrow money from banks (even at rates higher than those offered by the REA) to avoid restrictions and rules imposed by the REA, Cooke profoundly altered the way rural electrification advanced.[17] Instead of lending money primarily to power companies, the REA dispensed it to farmers' electric cooperatives, similar to the Alcorn co-op created by the TVA two years earlier; they would use the funds to string new distribution lines and to construct (in a few cases) generation and transmission facilities.[18] Of the $69.4 million that the REA loaned until 1 November 1937, 78.5 percent went to co-ops and just 2.2 percent to private utility companies.[19]

The Government's Exemplary Work

The TVA and REA did impressive rural electrification work, but in the 1930s, they employed tools that private utilities could not muster themselves. And they enjoyed the cooperation of agencies, committed to similar goals, that provided consumers with previously nonexistent resources. In an interesting manner, however, some of the approaches used by the government entities (such as constructing cheaper distribution lines and employing promotional rate structures) built on the work done earlier by the CREA state bodies and the private utility industry.

FINANCING AND LOW-COST APPLIANCES

Though recovering substantially from the recession of 1920–21, ruralites did not enjoy the prosperity that the "roaring twenties" decade delivered to their city cousins. And as the Depression tanked the previously upbeat urban economy, it sent the rural economy into another tailspin, diminishing farmers' purchasing power. To make utility companies interested in serving this market, so went the standard wisdom, farmers would need to make capital outlays for electrically operated motors, milkers, and other equipment at a time when they had little money to invest. The task of selling these "articles of electrical virtue . . . is a heartbreaking one under present farm conditions," observed *Electrical World* in 1931.[20] A New York farm electrification advocate reiterated this concern, stating that the "one obstacle in the path of still further use of electricity" in his state "is the inability of farmers to finance the purchase of equipment which they know will be wealth-producing."[21] Within the harsh business environment of the early 1930s, neither power companies nor equipment manufacturers could easily lend money to the distressed farmers.[22]

Government planners who created the TVA addressed the financing problem directly. In December 1933, they established the Electric Home and Farm Authority (EHFA), a subagency of the TVA, to spur purchases of electric appliances largely by offering low-interest loans.[23] Viewed as an experiment by David Lilienthal, codirector of the Authority and head of the EHFA, the new program first made loans just to customers receiving TVA power, but it later provided financing to people served by private utilities.[24] Through the Reconstruction Finance Corporation, a body established in 1932 and affiliated with the Federal Reserve, the government initially made available $10 million of credit for home wiring and appliance purchases at attractive rates. The EHFA required small down payments (of 5 percent) with the balance financed at 5 percent annually for between two and four years.[25] Making such outlays easier for customers to manage, the installment charges appeared as monthly additions to electric bills.[26] To buy an electric refrigerator, range, and water pump, whose cash price totaled $351.50, for example, a customer put down $18.50 and paid $8.33 for forty-eight months.[27] For comparison, firms such as General Electric offered consumer loans for purchases of their products at annual rates of between 8 and 30 percent, with full payment expected within one year.[28]

Beyond favorable financing, the TVA and EHFA worked with manufacturers to design inexpensive, but well-made, versions of appliances for sale to rural customers, especially in southern states.[29] Among the earliest such pieces of hardware, General Electric's specially engineered low-cost refrigerator won the sobriquet of a "Model T appliance" from *Business Week* magazine in June 1934. The four-cubic-foot cooler did not contain "gadgets and minor luxuries" appearing in more extravagant consumer models, but it remained clean, simple, and compact. At $74.50, the price "puts electric refrigeration in position to crash the mass market." Considered "revolutionary" by the magazine author, the price came in well under the cheapest similar model listed in the Sears, Roebuck catalog (at $94.50).[30] Soon thereafter, the EHFA announced a refrigerator built by the Crosley Radio Corporation, whose price undercut that of the GE machine by $2.00.[31] By the end of 1935—just in time for Christmas—the EHFA had broadened its reach; beyond lending 95 percent of the money needed for refrigerators, ranges, water heaters, and water pumps, the agency financed clothes washers, driers, milk separators, milk coolers, vacuum cleaners, and motors. The *Chicago Daily Tribune* announced this expanded program in an article titled "EHFA Will Lend Santa Funds to Buy Appliances."[32] As important as the price of these basic appliances was the implicit seal of approval offered by the TVA and EHFA. Even private utility companies, such as Georgia Power, advertised these inexpensive appliances and used the government's imprimatur as a marketing tactic.[33]

The EHFA further contributed to rural electrification through its promotional and educational activities. To overcome cultural hesitation in the use of electrical appliances and to showcase ways in which the new energy form could improve people's standards of living, the agency advertised in publications and on the radio. In doing so, the organization actively advanced New Deal notions of progress and modernity.[34] And to reach more women, the TVA hired Eloise Davison, the home economics professor at Iowa State College who also served on the Iowa CREA committee and national CREA. Becoming director of the TVA's Domestic Electric Service program, she orchestrated public demonstrations of electric equipment and promoted electrification through courses taught at high schools and with talks given to church groups, clubs, and other community organizations.[35]

EHFA's financing, work with manufacturers, and promotional efforts appeared to boost purchases of electric devices. By 1942, the EHFA had collaborated with more than 7,000 appliance dealers and contractors to finance more than 431,000 installations of wiring and equipment.[36] By transforming the marketplace for appliances, such as refrigerators, which previously could only be considered for purchase by wealthy homeowners, the EHFA spurred increased electricity consumption.[37]

LENDING MONEY TO CO-OPS AND OTHER BENEFITS

As a government lending agency, the REA had financial expectations that differed from those of central station firms. Even with more liberal policies toward building lines developed by some utilities in the late 1920s and early 1930s, private companies generally wanted to recover the cost of their high-line expenses within three to five years. The REA, however, initially loaned money to cooperatives at 3 percent a year, equaling the average rate of interest paid on long-term government bonds, for up to twenty years.[38] The Rural Electrification Act of 1936 extended the maturity of loans to twenty-five years.[39] Low lending rates, coupled with long loan periods, meant the co-ops could reduce monthly charges to customers by about two-thirds compared with financing arrangements having five-year terms.[40] For those suffering from the long farm recession and Depression, such a reduction in payments made a huge difference. Sweetening the situation, Congress in 1944 passed an amendment to the 1936 act that fixed a 2 percent borrowing rate with up to thirty-five years of financing, knocking down monthly charges more.[41] At the time, long-term interest rates for the government bonds reached 2.48 percent, and corporations typically obtained funds for about 3.00 to more than 3.50 percent annually.[42] Expressed simply, co-ops received extremely favorable borrowing terms.

In December 1935, the REA announced another financing initiative—this time to wire farmsteads. As utility managers observed earlier, for ruralites to exploit the value of electricity, they needed to have enough spare capital to pay for wiring their properties, an unlikely occurrence when farm income remained low. The REA acknowledged the utilities' concerns, noting that "[i]n the past, lack of adequate financing facilities has retarded the extension of electric service into rural areas." Paired with financing provided by the EHFA, the new REA loans enabled more farmers to modernize their properties and purchase electrical appliances. Additionally, the REA encouraged tradesmen to wire large numbers of farmsteads under one contract.[43] By 1939, the combination of REA loans and "group wiring plans" appeared successful, with reductions in wiring costs by as much as 35 percent, according to the REA administrator.[44]

In some states, co-ops received special treatment that permitted them to avoid costs in ways utilities could not. The Wisconsin legislature, for example, exempted REA co ops from various forms of regulation. (See more discussion on the regulation of co-ops in the next chapter.) Furthermore, co-ops for several years did not need to set aside funds for depreciation (to draw upon for replacing worn-out equipment), as did private utilities under conventional accounting procedures. Beyond that, a 1939 law exempted co-ops from a tax paid by utilities, then about 17 percent and rising to 25 percent in 1945.[45] The REA's model co-op legislation, adopted by numerous states, also freed the organizations from paying excise and income taxes, replacing them with an annual fee of $10 for each one hundred co-op members.[46] In the late 1930s, some states offered complete tax-free status permanently or for several years.[47]

As nonprofit entities, REA co-ops also won benefits such as exemption from federal taxes under the Revenue Act of 1916 (which obviously predated electric cooperatives but not farmers' cooperatives).[48] The REA's *Rural Electrification News* commented in 1936 that "cooperatives are especially favored by the Federal Government and by the laws of many States with respect to taxation," though they sometimes paid local taxes.[49] By contrast, throughout the 1920s, utility officials complained about what they saw as excessive taxation, which consumed 10 percent of gross revenue (and about 20 percent of net operating revenue) in 1930. Higher taxes resulted from passage of the federal 1932 Revenue Act (during the Hoover administration), which imposed a 3 percent sales tax on electricity consumers and 13.75 percent tax on utility corporations. These expenses, which raised the overall cost of selling electricity by commercial power companies, generally did not apply to co-ops.[50]

REA-funded cooperatives obtained other advantages that lowered the cost to construct distribution lines. For example, the REA allowed farmers to work on erection of poles and wires instead of paying co-op membership fees. The REA's 1942 report noted that the use of labor for "clearing rights-of-way, digging holes, driving trucks, and other ground work" are paid, not as cash, but credited to members' accounts.[51] Significant as well, many co-ops could construct lines on farmers' properties without paying for the rights of way. REA policy forbade using lent money to purchase such rights, a practice viewed as inconsistent with the organization's philosophy of encouraging collaborative work to achieve lower costs for members.[52] By 1941, the co-ops had collected more than one million easements.[53]

Perhaps as important, the REA developed engineering practices that had started to come into use by private companies (described in chapter 11) for reducing expenses. By employing high-strength wire conductors, the agency could extend the span between poles—from one every 150 to 200 feet to between 300 and 400 feet by 1937—and lessen costs accordingly. REA poles sometimes did not contain wooden cross arms, such that wires simply were attached to insulators on vertical shafts, diminishing costs further. The co-ops also often built long lines— one hundred to two hundred miles at a time—and they contracted with experienced crews (supervised by REA field engineers) using a "moving belt" or "assembly-line" process to dig holes, raise poles, string wires, and install transformers.[54] These approaches, along with standardization and volume purchases of equipment, facilitated low costs. As early as 1937, the REA reported that a recently approved group of lines cost only $850 per mile.[55] A job in Texas came in even cheaper—only $550 per mile, which included the expense of transformers.[56]

The REA additionally reduced costs for many rural customers by providing "area coverage." Instead of extending distribution wires from existing lines to farmers' residences on a piecemeal basis, the agency proposed to serve everyone within a certain region, all of whom paid the average cost to serve them. By doing so, it could take advantage of its assembly line method for building less expensive lines. The approach also meant that those most distant from existing lines, who would otherwise experience high costs and likely never get service from a private utility company, could still afford service.[57] The technique had the added advantage (no doubt well considered by REA managers) of preventing private power companies from stealing its most attractive rural customers. By offering electricity to an entire geographical region, a co-op could claim it as its service territory—a de facto franchise district—and therefore make it difficult for private companies to sell power to individual customers within it.

As suggested by these descriptions of ways the REA helped co-ops achieve their goals, the federal agency assumed several activities that would have been costly and perhaps impossible for local membership organizations to provide by themselves. In addition, REA staff in Washington, DC, provided legal services (such as the drafting of state co-op laws), technical directives, and management advice without charge to its clients. The government shouldered these administrative costs, paid for from congressional appropriations, and did not pass them on to co-ops.[58]

RATES

Government rural electrification efforts enjoyed success in part because they could perform experiments in ways that seemed impossible for private utilities. In other words, because the TVA and REA did not incur the same restrictions as companies—in particular, the requirement to make a profit for stockholders—they tried novel approaches to stimulate increased farm use of electricity. Innovative and low-cost rate structures constituted the most unusual—and most successful— of these innovations, and their use by the TVA in particular provided useful lessons for managers in the utility industry.

Standard accounts of rural electrification note that private utility companies may have taken some actions to increase power use on farms, but they did not do much to reduce the rates paid for electricity, thus forsaking one significant avenue for stimulating demand. Put simply in these narratives, high prices discouraged the widespread use of electricity that would have improved people's lives. Morris Cooke, the first REA administrator, observed in his writings that companies justified charging farmers steep rates to compensate for "the alleged extra costs" of stringing wires to sparsely populated areas.[59] Soon after becoming the REA administrator in 1935, he observed that utilities' unwillingness to change pricing policies, even when offered REA loans, drove him to enlist newly formed co-ops to bring electricity to ruralites.[60]

In his writings on the REA, D. Clayton Brown repeated similar concerns about the negative impact of high rates. The influential historian noted in his 1980 book that farm customers paid rates about twice as great as those incurred by city folk (who, significantly, did not pay the initial costs of installing distribution lines).[61] Brown continued with this criticism elsewhere by noting that private companies "never dealt directly with the cost of electric service, which was a critical obstacle in the electrification of the rural areas."[62] Even companies working with state CREA affiliates charged high prices, he asserted.[63]

Though not necessarily unfair, the criticism reflects knowledge that did not exist widely in the 1920s and early 1930s about the "science" of rate making. In particular, utility managers did not truly understand the costs involved in distributing electricity, which (in theory at least) constituted the basis for making rates to recover those costs.[64] Unquestionably, the cost to serve distant and dispersed customers exceeded that of selling power to urban users. As noted in chapter 2, utility managers often quantified the higher cost as justification for bypassing the rural market.

The appearance of numerical precision, however, hid the fact that large uncertainties existed in calculating costs. In 1926, mechanical engineer Lesher Wing acknowledged, "Rate making is not an exact science; it is an art" because of the difficulties in allocating costs accurately to different customer classes (to industrial, business, residential, and rural users, for example).[65] Utility officials and academics came to a similar verdict while attending a rate-making symposium convened by Morris Cooke in January 1933. Even after several discussions drawing on theory and practice, participants failed to reach a common understanding of how to assign companies' costs to rural customers. Perhaps the best summary of the proceedings appeared as a frontispiece drawing in a book containing the contributors' papers. Showing an "area of known costs," which included the space from generating plants to substations, the illustration also portrayed a literally darker "area of unknown costs" where distribution lines took power to small towns and farms. The caption noted, "Ample data as to costs are available up to the distribution sub-station. From this point to the retail customer's meter, however, nothing has been known about costs—until recently" (figure 12.2).[66] But that conclusion seems somewhat disingenuous, since freshly acquired knowledge did not markedly improve comprehension of distribution costs. One meeting participant referred to such costs as "the final mystery to be solved,"[67] while Joseph Swidler, a reviewer of the symposium's papers and later a chief TVA lawyer, described the conference's findings as "only tentative and inconclusive."[68]

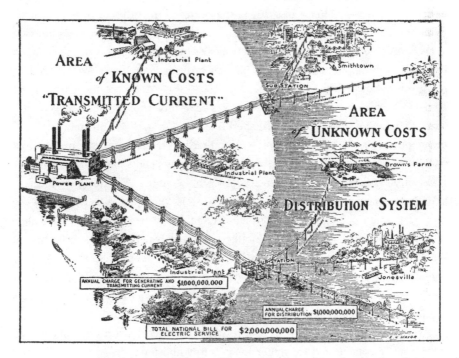

Fig. 12.2. Depiction of the vast area of unknown of costs in the electric distribution system. *Source*: Frontispiece of Morris Llewellyn Cooke, ed., *What Electricity Costs in the Home and on the Farm* (New York: New Republic, 1933). Used with permission of the *New Republic*.

In short, it appears that even the experts did not fully understand the economics of distribution lines. And without a good understanding of costs, how could managers accurately determine rates to be paid to recover those costs?

The conundrum did not inhibit TVA managers who developed a rate structure for their customers. In theory, ratemakers should have established per-kilowatt-hour prices based on calculations of the amortized cost of the acquired Wilson Dam, the Authority's first hydroelectric generating facility, and a prediction of the amount of electricity that would be consumed over a specified time period. (Roughly, the cost per kilowatt-hour was determined by dividing the annual cost of the plant, which included fixed and variable expenses, by the projected number of kilowatt-hours used in the year.) But according to historian Thomas McCraw, managers could not easily assess how much of the dam's cost should be paid for through electricity rates because of the facility's multi-use nature.[69] Deciding to influence public opinion favorably and to raise consumption through the use of low rates (which, according to TVA director David Lilienthal, would "add to the strength and richness of living of the people of the valley"), government accountants deliberately minimized the plant's worth, and they estimated (or guessed at) how much power customers might use.[70] By undervaluing the plant's cost and inflating consumption numbers, TVA officials could arrive at much lower rates than those offered by private companies.

McCraw documented how the managers also "took short cuts, employed arbitrary figures and methods, and finished their work in a ridiculously brief time," developing a simple, easily understood, and radical rate structure. Often known as the "3-2-1" schedule, the formula offered power at 3 cents per kWh for the first 50 kWh, 2 cents per kWh for the next 150 kWh, 1 cent per kWh for following 200 kWh (and 0.4 cent per kWh for subsequent usage).[71] A press release noted that the agency designed the rates "to encourage and make possible the widest use of electric service, with all the individual and community benefits which go with such wide use." The structure resulted in average prices of about half of what nearby private utility companies charged.[72] Academic economists in 1941 observed that the TVA "rates were not based on a cost-of-service theory but on a 'social-basis theory'" in which actual costs to provide service were disregarded.[73]

Private utilities had already discovered that increased consumption of electricity generally contributed to higher load factors and lower per-unit prices. And indeed, most firms reduced the price of electricity over the years as their costs declined. Even companies that served territories later taken over by the TVA had instituted rate cuts, planned before creation of the federal agency.[74] Overall, the average price per kWh for a residential (not rural) customer dropped from about 16 cents in 1902 to under 6 cents in 1932. Utility managers expected further price cuts as consumers bought more electricity and as production costs continued their downward trend (largely as a result of improvements in generation and transmission technologies).[75] Drawing on knowledge of these experiences, CREA state affiliates (as noted earlier) sought to identify and demonstrate viable applications of electricity that would significantly increase farm use of electricity. As western companies had already discovered with irrigation customers, heavy consumption combined with low prices yielded high revenues—enough to offset costs and ensure a reasonable (and regulator-approved) profit.

The TVA's use of questionably justified low prices would only make sense if farmers used extraordinary amounts of power.[76] Happily for the New Deal planners, the rate-making venture appeared to work. As historian Michelle Mock has documented, the EHFA helped customers purchase high-load appliances such as refrigerators and ranges, causing consumption to grow rapidly.[77] The agency first promoted its services in the TVA demonstration town of Tupelo, Mississippi, in 1934 and, in a one-month period, it spurred sales of 188 refrigerators and 43 ranges.[78] Continued promotion of appliances motivated Tupelo residential customers to double their individual consumption in the seven months after introduction of TVA's 3-2-1 rate schedule.[79] Employing the same pricing formula in TVA's first co-op in Alcorn County, customers went from consuming only 49 kWh per month (in May 1934, before the new rates took effect) to 152 kWh in May 1939. Moreover, the average price per kWh dropped from 5.37 cents to 1.73 cents, while the number of consumers jumped from 1,180 to 2,133.[80]

The TVA and EHFA accelerated thinking about lower priced electricity as a way to stimulate demand. Despite utility companies' denigration and protestations about the organization serving as a "yardstick" for comparing public to private electricity providers, the TVA demonstrated, in a more dramatic fashion than the CREA experiments, that low rates and inexpensive-to-finance appliances could motivate greater sales of electricity.

Undoubtedly, the TVA experience provided justification for Morris Cooke to argue that REA co-op customers would benefit from similar low-cost rate structures. Unlike the TVA-created co-ops (such as the one in Alcorn County, Mississippi), which obtained electricity from the government agency, however, the REA co-ops did not enjoy a single power supplier. Instead, they drew on a variety of sources. The most important consisted of public electricity authorities, such as the TVA, which incurred legal requirements to sell to co-ops and municipalities before offering excess power to private companies.[81] The TVA reported in 1941, for example, that it passed on its low-cost electricity to 114 municipal and co-op distributers at a rate of 4.23 mills per kWh hour. (A mill is one-tenth of a cent or a thousandth of a dollar.) During the following year, it sold power to 128 power distributors at 4.27 mills per kWh; after consideration of distribution and other expenses, the ultimate customer paid 1.25 cents per unit.[82]

Rural co-ops also obtained power from government agencies such as the Bonneville Power Administration, created by the federal government in 1937 in the Pacific Northwest, which also sold electricity preferentially to public entities and cooperatives.[83] Like federal power organizations, the state-government funded Santee-Cooper project (operated by the South Carolina Public Service Authority, created in 1934) sought to develop rivers for navigation, to produce and sell electricity, to reclaim swamplands, and to reforest watersheds.[84] Energy began flowing from the state's public power projects (after several lawsuits initiated by utilities) in 1942, with co-ops purchasing one-quarter of the authority's electricity by the early 1950s.[85] Other public power enterprises sprang up with similar goals and requirements. One consisted of the Buggs Island Lake hydroelectric plant, built in Virginia by the Army Corps of Engineers on the Roanoke River (authorized as part of the 1944 Flood Control Act). The US Department of the Interior distributed power from the facility with priority going to government entities and co-ops.[86] Additionally, the Southeastern Power Administration, established in 1950 (and also managed by the Department of Interior), sold electricity to REA co-ops from more than twenty water projects at favorable prices.[87]

When co-ops could not obtain power from public agencies, they often received cheap electricity from private companies. The original Rural Electrification Act of 1936 (with subsequent amendments) likely motivated this benevolent corporate behavior, since the law authorized inexpensive loans for construction of generation plants and transmission lines when co-ops could not obtain electricity from other sources.[88] Presumably worried that the REA organizations would build such facilities themselves, thus creating competitive entities, utility company managers at Carolina Light and Power, for example, sold wholesale power at 12 to 15 mills per kWh to some North Carolina co-ops. That rate dropped to 7.5 mills per kWh in 1945, where it stood until 1971.[89] In general, co-ops nationally saw wholesale rates at around one cent (10 mills) per kWh for decades after the REA's creation, enabling them to undercut the rates originally offered to rural customers by private utilities.[90] Simply stated, the co-ops usually obtained cheap power even when they could not get it from public agencies.

By the middle of 1940, about 27 percent of the 6.1 million American farms received high-line service, and it appeared that the government had made great progress toward achieving a desirable goal.[91] Undoubtedly, the TVA and REA brought electricity to tens of thousands of unserved farm families. But this narrative, as conventionally presented, should not be mistaken for good history. Only rarely does the standard account provide a context or understanding of the actors who, on one hand chose not to electrify every farm in America, and, on the other hand, those who actually accomplished a large amount of farm electrification within the political and economic framework of the times.

The success of the TVA and REA in pursuing rural electrification was more nuanced than traditional narratives suggest. Put differently, the failure of power companies to electrify a larger percentage of farms remains more complicated. As noted in this chapter, the TVA and REA co-ops enjoyed several advantages. Perhaps most important, the co-ops obtained federal loans at low rates and for long periods. Customers also received cheap money for wiring and appliances through the Electric Home and Farm Authority, while government encouragement of manufacturers made possible the supply of welcomed, inexpensive appliances. Co-op members further benefited from lower taxes, no payments for rights of way, and more affordable rates based on average costs. And by purchasing electricity from government facilities at low cost, becoming preferential customers to public power projects, the co-ops offered low prices to customers with the hope that consumption would increase.[92] Beyond receiving such favored treatment, REA co-ops obtained free or low-cost technical and management advice from headquarters and from REA field engineers who visited construction projects.[93]

These advantages had substantial value. Economists Carl Kitchens and Price Fishback recently estimated that, compared with other types of loans available at the time to homeowners and businesses, the government provided an average subsidy to REA customers of just under 19 percent of the loans' principal.[94] Robert Bradley, a political economist, drew on accounts from the late 1930s to suggest that "[t]he REA was engaged in loss economics," since it financed connections that appeared uneconomical, costing from 25 percent to 50 percent of the $222 million spent by the agency through May 1939.[95] Meanwhile, economist John Neufeld noted that the REA (rather than co-ops) assumed costs for loan processing and for technical and management services; these costs could become quite large, providing a hidden and unquantified government subsidy.[96]

Interestingly, even Morris Cooke acknowledged that perhaps the private utility industry could not do what the most ardent rural electrification advocates desired. In his 1934 argument for creation of a rural electrification agency, Cooke admitted that private companies obtained a poor return on investment on rural lines, requiring them to charge high rates, such that many farmers found it more economical to install isolated power plants. Moreover, he observed that the key to obtaining lower rates consisted of motivating farmers to use large amounts of power—the same argument made by utility managers. "Real rural electrification," he wrote, "implies large average use of current, for without large use[,] rates cannot be made low enough to effect the coveted social advantages." The chicken and egg ("high rate[,] low-use") situation stymied the private utility industry, Cooke noted, such that only one solution appeared practical: "Large average use, especially in the initial states, seemingly requires a planning and investment beyond the capacity of a private company to initiate. Perhaps only the power and force of the Government can master the initial problem."[97]

Despite the appearance of the REA as an innovative organization, whose leaders willingly took risks and experimented with the hope of achieving laudable economic and social goals, recent scholarship suggests it also contained a streak of conservativism. In fact, the agency's management did not stray totally from the financing attitudes held by private power company executives. To maintain support in Congress and among the public, for example, the REA strove (especially after earning its congressional mandate in 1936) to position itself as a responsible lending organization, not a relief agency, whose money would be returned to taxpayers with interest. To provide evidence of good management of taxpayer assets, administrators often pointed to co-ops that paid back loans early and to the low number of delinquencies. (Only twenty-seven installments of interest and principal on loans went unpaid as of 30 June 1939, putting them technically in default. Nevertheless, the total arrearages came to a relatively miniscule $65,616.)[98] As historian Abby Spinak has suggested, early REA leaders accomplished such feats by undertaking the practice of "cream skimming"—the wiring of the densest and likely the most profitable areas (and something the REA blamed private utilities for doing)—as a means of ensuring success.[99] A 1936 Works Progress Administration report on REA operations observed that the lending agency exercised "utmost care in judging the self-liquidity [the ability to pay back loans] of all projects." It also noted, "Loans are well secured."[100] Moreover, the REA made sure that farmers would use large quantities of electricity—more than for just a few lights—to ensure that the co-ops earned enough revenue to satisfy loan payment requirements.[101] In 1938, Judson King, a special consultant to the agency, commented on its less-than-revolutionary practices and remarked that the REA exercised "a banker's care" in making loans."[102]

Sounding very much like a commercial power company document, the Works Progress Administration report on the REA noted, "The use of electricity for lighting alone will not yield a sufficient return to warrant building a project and will not result in rural electrification."[103] Rather, customers needed to purchase appliances so they consumed about 100 kWh or more per month. To reach this goal, the REA facilitated easy financing and offered assistance to wire farmsteads. The agency also received help from the Electric Home and Farm Authority and Federal Housing Administration, the latter of which provided cheap loans for home wiring and for electrically operated water pumps that enabled indoor plumbing.[104]

This revisionist description of the government's rural electrification efforts offers some dissonance to the conventional account that portrays the REA as working selflessly to bring universal rural electrification against all odds and for society's overall benefit. As recent historians have shown, REA managers may have acted more like their for-profit utility cousins than previously portrayed. When selecting recipients of loans, government bureaucrats scrutinized the likelihood that co-ops could repay loans quickly, in a way that would make the agency look good as it sought to retain support from Congress and other stakeholders.

Such risk-averse behavior notwithstanding, government rural electrification offered important lessons to private company managers. Even as they read articles in *Electrical World* in 1934 about efforts to block the TVA (titled, for example, "Coal Men Fight Federal Projects" and "Edison Institute Opens War on Administration's Power Plan"),[105] utility leaders also learned about initial efforts in Tupelo and elsewhere suggesting the success of vigorous efforts to build load through appliance sales and reduced prices. To be sure, the TVA and the EHFA—and later the REA—employed resources (such as low-cost capital and long-term loans) that utilities did not possess. Nevertheless, the government's intense marketing activities and rate-making experiments dramatically demonstrated the value of promoting high consumption. Additionally, the initiatives illustrated (even more than CREA investigations) that once farmers obtained electricity, they purchased new electrical equipment and increased their usage.

The point of this chapter is not to argue that the TVA and REA co-ops enjoyed unfair advantages over utilities. Rather, it suggests that the government agencies took advantage of special conditions that permitted them to pursue a different objective than did utilities. Private power companies constituted profit-seeking enterprises whose managers remained accountable primarily to investors. The firms pursued a conservative approach toward rural electrification, one that did not threaten their financial integrity by taking high perceived risks. The TVA and REA co-ops, however, served (at least in part) a social purpose, established during a time in American history when the nature of public institutions changed. Not bound by normally accepted accounting techniques nor tethered to the need to make profits, they could undertake activities unthinkable to corporate managers. In the process, the government entities offered strong evidence that private companies could make good money by serving customers who had previously been considered unfavorable prospects.

Chapter Thirteen

Competition and Private Utilities in the REA Era

> Thus, the cooperative has not only itself brought electricity to the farm, but it has obviously stimulated extensions on the part of private companies.
> —C. WOODY THOMPSON AND WENDELL R. SMITH,
> ACADEMIC ECONOMISTS, 1941

By mid-1949, just fourteen years after the REA began operations, more than 78 percent of the nation's farms received central station service. The REA administrator noted in his annual report that the fiscal year (from 1 July 1948 to 30 June 1949) had witnessed "the greatest expansion of rural electrification" in the REA's history. "Rural people wanted electricity quickly," he exulted, "and the public and the Congress wanted them to have it."[1] On the organization's fifteenth anniversary in 1950, *Rural Electrification News* gloated that about 83 percent of farms had obtained high-line power. This accomplishment occurred despite the attitude of private power companies, which in 1935 "did not want to borrow Federal funds, and no one could force them to do so." The only feasible approach for rural citizens to acquire electric service was to organize co-ops—to " 'take the bull by the horns' and do the job themselves."[2]

This storyline of a benevolent federal agency assisting the neglected farm population resonates in standard narratives of rural electrification. But it only tells part of the story, one that highlights the REA's work while discounting efforts made by the utility industry. Most important, the narrative omits the fact that until 1950, nongovernmental power companies entered the rural electrification market aggressively. Though using morally and legally ambiguous tactics, the firms energized more farms than did the REA: 2,407,046 obtained power from private companies in 1949, while 2,343,738 received it from co-ops (figure 13.1).[3]

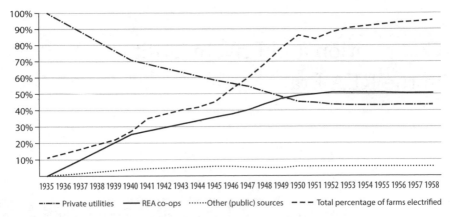

Fig. 13.1. Percentage of electrified farms and providers of electricity, 1935–58. *Sources:* Total percentage of electrified farms data from *Report of the Administrator of the Rural Electrification Administration*, various years, for fiscal years ending 30 June. Percentage of farms served by private companies, the REA, and other public entities calculated from data provided at the end of calendar years in US Department of Commerce, *Statistical Abstract of the United States* (Washington, DC: GPO, 1948, 1949, 1953, 1957, and 1960). The decline in the total farm electrification rate in 1951 likely resulted from the change in the US Census Bureau's definition of a farm.

The crushing economic consequences of the Great Depression, combined with assaults on the utility industry's management behavior and financial structure, put utility executives on the defensive during the New Deal era. Moreover, company leaders believed that the TVA, REA, EHFA, and other government agencies had impinged on the way they previously had controlled large elements of the electric utility system.[4] Despite these forces—or, more likely, because of them—the commercial power industry made remarkable advances in rural electrification in the period from 1935 to the 1950s and later. Of course, many of the gains resulted from the real and perceived competition from the REA. But electric company managers likely also realized that, based on the work done by the CREA and the experiences of organizations such as the TVA and REA, the farm market would ultimately prove eminently worthwhile.

As a historian reading trade literature and archival documents, I never encountered statements from utility managers admitting that they pursued the farm market in the mid-to-late 1930s because of the rivalry posed by government agencies. However, *Electrical World* often published congratulatory pieces about the industry's rural electrification work. The articles illustrated the means by which companies reduced costs so they could better serve farm customers, while others alerted managers of the TVA's and REA's activities. And naturally, the journal included editorials lamenting the fact that the public entities earned unfair advantages in ways that threatened the state takeover of private enterprise. In essence, the TVA and REA appear to have created a true sense of crisis among utility leaders in the 1930s (during a decade of other crises) that spurred competition without explicit mention. This chapter's analysis of the private utilities' efforts to serve more rural customers therefore constitutes a highly plausible account, but one that cannot draw on the type of authentication (in reports, oral history interviews, and so forth) that historians might prefer.

Fears of Encroachment

Utility officials likely believed the government's rural electrification program constituted an existential threat. Since the emergence of the electric power industry in the 1880s, managers developed a culture in which they portrayed themselves as stewards of an increasingly essential infrastructure that generated not just electricity, but technological and social progress.[5] Consequently, they felt vulnerable and vigorously opposed challenges to their control, such as the proposed Giant Power project in Pennsylvania and other public power initiatives in the 1920s. Though able to impede those efforts, the industry still endured the sting of accusations of questionable business activities disclosed by the Federal Trade Commission (FTC) investigations beginning in 1928. Revelations of unseemly practices, such as the use of ghostwritten editorials placed in local newspapers, the hiring of university professors to serve as "consultants," and the insertion of utility-favorable paragraphs into school textbooks, had stained the industry's image. The exposure of questionable (though technically legal) means of leveraging investments through the holding company structure, exploited by Samuel Insull and others, further tarnished the reputation of the business.

As one approach to improve its appearance, the utility industry disbanded its trade organization, the National Electric Light Association, and replaced it with the Edison Electric Institute (EEI). A *Washington Post* article on the new organization's formation, announced on 12 January 1933, observed that NELA had suffered attacks in the FTC hearings "for alleged propaganda activities prejudicial to the public good." But the new group, even though headed by George Cortelyou, a former NELA president, would act differently, assuming "an attitude of frankness and ready cooperation in its dealings with the public and with regulatory bodies."[6] The *New York Times* wrote positively about the industry's promises to behave more ethically, and it highlighted a "purging process" that had (since the beginning of the Depression) eliminated many holding company abuses.[7]

Of course, creation of the defensive-acting EEI did not forestall continuing attacks on the utility industry's autonomy. By the time of the new organization's creation, Franklin Roosevelt had already been elected (but not yet inaugurated), and many utility officials worried about increased government encroachment in the power sector. As New York governor, FDR had signed legislation in 1931 creating the State Power Authority, which offered hope for more rural electrification and for construction of municipal power plants as a means to undermine the allegedly high rates charged by urban utility companies.[8]

Fears of greater government involvement in the power industry materialized soon after FDR assumed the presidency. The establishment of the Tennessee Valley Authority in May 1933 created a perhaps unrealistic, but publicly appealing "yardstick" for providing inexpensive electricity to rural areas. Formation of the federal Public Works Administration in 1933 also threatened utilities' near-monopoly in producing power, as the agency received authority to finance municipal power plants and hydroelectric projects. Soon thereafter, during the spring and summer of 1935, Congress debated the financial structure of utility firms, ultimately posing (with the Public Utility Holding Company Act) a "death sentence" that eliminated multilevel holding companies starting in 1938.[9] In this environment, President Roosevelt signed the executive order in May 1935 establishing the REA as a temporary relief measure, with Congress making it a permanent lending agency a year later. Despite Morris Cooke's frequent statements soon after the REA's founding that "there'll be no competition . . . no butting in on private utilities," the leaders of the power industry correctly saw the situation otherwise.[10]

Among the measures taken to thwart the REA's challenges, utility managers employed the legal mechanisms of regulatory commissions. Though at first seeking to work with power companies to spend the initial emergency funds to electrify rural areas, Cooke ultimately depended on farmers' cooperatives to pursue the erection of distribution lines. In some cases, private firms argued that the new entities should be forced to obtain certificates of convenience and necessity from regulatory bodies before they built lines.[11] In theory, application for the certificates guaranteed that utilities did not duplicate resources or contribute to destructive competition in a monopolistic enterprise.[12] But the Raleigh (NC) *News and Observer* saw through the tactic, opining in 1937 that imposing such an obligation on co-ops merely benefited power companies. "The only real purpose in requiring farmers to get a certificate of convenience" the newspaper argued, "is to delay every farmer effort to provide themselves with electricity, . . . and to throw an obstacle in the way . . ." of the co-ops.[13]

REA officials perceived the menace posed by utilities' use of regulatory procedures. Arguing against proposed state oversight in Illinois, for example, Cooke wrote in 1937 that private companies would hold great advantages over co-ops under regulation, since the latter "would be weighted down in ninety-nine cases out of one hundred by the talent, engineering and legal, of the private companies."[14] Reporting on a North Carolina bill that sought to subject the new entities to regulation in 1937, a newspaper article observed that the disadvantaged co-ops "would have to oppose the great power companies of the country."[15] Put simply by Cooke, if commissions exerted control over co-ops, attempts by farmers to serve themselves would be "doomed to failure."[16]

The REA also fought against state regulation of co-ops because of their unusual organizational arrangement. Government oversight of utilities appeared necessary to prevent monopoly power companies from earning excess profits when serving the public and for protecting investors against managerial abuses. But since they only offered electricity to their own members, co-ops should not be subject to the same rules that governed companies dealing with the general population, the REA contended. Moreover, co-ops established themselves as nonprofit entities, with excess funds used to retire loans or to reduce rates, not to pay dividends to financial backers. And unlike utility firms, co-ops did not issue equity securities or bonds, so they did not require supervision of financial activities.[17]

Some of these assertions carried weight in various jurisdictions, thus defeating early utility efforts to discredit co-ops, but not everywhere. In passing the Electric Cooperatives Act in March 1936, for example, the Virginia Assembly established the legal basis for co-ops to operate, but it also subjected them to regulation "in the same manner and to the same extent as are other similar utilities."[18] The legislation's requirement therefore enabled power companies to employ, at least in theory, means to hinder the co-ops' work through the regulatory process. As comparable measures appeared in other state legislatures, it remained unclear how the new co-ops would be treated. Observers noted in 1936 that the question of regulation remained in a "twilight zone" of uncertainty.[19]

In large part to facilitate the lawful existence of co-ops and to gain exemption from traditional regulation, the REA Legal Division devised a model statute that won enactment in fourteen states by the end of 1937.[20] Two years later, the REA happily reported that state legislatures defeated several "injurious laws" that would have inhibited the co-ops' activities; lawmakers released the organizations from government oversight in Alabama, Montana, New Mexico, Pennsylvania, Oklahoma, South Carolina, and Tennessee.[21] The early exemption of co-ops from regulation in Alabama and Tennessee, noted TVA lawyer Joseph Swidler in June 1935, "will result in a considerable saving to cooperatives, which can ill afford the expense attendant upon commission regulation."[22] In some states, such as Utah, courts (rather than legislatures) took the lead, holding that co-ops did not constitute public utilities, thus relieving them of supervision.[23] By 1940, twenty-five states had exempted co-ops from regulation or claimed no authority over them, according to REA administrator Harry Slattery. Six states exercised some control, and five retained complete oversight by commissions.[24]

But not all state regulation impeded co-op activities. The Wisconsin Public Service Commission in late 1935, for example, organized five conferences between representatives of utility companies and potential REA co-ops to allocate service areas (though not always successfully).[25] And the Public Service Commission of Kentucky—one that Slattery listed as having "complete jurisdiction" over the co-ops' activities,[26] an implicit negative toward rural electrification—claimed in 1940 to be "proud of the record it has made in encouraging the spread of rural electrification throughout the State." Through favorable administrative orders, the body helped co-ops obtain low wholesale rates from existing companies.[27] Additionally, the commission processed 123 applications for franchises in rural areas and approved loans enabling 23 co-ops to serve 19,311 homes by the end of 1939.[28]

At the same time that they fought co-ops by insisting on their need to endure traditional regulation, power companies sought to preempt the new public power entities in another fashion: they often employed a technique, vilified by the REA, known as "spite line" construction. A 1937 agency report explained that the term "came to stand for the construction which a private utility hurriedly undertook after a farmers' cooperative project had received an allotment of funds from REA."[29] Because creation of co-ops and plans for erection of distribution lines quickly became common knowledge through newspaper reports and public notices about organizing meetings, utilities sometimes built lines hastily, thus making area coverage by the REA—service to entire expanses within a region—impractical. As some customers received service in this manner, utilities argued that the co-ops should not be permitted to duplicate the lines.

Related to spite line construction, "cream skimming" had the effect of reducing the REA's interest in an area. Using this tactic, utilities quickly provided power to the most prosperous customers and therefore made service to the remaining customers less economically feasible, even for co-ops. The use of spite lines and cream skimming effectively gave private utilities an advantage over the membership organizations that had planned, but not yet begun, construction of lines. As even the REA noted, "the express terms of the Rural Electrification Act . . . limited [the agency] to the serving of persons who are not already receiving service."[30] Once private companies supplied electricity to farmers, in short, REA co-ops could not offer alternative service. Of course, the ploy did not go unnoticed. The West Virginia regulatory commission commented on the irony that within two weeks of a co-op receiving a charter in June 1937, a private company started building twenty-eight new power line extensions totaling about 124 miles in the same territory, which contrasted with the 19 miles constructed in the previous two years. And whereas the company earlier required rural customers to guarantee minimum purchases of electricity, it ceased making that request after the co-op's formation.[31]

Occasionally, central station companies simply bought out co-ops as they began establishing themselves. Carolina Power and Light Company (CP&L), for example, stopped the emerging Johnston County, North Carolina, co-op by using this tactic. In May 1937, the company offered $6,750 to assume the organization's expenses, raising its bid in July to $15,000. REA administrator John Carmody told the co-op that abandoning the project would violate its contract and would have a negative impact on the creation of REA-funded co-ops elsewhere.[32] He further sought to get the state's governor involved, though (according to a news report) he may have poisoned the waters because Carmody's letter to him implied that state government was "dominated by the power companies."[33]

In a surprise to most observers, four of the six co-op directors accepted the power company's offer, without notice to the REA or the organization's members.[34] Some Johnston County residents (though not co-op members) tried to overturn this action by suing CP&L. As a finding of fact, a lower judge noted, "The purpose of the Carolina Power & Light Company in making said offer and in taking over the entire project under contemplation by the Johnston County Electric Membership Corporation, was to acquire a monopoly of the business in which it was engaged."[35] The justice also assailed the directors for a "violation of the trust imposed in them" by the co-ops members, who were not consulted, making the whole transaction "tainted with mala fides."[36] Still, the state's Supreme Court dismissed the case because the plaintiffs did not belong to the co-op and therefore had no right to sue.[37] A news article observed that the Johnston County farmers simply got tired of the "endless red tape" bureaucracy, then going on for a year and a half, and opted for CP&L, which "is ready to go to work and build 155 miles of additional rural lines immediately and will probably get them finished before the cooperative could get its loan." After all, the article noted, "what the farmers want are rural lines and electricity now, not sometimes [sic] in the future."[38]

Though farmers obtained service with such ploys, the REA did not appreciate the power companies' efforts to hinder creation of co-ops. Its administrator, John Carmody, wrote to the West Virginia governor in May 1937 to seek help in stopping utilities from frustrating his agency's efforts to fund cooperatives. He noted that a private company had assumed an approved REA project in August 1936; a few months later, another endeavor had to be "rendered inactive" since power companies "had taken so much of the territory as to make the remainder of the project unfeasible." Utility managers argued, Carmody suggested, that it did not matter who built the lines as long as farmers got power. Disagreeing, he noted that "experience indicates that where cooperatives are defeated[,] the aggressive spirit of the farmers is killed[,] and the utilities promptly cease to build lines into less profitable rural areas."[39]

Despite similar skirmishes that occurred throughout much of the country, REA co-ops eventually emerged from an initial "legal wilderness" and obtained the right to operate.[40] As popular support of the REA grew (no doubt, in part, because of the effective self-promotion of the agency, described in chapter 1), state legislatures, courts, and regulatory bodies often took substantive action to frustrate utilities' efforts to emasculate the new organizations in the late 1930s: the Tennessee commission, for example, issued orders to prevent cream skimming by utilities. The Wisconsin and Pennsylvania legislatures also prohibited utilities' construction of distribution lines in territories in which co-ops demonstrated an interest by filing application materials with commissions.[41] The initiatives therefore prevented further incursions, such as those in North Carolina, in which a utility strung up lines soon after co-ops publicly announced their intention to expand in a particular region. By 1941, all but three states had established the statutory machinery for permitting co-ops to operate, having mitigated (to a large extent) the use of legal and regulatory means to inhibit the new entities.[42] In the same year, Administrator Claude Wickard reported that the "number of power companies still actively engaged in obstructive tactics is small in comparison with the number which were actively hostile in the earlier stages of the REA program."[43]

Better Market Conditions Lessen Hesitation

While the government's competitive threat likely spurred most rural electrification efforts among utilities, managers may have felt more disposed to take risks on rural customers because of improving economic conditions. After net farm income bottomed out at a little under $2.3 billion in 1932, it grew to almost $3 billion in 1933, $3.5 billion in 1934, and just over $5 billion in 1935.[44] The farm economy had not yet returned to its 1929 level (with about $6.7 billion in net income), of course, but it clearly showed a rebound. Even the REA observed that increased farm income had already translated into sharp increases in sales of machinery and automobiles.[45] Perhaps farmers might have enough money to purchase electricity and productivity-enhancing equipment, especially after receiving aid from government agencies.

As the farm economy strengthened in 1935, so did the environment for utility stock and bonds, both of which augured well for the prospect of returning prosperity.[46] Sales of securities for purchasing power plants and other equipment (that is, capital expansion) had plummeted in the early 1930s: as late as 1931, the industry raised more than $1.5 billion, but that amount dropped to about $470 million in 1932 and just $56 million in 1933.[47] Securities sales more than doubled in 1934 from the 1933 low point, however, and then exploded eightfold to more than $1 billion in 1935. A huge consumer of capital, the utility industry also started to obtain funds from investors at favorable borrowing rates, dropping to about 3.5 percent in 1935, below the 4–5 percent range of 1934 for long-term mortgage bonds. Evidence of the improving circumstances also appeared in bond prices in the secondary market: prices jumped from under $84 per hundred dollar issuances in early 1934 to more than $104 in late 1935.[48] The overall stock market, moreover, came back to life, and power companies' share prices surged 66 percent in value during 1935 as measured by the Dow Jones Utility Average.[49]

The improving financial conditions likely gave power company managers confidence in their own ability to pursue rural electrification (and other endeavors) without government help. In other words, their inability to obtain REA loans during 1935 at a rate of 3 percent for twenty years may not have appeared as a lost opportunity.[50] The better environment meant that utilities could borrow money at terms comparable to those offered by the REA—around 3.5 to 3.8 percent, though usually for bonds having longer, and more attractive, maturities of twenty to thirty years.[51] Additionally, the power firms would not need to abide by REA rules and procedures.

Morris Cooke took note in September 1935 of the healthier financial conditions when he publicly congratulated a New York utility company, which had withdrawn its application for REA funds to add five hundred miles of rural lines. "It is particularly gratifying," commented Cooke, "that the good credit of your company, the better market for investment securities and the low interest rates prevailing should have made it possible for you to proceed with this construction with private financing without the necessity of borrowing from this Administration." He further hoped "other utility companies will show similar enterprise and follow your excellent example." As important, he assured the power firm's president that the government, through the Electric Home and Farm Authority, would provide low-cost financing for wiring farms and for appliance purchases.[52] It therefore seemed that private firms and the new agency would provide valuable complementary services to overcome frequently discussed impediments to widespread electrification.

Using Rural Electrification as a Path toward Recovery

Some utility managers may have also seen rural electrification as a way to gain new business during the depths of the Depression. With manufacturing and industrial activity plummeting, power company executives realized that they needed to do something to buttress electricity sales. Initial attempts to promote increased usage by industrial customers, often by lowering rates, did not have much impact since manufacturers could not sell their products as easily as in the 1920s, and they had little choice but to curtail production—and with it, electricity consumption.[53] As one response to declining sales to industrial clients, managers began offering lower rates to other customers, such as residential users. The Milwaukee Electric Railway and Light Company saw its industrial load sink and turned to selling appliances and using cheaper, experimental rates for home consumers; the strategy pushed up average residential consumption by 60 percent between 1933 and 1938.[54]

Power companies elsewhere pursued similar programs to boost domestic (residential) use of power. In early 1932, *Electrical World* observed that electricity purchases by these customers had been increasing and that, despite lower unit prices, the amount of revenue per customer continued to grow, largely owing to their burgeoning use of electric appliances, such as refrigerators, ranges, and water heaters.[55] In the five years after 1926, domestic energy sales almost doubled, and the future looked bright, especially since the refrigerator—a relatively new appliance—only found a place in about 11 percent of electrified homes in 1931.[56] Drawing about 600 kWh per year, the appliance could do wonders for utilities seeking increased sales once more homeowners bought them.[57] Economists in 1941 observed that the focus on domestic customers "was so successful as almost to offset the sharp loss of the industrial load during the worst days of the depression of 1929."[58]

Even with a strategy that sought to enhance consumption by residential customers, growth of the domestic market had limits, since (as Toledo Edison manager W. E. Trimble observed in 1932) about 92 percent of urban homes had been electrified. But the rural market beckoned. "[W]e see that the field for rural electrification"—in which approximately 12 percent of farms used central station electricity—"is very large," he commented. And because of the near saturation of the urban market, utility company growth "is largely dependent on rural electrification."[59] Other industry stakeholders expressed like-minded views. Anticipating an untapped potential of 3 million customers (about half of all farms at the time), *Electrical World* in January 1932 observed that the almost seven hundred thousand already-connected farms still needed to purchase hundreds of thousands of appliances, such as electric pumps, general-purpose motors, milking machines, refrigerators, incubators, and water heaters. And in its preview of 1933 business conditions, *Electrical World* noted that the power industry lost customers in the previous year. To regain some business, companies needed to look to rural regions, farms, and villages.[60] More explicitly, the trade magazine explained that "there's the farm market eager for the things that electricity can do and the leisure and economies it can bring."[61] Expressed simply, farm customers could help utilities lift revenues during a difficult time.[62]

Perhaps unexpectedly, the TVA furthered this line of thinking, by suggesting that farmers might have potential to become better customers than previously believed. The public agency did so by demonstrating that markedly lower prices stimulated demand substantially—enough to yield higher-than-expected revenue for utilities. This elasticity in demand (such that consumption increased as prices decreased) had already been experienced in many state CREA studies and as utilities in the late 1920s and early 1930s saw increasing sales among existing farm customers. But the low TVA rates, publicized as a yardstick for comparing private companies' prices, drove home the point more effectively. The work of the EHFA in providing inexpensive loans to customers for appliances—along with well-received endorsements of cheap appliances—did much to spur purchases of the demand-increasing devices that dramatically improved the economics of selling to farmers.

Using the Tennessee Electric Power Company as an example of the value of low-cost pricing, the TVA noted in its 1939 *Annual Report* that residential (not rural) customers in 1933 paid an average of 5.77 cents per kWh for power and consumed about 600 kWh per year. When the company began reducing prices after obtaining TVA-produced power, consumption jumped impressively, such that in 1938, customers paid only 2.75 cents per kWh and used 1,461 kWh.[63] (Interestingly, a graph in the TVA report shows that from 1926 to 1933, the company reduced prices and also saw increased usage, though the growth rate jumped after 1933 when prices declined more markedly.) Even with lower rates, greater consumption translated into more revenue (by about 14 percent, with each customer paying $40 per year, up from $35). The lesson seemed obvious to TVA managers: "Dependence for sufficient revenues to meet the costs of power and its distribution is placed on volume sales of electricity rather than upon a high-rate level."[64] Put differently, the pricing schedule instituted by the TVA planners in 1933, made on the basis of faith and hope rather than on existing economic principles, had proven successful, increasing overall revenue. So potent was this early demonstration that in August 1939, Alcorn County co-op members saw their already low rates decrease further: the easy-to-remember 3-2-1 rate schedule became less memorable but cheaper. For the first 50 kWh, customers paid 2.5 cents per unit; for the next 100 kWh, they spent 2 cents per kWh; the next 250 kWh cost a penny each; consumption beyond 400 kWh pushed the rate below 1 cent per unit.[65]

Georgia Power executives learned much from the TVA experiences. As historian Brent Cebul has recently documented, the company (a part of the Commonwealth and Southern holding company that ceded some of its service territory to the TVA) initially saw its investment ratings, along with its stock and bond prices, decline soon after the federal organization began work. But its leaders quickly took advantage of the EHFA's stimulation of appliance sales to build up its own rural market, pursuing a process that Cebul termed "creative competition." In response to the TVA's yardstick prices, the company advertised EHFA-approved appliances and lowered its rates. Sales increased significantly after bottoming out in 1932, providing greater revenue to compensate for lower unit prices.[66] The firm continued an emphasis on offering consumer credit, as did the EHFA, and it expanded rural service in ways that emulated the New Deal programs—with positive effects. By 1936, Moody's Investors Service remarked on the improved financial health of Georgia Power, in large part because of its new business model. ("Satisfactory rate of growth recorded despite substantial rate reduction," Moody's reported.)[67] As business conditions continued improving in the 1930s and later, observed Cebul, "rural Georgians enjoyed a modern standard of living thanks largely to Georgia Power's competition with and emulation of the New Deal's TVA, EHFA, and REA." In an ironic sense, "While Georgia Power had its creative competition with the New Deal to thank, thousands of rural Americans directed their gratitude to private companies."[68] The New Deal agencies, Cebul argued, offered "public models and motivation" that benefited the for-profit utilities.[69]

Viewing the rural market as partial salvation for power companies' woes during the Depression did not seem odd to managers like Grover Neff. As he pointed out in a 1934 speech, the number of farmers who purchased electricity from central station firms continued to grow, even during the economic crisis. Only 130,000 farms nationally obtained high-line service in 1920, he observed, but at the end of 1932 the number had risen more than 400 percent to 709,449. The growth rate moderated during the slowdown, but "contrary to almost every other development," he remarked, an increasing number of ruralites continued to obtain central station service. At the end of 1933, 713,558 farms had gained high-line connections, and another 10,809 obtained service by the end of May 1934.[70] As further evidence of this encouraging trend, *Electrical World* reported in early 1935 that "[f]arm customers are the only group revealing a steady gain throughout the depression."[71]

Even the creation of REA co-ops seemed to offer modest financial benefits to utilities. A 1936 Works Progress Administration report noted, "Many private companies have been content to permit farm cooperatives and public bodies to construct distribution lines because they see in them an increased market to absorb their generated power." Except in rare cases, the co-ops served as a new market for the power firms, the report added, because the REA planned only to "take current from existing power generating plants," most of which were operated by private companies.[72] In 1937, the REA administrator noted that about two-thirds of the agency's wholesale power came from utilities—"from their existing sources and at a profit to them."[73]

A slight recovery in the farm economy and securities markets may have provided some encouragement for power company managers to move more forcefully into the rural electrification business. But the more significant impetus probably came from the competitive threat of REA co-ops and other public agencies. Leaders of the private utility industry remained defensive and, as they had done in earlier fights against public power incursions, they sought to retain control of a system that they believed they (and their managerial ancestors) created and operated well. An observer sympathetic to the public power movement in 1944 noted (accurately) that the government's creation of the TVA, REA, and other public projects "demonstrated the feasibility of a new approach and has stimulated activities which have released previous latent energies in the electric power industry." Most important for this study, the author commented that, in the newly "competitive situation," the private companies provided power to two-thirds of all electrified farms in 1944.[74]

Economist John Neufeld recently provided nuance for understanding the utility companies' forceful response to the REA menace by focusing on the notion of lost profit opportunities. He argued that in the years before the existence of government-supported co-ops, utility managers could avoid making what they viewed as speculative investments to serve rural customers adjacent to their existing bases of operation, mostly in cities or towns. (Recall that utilities obtained franchise territories; farmers generally resided outside of these regions, though they could often obtain power by satisfying companies' requirements.) Even with more liberal utility policies for rural service in the years before the New Deal programs began, however, most farmers could not afford guaranteed monthly payments or up-front outlays for distribution lines. The firms therefore could justifiably bypass ruralites, focusing on industrial, commercial, and urban customers who would likely prove profitable. Best of all, they could simply wait until the farm market became demonstrably more promising before addressing its needs, expanding later into previously unassigned service areas.[75]

But the REA's establishment eliminated the power companies' freedom to delay making decisions about service to farms. The Rural Electrification Act of 1936 explicitly authorized loans to erect high lines to farmsteads that had not yet received central station power.[76] In the process, the co-ops could lay claim to territories that utilities had abrogated—or thought they could patiently ignore until another time. Pennsylvania's Electric Cooperative Corporation Act of 1937, for example, exempted co-ops from regulation and gave them the right of eminent domain. Once chartered by the state and given authorized service districts, the co-ops could erect high lines to previously unserved ruralites, thereby excluding power companies forever.[77]

As other states adopted similar laws and administrative procedures, utility leaders needed to expedite their thinking about when to enter the rural market. They could continue waiting and likely lose large numbers of potential customers to REA co-ops in regions that remained outside the companies' existing service territories. Alternatively, they could make investments immediately, effectively taking control of new areas that they hoped would eventually yield sufficient revenues. As Neufeld argued, utility leaders knew that "[r]ural electrification was widely expected to come eventually," a realization that I claim arose because of the work of Grover Neff, the national and state CREA organizations, and industry partners starting in the 1920s.[78] The critical factor in utilities' zealous pursuit of rural electrification in the 1930s, in other words, consisted of more than a decade of experiences that gave company executives confidence in the farm market's ultimate feasibility.

Utility managers responded to the competitive threat at a pace that seemed improbable a few years earlier. REA officials took note of the accelerated activity, observing in 1937 that while the agency brought power to about 250,000 farms in 1937, "[p]rivate utility construction, stimulated by REA, raised the living standards of hundreds of thousands more." Power companies, in fact, established new construction records, building about forty-one thousand miles of rural lines during 1937, up from about nine thousand miles in 1936 and double the number of the previous year.[79] REA managers saw that most of the utility expansion occurred where co-ops had begun projects, such as in Illinois in 1937, but none where the REA remained inactive.[80] In a philosophical manner (at least in this case), government administrators observed that while such competitive "tactics have damaged some REA projects and destroyed others, they have at least increased the amount of electrification."[81]

Relations between private utilities and public agencies did not become a love fest after rural electrification demonstrated its profitability. Even though Georgia Power prospered by emulating New Deal agency practices, for example, the president of its Commonwealth and Southern holding company, Wendell Willkie, still demonized the trend toward increasing government control of the economy as a form of "state socialism."[82] (Willkie had reluctantly sold assets to the TVA so it could transmit power, and he unsuccessfully led fights against the agency in Supreme Court cases decided in 1938 and 1939.)[83] Becoming recognized as an articulate spokesman for pro-business interests during the Roosevelt era, Willkie won the Republican presidential nomination in 1940.[84]

After World War II, private companies maintained their onslaught against public power entities. REA administrator Claude Wickard complained in 1946 that utilities persisted in blocking farmers' efforts to win government funding through "pressures of every sort."[85] Carolina Power and Light, for example, skirmished with REA co-ops when negotiating sales of wholesale power. In defense, the co-ops sued the company in 1977, with a resolution emerging in 1991.[86] The industry also continued its public relations campaign to denounce government incursions into the business realm. A vocal exponent for private ownership of the electricity business, Edwin Vennard, the managing director of the Edison Electric Institute, argued in a 1968 book that government should not usurp free enterprise. In ways reminiscent of 1920s rhetoric (but also consistent with Cold War discourse), he asserted that "the individual citizen, as taxpayer, as power consumer, and as investor, gains economically and in . . . individual freedom" through the application of "the free-market approach."[87]

When Vennard penned his book, the job of rural electrification had largely been completed. In its 1968 annual report, the REA proudly observed that 98.4 percent of farms obtained central station service ("as contrasted to 10.9 percent when REA was created in 1935").[88] Not unexpectedly, the administrator failed to give credit to the private companies that had wired up about 40 percent of those rural customers.[89]

Conclusion

The success in providing central station, high-line service to almost all farms suggests that the rural electrification subsystem had become mature and closed—using terms coined by historian Thomas Hughes. The closure likely occurred within about fifteen to twenty-five years after the founding of the REA, as electrification rates zoomed to 45.0 percent in 1945, 86.3 percent in 1950, 94.2 percent in 1956, and 96.5 percent in 1960.[1] During this relatively brief period, the subsystem integrated a collection of actors into a stable sociotechnical framework that realized a valued goal.

The newly mature subsystem did not look like the one envisioned by rural electrification advocates in the 1920s and early 1930s, however. Most significantly, it did not remain under the control of utility managers, who failed to resist radical inventions from outside the subsystem. Undeniably, promoters made tremendous headway in the 1920s to electrify farms through their establishment of social institutions and elimination of many technical and regulatory impediments. But by creating the TVA and especially the REA, President Roosevelt and Congress introduced dynamic actors into the subsystem that challenged the work done by private companies. These stakeholders ultimately acquired the majority of farm customers and retained political and rhetorical superiority for decades thereafter. They undercut the building momentum within the subsystem and destroyed the prospect—pursued so ardently by undercurrent champions such as Grover Neff—that utility managers would preside over an almost totally electrified rural America.

An Unanticipated Subsystem

Private industry ceded control of the rural electrification subsystem to public power organizations through the introduction of new, vigorous participants and the loss of existing supporters. Of course, the most important of the former consisted of the federal agencies—the TVA and REA—and the cooperative electric organizations they spawned. Given technical and managerial support from the government, the co-ops must have riled the psyches of company leaders. Generally harboring condescending attitudes toward rural folk, these executives came to realize that unsophisticated farmers had begun serving their brethren in ways that had eluded the power firms for a half-century. The supposedly naive agrarians now constituted a competitive threat to parts of the utility managers' business, adding to the other menace—municipal systems—that had largely been tamed.

Likewise, the federal government stepped in as the primary financier of the new rural cooperatives, thus upending the more traditional actors within the private securities markets. Besides the TVA and REA, the government's Public Works Administration and Electric Home and Farm Authority offered several hundred millions of dollars in loans at unprecedentedly attractive terms for rural electrification.[2] Such funding swamped the resources furnished by banks and investment brokerage firms for farm work, especially in the early 1930s, when utilities had difficulty obtaining capital for almost any activity. (Recall that in 1933, the industry could raise only $56 million from securities sales, less than 4 percent of the amount it procured two years earlier.)[3]

As the new financial stakeholders took leading positions in the rural electrification subsystem, existing supporters shifted allegiances. Especially significant, university-based academic agricultural engineers found that working with federal agencies offered new opportunities for their profession and constituencies, such as students, extension agents, and farmers. Soon after 1933, for example, the TVA and the USDA established agreements with land-grant colleges in seven states to pursue electrification research and education. The TVA also provided funds for training extension service agents who instructed farmers about the best electrical practices.[4] Additionally, REA officials regularly met with agricultural engineers at universities and at professional conferences.[5] Separately, the REA established a junior training program that brought recent agricultural engineering graduates into the agency and then hired several into higher-level managerial positions.[6] In creating its Utilization Division to expand load-building work and to staff agency offices, the REA employed many of these college-educated professionals.[7]

Agricultural engineering programs at several land-grant colleges expanded activities after the New Deal agencies ramped up. At North Carolina State College, for example, Professor David Weaver began rural electrification research with money provided by the Federal Housing Administration in 1934. Creation of the REA accelerated work, with Weaver serving as the liaison between the federal agency and the state beginning in 1936. To help farmers after they created REA co-ops, the college's agricultural extension service hired new personnel to teach about proper home wiring techniques.[8] The REA also spurred Alabama Polytechnic Institute's agricultural engineers and extension agents to emphasize instruction on safe wiring techniques, electric water pumping, and refrigeration.[9] Virginia Polytechnic Institute pursued similar research and educational activities, with fifteen REA-financed projects established in its state through 1943.[10]

Once a supremely important stakeholder as the utility industry's research arm and key information disseminator, the Committee on the Relation of Electricity to Agriculture saw its influence wane as government actors gained prominence. Within a year of creation of the TVA in 1933, for example, the American Farm Bureau Federation and the National Grange pulled their representatives from the CREA, as these institutional supporters allied themselves instead with the federal government.[11] And quickly after President Roosevelt established the REA, AFBF directors convened a Rural Electrification Committee and "voted unanimously to give fullest cooperation" to the new agency.[12] Demonstrating this shift in partnerships, the AFBF invited President Roosevelt to speak at its annual convention in December 1935, with the organization's leader expressing support for New Deal programs.[13] The December issue of the AFBF's newsletter observed that its members commended the work done by the TVA and REA, urging the government's rural electrification efforts "be expedited in every possible way."[14]

The involvement of the national CREA as an important stakeholder in the rural electrification subsystem experienced another setback when representatives from the Departments of Interior, Commerce, and Agriculture withdrew their memberships in October 1935.[15] According to newspaper columnist Rodney Dutcher, head of the Washington, DC, bureau of the Newspaper Enterprise Association, the government agencies distanced themselves from what appeared to be the industry's attempts to sabotage the REA's program. Clearly no admirer of private utilities—he referred to the "Power trust" companies that sought to maintain high electricity rates for farmers—Dutcher gave examples of state committee members who spread false information about the newly formed REA. He also noted that the American Farm Bureau Federation quit participating in CREA work a few years earlier because of the committee's supposed "friendliness with power interests."[16] More generously, a 1936 REA publication noted that while the CREA had done "much valuable work in acquainting the farmers with the use of electricity," the government agencies decided that more productive work on costs and rates could occur "from an independent position."[17]

As the national CREA lost influence, it pared its regular printing of issues of the *CREA Bulletin* and *CREA Newsletter.* The authoritative 332-page compendium of research, titled *Electricity on the Farm and Rural Communities,* appeared as the first number of the *Bulletin's* seventh volume in 1931, with the second (and last-ever) number seeing publication in July 1937. Perhaps consolidating resources during the Depression, CREA still produced its *Newsletter,* circulated since 1928, with annual editions appearing in 1935, 1936, and 1937. Its final issue appeared in June 1938, with no hint that none would follow. But the work of the national CREA had almost reached an end. In August 1939, director E. A. White wrote letters to heads of state affiliates indicating (without providing an explanation) that the national committee would discontinue activities by October of the same year.[18]

Alternative Narratives

Though the private power industry lost dominance over the rural electrification subsystem, it nevertheless made significant strides to bring electricity to farmers in the 1920s and early 1930s. It did so by engaging partners in educational institutions—most importantly the agricultural engineers in state land-grant colleges—and by building a social and technical infrastructure that brought electricity to more than 11 percent of farms by 1933. Rising from under 3 percent a dozen years earlier, this much higher level provides testimony to the stakeholders' work that built a foundation for even greater availability of electricity in subsequent years. These efforts and accomplishments, which I have described in this book, argue against the assertion that private power companies purposely neglected farmers. Rather, my interpretation suggests that leaders within an undercurrent movement of the utility business worked in a fashion consistent with the tenor of the times to boost power delivery to farms. In the strained political and economic environment of the 1930s, however, they could not prevent the rise of forces that gave control of the rural electrification subsystem to new players.

Of course, commercial utilities and their allies did not endorse a narrative that illustrated how intruders deprived industry leaders of their omnipotence. In their own attempt to present a positive account of the business they covered, the editors of a special issue of *Electrical World* in 1949 (commemorating the journal's seventy-fifth year of existence) celebrated the rapid strides made by the industry in various realms. But the publication too leniently explained away the industry's problems. The holding company abuses and security manipulations of the 1920s, for example, occurred during a "period of business prosperity" when "[e]veryone dealt in futures, saw profits now and greater gains ahead." Admittedly, "the power companies were front and center" in a decade of speculative machinations, with firms combining, changing ownership, and interconnecting; each event resulted in issuance of more stock that "the public bought . . . in eager, greedy, hope of doubled value overnight."[19] Implicitly, it appears, the editors shifted part of the blame for the financial excesses away from the utility magnates and onto the avaricious public.

In their description of the progress of farm electrification, the *Electrical World* editors acknowledged that companies focused initially on the urban market before devoting attention to the "thinner rural field."[20] But once the industry established the Committee on the Relation of Electricity to Agriculture in 1923, new knowledge emerged rapidly that helped farmers increase production—the most critical indicator (according to the editors) of farm electrification, though politicians often preferred to look at the number of farms served. But even that metric looked impressive: during the period between the end of 1923 and the end of 1929, the number of industry-served rural customers grew by a factor of 3.2.[21] Most businesses (even those of today) would win accolades for enlarging their customer base at an average compounded rate of 21.7 percent per year, especially as the country emerged from the agricultural recession of the 1920s. This rate of growth, the editors noted, if continued through the 1930s, would have "resulted in service to every farm in the nation by 1941"—and without government intervention.[22]

The *Electrical World* narrative included other demons who sidetracked the industry. Politicians, for example, perpetuated a "hate of the 'Power Trust' bogey," and they motivated creation of the REA, which "set out to prove that the power company approach to rural electrification was entirely wrong."[23] Not portrayed as a totally evil organization, the REA spurred the power industry to continue its pre-Depression efforts to electrify rural areas, such that almost 4.4 million farms received central station electric service in 1948. But the editors made sure to point out that, of these farms (then making up about 75 percent of those in the country) 51.3 percent obtained power from private firms, while only 43.7 percent received it from REA co-ops. (The balance drew power from municipal and other government agencies.) Better yet, by the late 1940s, the co-ops had abandoned some unwise business and technical practices, such that "the feeling of animosity [from the REA to the private industry] also belongs to the past, in spite of strenuous efforts of some men to keep it alive."[24]

A Historiographic Challenge

Clearly not as sympathetic to the power industry as the trade journal's self-serving storyline, this book has nevertheless presented a revisionist narrative for understanding rural electrification before the federal government's involvement began in the 1930s. This alternative account, however, may encounter impediments in gaining adherents because of the attractiveness of the established view, which held (in its simplest form) that money-grubbing utility companies deliberately disregarded the financially unappealing farmers. Beyond that, readers of the history of rural electrification have difficulty in putting aside their knowledge of the REA's rapid success in spurring construction of power lines to farmsteads. As noted, in the fifteen years after creation of the government agency in 1935, farm electrification from central station power plants jumped from about 11 percent to 86 percent. By contrast, the work in the 1920s appears distinctly unimpressive, especially when examining truly destitute states, such as Mississippi, which saw only 1 percent of its farms electrified before the advent of the TVA and REA. I have argued, however, that the perception of the utility industry's efforts suffers partly because of the unfair comparison with events that historians know occurred later. In other words, scholars have generally failed to view rural electrification from the perspective of people living in the 1920s and 1930s.

The challenge to the mainstream narrative provides a cautionary lesson for historians, though not necessarily a novel one. When telling other stories, professionals consciously try to avoid elements of presentism in accounts—that is, the tendency to interpret the past using current perspectives and knowledge.[25] One form of presentism, known as "whiggism," takes this notion further, by praising "principles of progress" in past acts that lead us "to produce a story which is the ratification if not the glorification of the present."[26] Others have interpreted presentism as "hindsight bias"—the tendency to interpret historical events as predictable after people know their ultimate outcome.[27]

I agree that the story of the government's rural electrification efforts excites strong emotions, serving as a exhilarating example of how federal authorities, during a dark period in American history, took bold actions that had long-lasting and positively viewed impacts. But it is simply bad history to describe past events using presentist approaches. The government could act so dauntlessly in part because it enjoyed the support of a populace willing to try almost anything to mitigate economic and social turmoil. And unlike utility companies, which in the early 1930s found themselves in dire financial and political straits (largely a result of their own doing), the federal government could exploit its ample resources, such as hundreds of millions of dollars that it loaned at favorable terms to farmers' cooperative organizations while also providing low- or no-cost managerial and technical support.

In another form of hindsight bias, people look upon the early days of electrification with today's understanding of electricity as an amazingly versatile and practical form of energy. They appreciate that the electric grid serves as the infrastructural backbone of modern society—so much so that the National Academy of Engineering proclaimed electrification as the most significant of the top "twenty engineering achievements that transformed our lives" in the twentieth century.[28] With this knowledge, people imagine that farmers one hundred years ago must have understood electricity in the same way. To be sure, by the 1910s and early 1920s, city and rural folk knew electricity provided modern services in the home—for lighting, ironing, cooking, ventilation, cleaning, entertainment, and more—and they wanted it. But we do not generally recognize that electricity competed with other forms of technology that farmers sought to incorporate into their lives. At the same time that electricity became available, farmers had also begun adopting devices such as steam engines, gasoline motors, and tractors into their work environments. These technologies had already enabled fewer farmers to produce more food for an increasingly urban and industrialized population, and the machines became leading wedges in an agricultural revolution. They made it possible for farmers to depend less on animal power, an ancient source of effort, and more on inorganic forms of energy. Furthermore, in the 1910s, ruralites began embracing the automobile in large numbers, seeing in this form of hardware a way to bridge the gap between farms and cities in a more direct fashion, and they often seemed more interested in lobbying for better roads than electric lines.[29] Likewise, farmers rapidly accepted (and modified) telephone technology as the means to reduce their feelings of isolation. In other words, rural folk of the era did not necessarily put electricity at the top of the list of critically necessary and desirable technologies.[30]

Other evidence suggests that electricity perhaps did not mean the same to early-twentieth-century farmers as it does to twenty-first-century readers. For example, at the first conference on National Country Life in 1919, convened "for the study and discussion of the *social problems of rural life*," speakers addressed several issues dealing with education, health, religious practices, and government.[31] A computer search for the word "electric" (and its variants) resulted in no hits in conference documents. The same null result occurred in an examination of the second conference's proceedings, though the word showed up in the records of the third conference (held in 1920) when describing electrified railroads.[32] It also appeared in a discussion of the need for good roads; a US Department of Agriculture representative assumed that farm electrification would come after the installation of highways, likely the result of cleared rights of way that power companies could use for stringing distribution lines.[33] At the 1921 conference, one speaker mentioned electric lighting, but only as a way of differentiating between people living in the country and the city.[34] Another noted that that "the utilization of electricity in rural communities is still in its infancy," but the potential of electricity generation from flowing water remained "enormous." To exploit the resource, the author suggested, communities should create nonprofit corporations for the production or distribution of power.[35] In the fifth conference, held in 1922, the only mention of electricity appeared in a list of desired outcomes for rural residents. It came after the need to "emphasize the importance of the rural family from the economic and social point of view," the encouragement of ownership of land instead of tenancy, the need for better housing, and the desire to "establish experiment stations for simplification of labor conditions in the farm home." As a fifth goal, the commission aimed to "familiarize the country people, and particularly the farm women, with the use of the modern mechanical and electric devices for reducing manual labor to a minimum."[36]

In a similar manner, the American Farm Bureau Federation, which sought to address pressing issues facing its constituents, did not view electricity as a top priority during its early history. A search for the term "electricity" within reports produced by the federation's executive secretary in 1921 or 1922, its first two years of operation, turned up nothing. The documents included discussions of the AFBF's major programs, research efforts, and legislative campaigns as well as accounts of annual meetings, which highlighted the severe recession of 1920–21. ("The past year has been a very troubulous [sic] one for the farmer," asserted AFBF president J. R. Howard at the November 1921 meeting.)[37] Significant concerns included the host of taxes that added to farmers' woes, difficult-to-obtain bank credit, high freight rates charged by railroads, and the need for greater cooperation in marketing farmers' output.[38] Not until the AFBF partnered with the national CREA in 1923, however, did a report specifically mention electricity, noting that "agriculture is on the verge of an electrical age."[39]

Even rural electrification promoters strove to temper some of the hype concerning their efforts. In 1927, for example, W. C. Krueger, a Wisconsin Utilities Association research fellow at the University of Wisconsin, chided a power company vice president about making an exaggerated claim for the economic value of electricity on the farm. The official wanted Krueger's advice on a draft version of a booklet for distribution to farmers that explained the "uses and helps which electricity will be to them on the farm."[40] But Krueger cautioned that the pamphlet likely overstated the case: "The question as to whether the use of electricity will make money for the farmer has come up in many a discussion, and the general opinion is that we cannot substantiate the introduction of electric service on this basis, at least in so far as our present observations hold good. Rather than 'make more money[,]' electricity saves work, lightens the drudgery, releases time for participation in the pleasures of living, and in general raises the standard of living."[41]

Perhaps most telling, Earl White, head of the national CREA and obviously a strong promoter of rural electrification, admitted in June 1931 that the future success of farming depends on many factors. Though highlighting the benefits of electricity in a presentation to agricultural engineers, he nevertheless showed great candor when predicting the main productivity enhancers on the farm: "There are many forces shaping the destiny of American agriculture," he observed, "and it is still an open question as to where electricity will rank on the list." The increased use of mechanical horsepower (from tractors, for example) constitutes perhaps the most important factor in augmenting farm productivity. Consequently, he continued, "electricity will be considered simply as one [form of farm life enhancement] among many."[42]

Less than a year later, agricultural educators at the University of Wisconsin seem to have reached a similar conclusion. In justifying expenses during the harsh early years of the Depression, Agricultural Engineering Department chairman Edward Jones wrote that "reduction of the cost of producing farm products is the chief answer to the farmers' present difficulty." The top means to help farmers achieve this goal, he noted, included—not farm electrification, which perhaps still remained in the experimental phase—but "[m]ore economical designs of farm buildings, the cheap and efficient remodeling of old buildings, [and] the stopping of extravagant expenditures in dairy barn ventilating systems."[43] When describing research, the department head listed projects dealing with the efficient use of power machinery, such as harvesting equipment loaned by manufacturers.[44] Of the eight major investigatory efforts pursued by the department, none focused specifically on electrification, though Jones observed that work on the subject continued by two Wisconsin Utilities Association fellows.[45]

In short, it appeared that other concerns predominated over electrification as means to improve farmers' welfare. While electricity constituted one of several new technologies that would alter life in rural areas, to many agrarians (and even to a large number of unwired urban customers), electricity remained a luxury rather than a necessity. It did not seem as essential as other improvements in the rural environment at the time, despite the great value we put on the energy form in the twenty-first century. Our histories of rural electrification should not reflect today's appreciation of electricity but rather the attitudes of people living almost a century ago.

A Greek Tragedy

While critical of presentist interpretations of rural electrification and the widespread use of hindsight bias, I also acknowledge that the standard narrative gained popularity for a more mundane reason. In a rather nonscholarly manner, the customary storyline attracted attention because it includes a wonderfully dramatic cast of characters with the attributes of angels and devils, making for a Greek tragedy of epic proportions.

Prominently, the account includes utility magnate Samuel Insull, who had been portrayed as a public hero in the 1920s. When *Time* magazine highlighted the president of the Middle West Utilities holding company in its 29 November 1926 issue—and put his portrait on its cover—the businessman had just secured a second guarantee of a half-million dollars to support the Chicago opera.[46] *Time's* article recorded the former Englishman's rise from obscurity to become head of a regional utility company and other organizations that provided energy services to millions of people. "Now, in a city of 200 square miles, no one can switch on an electric light without switching on Samuel Insull," the periodical gushed. Illustrating the great man's humility and ability to interact with the artists he sponsored, the magazine observed that between acts at opera performances, he frequently complimented the singers, who referred to him as "Papa Insull" and who gave him money to invest on their behalf. Happily for them, the article declared, "he has never lost a penny for himself or for anyone else."[47]

In the years after this flattering piece, however, the popular image of the electric power industry—and Insull in particular—took a beating. Like other utility managers, Insull had exploited the use of holding companies as management and investment tools, such that in 1930, his pyramid-like firms allegedly controlled assets in utility and manufacturing firms of more than a half-billion dollars with an investment of just $27 million.[48] Fears about financial abuses of holding companies arose earlier, however, as advocates of public power, such as Senator George Norris of Nebraska, spurred the Federal Trade Commission to begin investigations in 1928 into possible utility industry excesses. The government agency also continued its earlier line of questioning about whether companies, including those run by Insull, had unduly influenced elections of legislators and selection of regulators.[49]

While these practices attracted some immediate attention from utility critics, they became fodder for the general public after the stock market decline began in October 1929, followed by the Great Depression. The financial leveraging that made holding companies such valuable assets during good times unraveled when investors sought to sell securities at signs of distress. As several holding companies collapsed—perhaps most dramatically those managed by Insull—newspapers detailed the evisceration of savings among people who previously deemed utility securities to be safe enough for "widows and orphans."[50] Opera soprano Rosa Raisa complained in 1933, for example, that she had once regarded Insull as "the god of American business," and had given him several hundred thousand dollars to invest, an action she later regretted.[51] *Time* magazine reported in 1938 that Insull controlled securities having a market value of more than $3 billion at one time, with a good chunk owned by individuals, such as the opera star.[52] Journalist Ernest Gruening published scathing pieces highlighting the use of propaganda by Insull and others (including his brother, Martin J. Insull, president of the Middle West Utilities holding company until 1932) to acquire power and control.[53]

As the public image of utility companies soured, so did that of Samuel Insull, whose image appeared on *Time*'s cover again on the 14 May 1934 issue. But this time, the magazine did not offer a dignified portrait of a benevolent industrial saint. Rather, it showed a humiliated villain hiding his face with a hat as he readied himself for state and federal trials. The article detailed Insull's flight to Europe and Turkey to evade charges of embezzlement and securities fraud, after a life of success that made him worth $100 million or more. But his "towering corporate pyramids" eventually "cracked and crumbled," observed the piece, costing investors three-quarters of a billion dollars.[54]

Despite being acquitted of all charges in 1934 and 1935—a fact rarely publicized as often as his indictments—Insull died ignominiously in 1938; he moved to Paris after his trials and suffered a fatal heart attack in a metro station, supposedly carrying only 20 cents in his pocket. At his death, he reportedly held assets of only $1,000 and debts of $16 million. "A man of real and solid abilities," the *New York Times* commented, Insull "was the victim not merely of the period in which he lived but of his own defects."[55] *Time* magazine remarked that the "cocky, onetime clerk" of Thomas Edison and overly self-assured "operating genius" had decimated the worth of financial holdings owned by "smalltime investors," even though he won legal exoneration.[56] As these popular accounts suggest, Insull went from hero to villain.[57] Once a storied and beloved philanthropist, a successful businessperson who benefited society at large, he became "a perfect scapegoat for the depression," according to historian Harold Platt.[58]

Insull's antagonist in the real-life drama was Franklin Delano Roosevelt, the man who (according to a common interpretation) courageously brought America out of the Depression and saved capitalism and democracy. As New York's governor, Roosevelt had already acquired credentials as a backer of public power and disadvantaged farmers. When he campaigned for the presidency in 1932, he portrayed utility executives as contributors to the stock market crash and securities crises of the 1930s. In particular, he lit into Samuel Insull, describing the utility magnate with biblical references, such as "the lone wolf, the unethical competitor, the reckless promoter, the Ishmael or Insull, whose hand is against everyman's."[59] "The Insull failure," FDR noted in another speech, "has opened our eyes. It shows us that the development of these financial monstrosities [utility holding companies] was such as to compel ultimate ruin; that practices had been indulged in that suggest the old days of railroad wild-catting; that private manipulating had outsmarted the slow-moving power of government. As always, the public paid and paid dearly."[60]

Once elected, Roosevelt made good on his promises to alter the nation's power system. In 1933, he enlarged the scope of earlier efforts for the federal government to take control of the Muscle Shoals dam by establishing the Tennessee Valley Authority. Creation of the REA in 1935 and the Bonneville Power Administration in 1937 further extended the federal government's reach in the power sector. And passage of the Public Utility Holding Company Act in 1935, despite great opposition from the power industry, did much to eliminate corporate abuses, such as those linked in the public mind with Samuel Insull.[61] As the Chicagoan's reputation disintegrated, FDR's swelled, allowing for a narrative to arise that shamed the utility industry and praised the government's efforts to bestow a valuable gift— electricity—upon the nation's forgotten farmers.

The vivid contrast between Insull and Roosevelt in the standard narrative of rural electrification helps explain why historians have focused on the evils of the for-profit power industry. By perpetuating the conventional account, however, academics have overlooked the context of the 1920s and early 1930s, a period when commercial utilities sought to power up rural America in a conservative manner—and without the need for corporate or government subsidies. Though not holding the considerable capabilities of the federal government during a national crisis, the utilities and their partners nevertheless brought electricity to farmers at an unprecedented rate. In the process, they laid the groundwork, with new knowledge and creation of significant social and technical elements, for the more impressive expansion of rural electrification in the decades that followed.

Notes

Introduction

1. Edison Electric Institute, data presented in "Add 533,000 New Customers," *Electrical World* 106, no. 1 (4 January 1936): 62. Similar data appear in George W. Kable and R. B. Gray, *Report on C.W.A. National Survey of Rural Electrification* (Washington, DC: USDA, 1934), 53, using information provided by NELA, *Progress in Rural and Farm Electrification for the 10 Year Period 1921–1931* (New York: NELA, 1932): 5. For 1933, Kable and Gray report a slightly lower figure (of 710,642) than does the 1936 *Electrical World* report (of 713,558) for the number of farms served by electric utilities.

2. A note about terminology: "rural" and "farm" are not synonymous, though for many discussions of electrification in nonurban areas, they are viewed as such. The US

Census Bureau, for example, distinguished between rural-farm and rural-nonfarm populations in *Fifteenth Census: 1930, Population*, vol. 3, pt. 1 (Washington, DC: GPO, 1932), 5–6. A 1929 utility industry report observed, however, "It should be clearly recognized that, primarily, rural electrification means electricity at work on the farm." G. C. Neff, "Essentials for Development of Rural Electric Service," NELA, *Proceedings* 86 (1929): 228.

3. J. C. Martin, "Rural Line Development," *NELA Bulletin* 10 (October 1923): 609; and "Notes on Progress of Farm Electrification," *NELA Bulletin* 11 (January 1924): 24.

4. NELA, *Progress in Rural and Farm Electrification*, 8.

5. Comments by G. C. Neff after presentation of "Report of the Overhead Systems Committee," NELA, *Proceedings* 76 (1921): 851.

6. USDA, *Report of the Administrator of the Rural Electrification Administration, 1950* (Washington, DC: GPO, 1950), 2.

7. Thomas P. Hughes, *Networks of Power: Electrification in Western Society, 1880–1930* (Baltimore, MD: Johns Hopkins University Press, 1983).

8. Richard F. Hirsh, *Technology and Transformation in the American Electric Utility Industry* (New York: Cambridge University Press, 1989), chap. 3, "Manufacturers and Technological Progress before World War II."

9. Richard F. Hirsh, *Power Loss: Deregulation and Restructuring in the American Electric Utility System* (Cambridge, MA: MIT Press, 1999), 51; and Thomas P. Hughes, "The Evolution of Large Technological Systems," in Wiebe E. Bijker, Thomas P. Hughes and Trevor J. Pinch, eds., *The Social Construction of Technological Systems: New Directions in the Sociology and History of Technology* (Cambridge, MA: MIT Press, 1987), 52–53.

10. National Rural Electric Cooperative Association, "Electric Co-op Facts & Figures," https://www.electric.coop/electric-cooperative-fact-sheet, accessed 15 July 2021.

11. Bert Klandermans and Conny Roggeband, eds., *Handbook of Social Movements across Disciplines* (New York: Springer, 2010), 5.

12. Aldon Morris, "Reflections on Social Movement Theory: Criticisms and Proposals," *Contemporary Sociology* 29, no. 3 (May 2000): 445.

13. For discussions of narratives and counternarratives, see William Cronon, "A Place for Stories: Nature, History, and Narrative," *Journal of American History* 78, no. 4 (March 1992): 1347–76; and David E. Nye, *America as Second Creation: Technology and Narratives of New Beginnings* (Cambridge, MA: MIT Press, 2003).

14. Claire Turenne Sjolander, "Through the Looking Glass: Canadian Identity and the War of 1812," *International Journal* 69, no. 2 (2014): 152–67.

15. Different accounts of Israel's creation in 1948 are discussed in Avi Shlaim, "The Debate about 1948," *International Journal of Middle East Studies* 27, no. 3 (1995): 287–304.

16. The phrase "history written in reverse" is used in a book review by Michael Hiltzik, "Keeping Profits on Track," *Wall Street Journal*, 28 July 2020, A15.

CHAPTER ONE: **The Standard Narrative and Its Defects**
Epigraph: Morris L. Cooke, "The Early Days of the Rural Electrification Idea: 1914–
    1936," *American Political Science Review* 42, no. 3 (June 1948): 434.

1. Arthur M. Schlesinger Jr., *The Politics of Upheaval*, vol. 3 of *The Age of Roosevelt* (Boston: Houghton Mifflin, 1960).

2. Robert Kanigel, *The One Best Way: Frederick Winslow Taylor and the Enigma of Efficiency* (New York: Penguin, 1997).

3. Jean Christie, "New Deal Resources Planning: The Proposals of Morris L. Cooke," *Agricultural History* 53, no. 3 (July 1979): 597–606; and Jean Christie, *Morris Llewellyn Cooke: Progressive Engineer* (New York: Garland, 1983).

4. W. S. Murray, "Mr. Murray on the Giant Power Survey Report," *Electrical World* 85, no. 10 (7 March 1925): 512.

5. A frequent critic of utility companies, which allegedly enjoyed "astounding prosperity," Cooke urged the firms in 1928 to reduce rates for residential customers. Morris Llewellyn Cooke, *What Price Electricity for Our Homes: An Open Letter to the Electrical Industry* (Philadelphia: Cooke, 1928).

6. Norman Wengert, "Antecedents of TVA: The Legislative History of Muscle Shoals," *Agricultural History* 26, no. 4 (October 1952): 141–47.

7. Christie, *Morris Llewellyn Cooke*, 100; and Wengert, "Antecedents of TVA," 146.

8. Morris Llewellyn Cooke, *Report of the Mississippi Valley Committee of the Public Works Administration* (Washington, DC: US Federal Emergency Administration of Public Works, 1934).

9. Morris Llewellyn Cooke, "National Plan for the Advancement of Rural Electrification under Federal Leadership and Control with State and Local Cooperation and as a Wholly Public Enterprise," typescript document, undated but likely 1934, from Franklin D. Roosevelt Presidential Library, Hyde Park, NY.

10. President Franklin D. Roosevelt, Exec. Order No. 7037, "Establishing the Rural Electrification Administration," 11 May 1935; and "Cheap Electricity Planned for Farm," *New York Times*, 14 May 1935, 27.

11. Rural Electrification Act of 1936, Pub. Law No. 74-605, chap. 432, 49 Stat. 1363, 20 May 1936; and "Roosevelt Signs REA Bill," *New York Times*, 22 May 1936, 3.

12. Morris L. Cooke, "A New Viewpoint," *Rural Electrification News* 1, no. 2 (October 1935): 1.

13. "Senator Norris Proposes Federal Program to Electrify All Farms," *Rural Electrification News* 1, no. 3 (November 1935): 3, 7.

14. "With the Editors," *Rural Electrification News* 1, no. 3 (November 1935): 26.

15. Morris L. Cooke, "Electricity Goes to the Country," *Survey Graphic* 25, no. 9 (September 1936): 506.

16. "Cooke Quits Post as Roosevelt Aide," *New York Times*, 7 February 1937, 18.

17. USDA, *Report of the Administrator of the Rural Electrification Administration, 1948* (Washington, DC: GPO, 1948), 2.

18. Morris Llewellyn Cooke, "The Early Days of the Rural Electrification Idea: 1914–1936," *American Political Science Review* 42, no. 3 (June 1948): 438.

19. Ibid., 437, 444, 446, 447.

20. "Slattery Named REA Chief," *Washington Post*, 8 September 1939, 2; "Harry Slattery to Leave Interior Post for REA Job," *Chicago Daily Tribune*, 26 September 1939, 5; and "Slattery Resigns to Air REA Dispute," *New York Times*, 12 December 1944, 1.

21. Harry Slattery, *Rural America Lights Up* (Washington, DC: National Home Library Foundation, 1940), 23. Slattery's representation of Neff's projection may have been

a little inaccurate as well. In 1927—later than "[s]hortly after the CREA was organized"— Neff indicated that if the gains of the previous three years continued unabated, "there will be approximately 1,000,000 electrified farms in this country by the end of 1932 and by 1938, this number will be increased to approximately 3,000,000." "Rural Electric Service Committee," NELA, *Proceedings* 84 (1927): 81.

22. USDA, *Report of the Administrator of the Rural Electrification Administration, 1940* (Washington, DC: GPO, 1941), 2.

23. Slattery, *Rural America Lights Up*, 26.

24. Ibid., 27–37.

25. Claude R. Wickard, "Power Revolution on the Farm," *New York Times Magazine*, 9 September 1951, 15.

26. Ibid., 38.

27. Ibid., 15.

28. REA, *Power and the Land*, film, 1940, YouTube, https://www.youtube.com/watch ?v=-KVwWAJBJUA, accessed 9 August 2016. Ronald R. Kline analyzed the film in "Ideology and the New Deal 'Fact Film' *Power and the Land*," *Public Understanding of Science* 6, no. 1 (1997): 19–30.

29. "Pamela Popeson, "Lester Beall and the Rural Electrification Administration," Museum of Modern Art, http://www.moma.org/explore/inside_out/2012/03/22/lester -beall-and-the-rural-electrification-administration, accessed 9 August 2016.

30. One of Sekaer's famous photographs can be seen at National Archives, http:// www.archives.gov/exhibits/new_deal_for_the_arts/images/work_pays_america/rural _electrification.html, accessed 9 August 2016.

31. Aaron D. Purcell, *White Collar Radicals: TVA's Knoxville Fifteen, the New Deal, and the McCarthy Era* (Knoxville: University of Tennessee Press, 2009), 70–71.

32. A flyer and essay on the play are available at New Deal Network, "Electricity in the Limelight: The Federal Theatre Project Takes on the Power Industry," Internet Archive, https://web.archive.org/web/20160720134942/http://newdeal.feri.org/power/essay01 .htm, accessed 8 July 2021.

33. Brooks Atkinson, "The Play: 'Power' Produced by the Living Newspaper under Federal Theatre Auspices," *New York Times*, 24 February 1937, 18. See chapter 7 for more on the FTC investigations.

34. USDA, REA, *Rural Lines-USA: The Story of Cooperative Rural Electrification* (Washington, DC: GPO, 1966), appendix, "REA Electric Borrowers; Date of First Loans," 39.

35. Arthur Arent, *Power: A Living Newspaper*, act 2, scene 2, Internet Archive, https:// web.archive.org/web/20160813225559/http://newdeal.feri.org/power/pwr2-01.htm, accessed 8 July 2021.

36. Ibid., act 2, scene 6. The play was written before Supreme Court challenges had been resolved in the government's favor.

37. The head of the Works Progress Administration, Harry Hopkins, noted that some people would call the play propaganda. "Well, I say, what of it? It's propaganda to educate the consumer who's paying for power. It's about time someone had some propaganda for him. The big power companies have spent millions on propaganda for the utilities. It's about time the consumer had a mouthpiece. I say more plays like *Power* and more power

to you." Quotation cited in Hallie Flanagan, *Arena* (New York: Duell, Sloan and Pearce, 1940), 185. Also see Jane de Hart Mathews, *The Federal Theatre, 1935–1939: Plays, Relief, and Politics* (New York: Octagon Books, 1980), 115.

38. David Cushman Coyle, ed., *Electric Power on the Farm* (Washington, DC: GPO, 1936), 11–12.

39. Ibid., 15.

40. Ibid., 69–70.

41. REA, *The Electrified Farm of Tomorrow* (Washington, DC: GPO, 1939), 6–7.

42. REA, *A Guide for Members of Rural Electric Co-Ops* (Washington, DC: GPO, 1950), 4–5.

43. Clyde T. Ellis, *A Giant Step* (New York: Random House, 1966), 28.

44. Ibid., 29, 33.

45. Ibid., 33.

46. Ibid., 33–34.

47. NRECA, "History," http://www.electric.coop/our-organization/history, accessed 14 July 2020.

48. Richard A. Pence, ed., *The Next Greatest Thing* (Washington, DC: NRECA, 1984), 10–11.

49. Ibid., 11.

50. Ibid.

51. The author gave credit for using excerpts from Robert A. Caro, *The Years of Lyndon Johnson: The Path to Power* (New York: Alfred A. Knopf, 1982), in Pence, *The Next Greatest Thing*, 15.

52. Ibid., 39, 65.

53. Ibid., 70

54. Ibid., 118.

55. Schlesinger, *The Politics of Upheaval*, 380.

56. Ibid.

57. Ibid., 384.

58. Ibid.; see notes on pages 689–90. Resources cited included a 1944 book on the REA (Frederick Muller's sympathetic *Public Rural Electrification*); Cooke's 1948 "The Early Days of the Rural Electrification Idea"; a 1950 article written by an REA consultant; REA administrator Claude Wickard's 1951 *New York Times Magazine* article, "Power Revolution on the Farm"; a 1951 book written by an employee of the government's Reconstruction Finance Corporation; Marquis Childs's *The Farmer Takes a Hand*, published in 1952; K. E. Trombley's 1954 biography of Cooke, *Life and Times of a Happy Liberal*; and REA administrator John Carmody's "Public Power" article written in 1956.

59. William E. Leuchtenburg, *Franklin D. Roosevelt and the New Deal, 1932–1940* (New York: Harper and Row, 1963).

60. Alfred B. Rollins Jr., review of *Franklin D. Roosevelt and the New Deal, 1932–1940*, by William E. Leuchtenburg, *Journal of American History* 51, no. 1 (June 1964): 124–25.

61. Reviewer Robert A. Divine described the book as "the best single volume on the New Deal." *Political Science Quarterly* 80, no. 1 (March 1965): 137–38.

62. Leuchtenburg, *Franklin D. Roosevelt and the New Deal*, 157–58.

63. Ibid.; and "Electrified Thumb," *Time* 32 (4 July 1938): 14.

64. John D. Garwood and W. C. Tuthill, *The Rural Electrification Administration: An Evaluation* (Washington, DC: American Enterprise Institute, 1963), 2.

65. Ibid., 4.

66. Philip J. Funigiello, *Toward a National Power Policy: The New Deal and the Electric Utility Industry, 1933–1941* (Pittsburgh, PA: University of Pittsburgh Press, 1973), 123; and Sidney I. Simon, review of Funigiello's book, *Annals of the American Academy of Policy and Social Science* 147, no. 1 (January 1975): 210. Other generally positive reviews came from Ellis W. Hawley in *American Historical Review* 80, no. 3 (June 1975): 735–36; and Robert D. Cuff in *Journal of American History* 61, no. 3 (December 1974): 822–23.

67. Funigiello, *Toward a National Power Policy*, 124.

68. Ibid., 124–25. In addition, Funigiello drew on Forrest McDonald, *Let There Be Light: The Electric Utility Industry in Wisconsin, 1881–1955* (Madison, WI: American History Research Center, 1957) to describe impediments to rural electrification.

69. Jack Doyle, *Lines across the Land: Rural Electric Cooperatives; The Changing Politics of Energy in Rural America* (Washington, DC: Environmental Policy Institute, 1979), 1.

70. Ibid., 2.

71. D. Clayton Brown, *Electricity for Rural America: The Fight for the REA* (Westport, CT: Greenwood Press, 1980). Reviewer Donald Holley called the book "the first scholarly study of the Rural Electrification Administration to appear in book form" in *American Historical Review* 86, no. 2 (April 1981): 476–77.

72. Brown, *Electricity for Rural America*, 3, 10. As I indicate in chapter 9, some of CREA's most significant contributions, such as publication of an encyclopedic compilation of rural electrification studies, occurred after 1930. Brown's book received a generally good reception, though a few reviewers commented on its focus on rural electrification in southern states and its sometimes-superficial evaluations of the political climate and business opposition. Philip J. Funigiello, *Journal of American History* 67, no. 4 (March 1981): 959–60; Charles Johnson, *Tennessee Historical Quarterly* 40, no. 2 (Summer 1981): 209–10; Albert N. Sanders, *Journal of Southern History* 47, no. 1 (February 1981): 135–36; James F. Doster, *North Carolina Historical Review* 57, no. 4 (October 1980): 484; Douglas D. Anderson, *Business History Review* 55, no. 4 (Winter 1981): 582–83; James W. Whitaker, *Technology and Culture* 22, no. 2 (April 1981): 315–16; and David L. Nass, *Wisconsin Magazine of History* 65, no. 3 (Spring 1982): 222–24. Anderson's review criticized Brown for failing to provide deep analyses of events, noting that the book "too often skims along the surface and at times, stretches to find design where there probably was not any."

73. Brown, *Electricity for Rural America*, 133n24.

74. Charles Johnson, review of *Electricity for Rural America*, by D. Clayton Brown, *Tennessee Historical Quarterly* 40, no. 2 (Summer 1981): 209–10.

75. This calculation is based on the 21.7 percent annual compounded growth rate of electrification from 31 December 1923 to 31 December 1929 as reported by NELA, *Progress in Rural and Farm Electrification for the 10 Year Period 1921–1931* (New York: NELA, 1932): 5, and extended beyond 1930. Of course, with the Depression hitting farmers especially hard and World War II diverting resources for other purposes, the private utility industry could not have maintained such a high growth rate. But I simply point out that

Brown's assessment, based on an unattributed source, showed a bias against the utility industry.

76. D. Clayton Brown, "Rural Electrification Administration (REA)," in Robert S. McElvaine, ed., *Encyclopedia of the Great Depression* (New York: Macmillan Reference USA, 2004), 855, 856; my emphasis.

77. Ibid., 855.

78. Robert A. Caro, *The Years of Lyndon Johnson: The Path to Power* (New York: Alfred A. Knopf, 1983), chap. 27. Caro repeats the phrase (as did Pence in *The Next Greatest Thing*, described above) on 504, 505, 507, 511.

79. Caro, *The Years of Lyndon Johnson*, 528.

80. Ibid., 516. Some lines cost that much, but most utilities experienced (and reported) costs closer to $2,000 or less in the early 1930s. See my discussion in chapter 9.

81. The dramatic style continued throughout the book; one reviewer commented on the author's frequent use of "hyperbole" such that the portrait one gets of Johnson "is unbalanced and overstated," owing, perhaps, to the lack of "subtlety and analytic perspective." Robert Dallek, "Caro versus Johnson," review of *The Years of Lyndon Johnson: The Path to Power*, by Robert Caro, *Reviews in American History* 12, no. 1 (March 1984): 148–53.

82. Laurence Malone, "Rural Electrification Administration," edited by Robert Whaples, EH.Net Encyclopedia, March 16, 2008, http://eh.net/encyclopedia/rural-electrification -administration, accessed 14 October 2014; my emphasis.

83. Gary Alan Donaldson, "A History of Louisiana's Rural Electric Cooperatives, 1937–1983" (PhD diss., Louisiana State University, 1983), 11, 14.

84. W. G. Beecher, "Is It Time to Revoke the Tax-Exempt Status of Rural Electric Cooperatives?," *Washington and Lee Journal of Energy, Climate, and the Environment* 5, no. 1 (1 September 2013): 234. For another law article drawing on Brown's work, see Debra C. Jeter, Randall S. Thomas, and Harwell Wells, "Democracy and Dysfunction: Rural Electric Cooperatives and the Surprising Persistence of the Separation of Ownership and Control," *Alabama Law Review* 70, no. 2 (2018): 361–454.

85. "Rural Electrification Administration 1934–1941," part of "Historic Events for Students: The Great Depression," Gale Group, 2002, http://www.encyclopedia.com/educa tion/news-and-education-magazines/rural-electrification-administration-1934-1941, accessed 19 April 2017.

86. Katherine Jellison, "Women and Technology on the Great Plains, 1910–1940," *Great Plains Quarterly*, paper 432 (1988): 148, http://digitalcommons.unl.edu/greatplains quarterly/432, accessed 12 February 2013. The original source of this citation is USDA, *Domestic Needs of Farm Women*, Report no. 104 (Washington, DC: GPO, 1915), 33. Also see Jellison, *Entitled to Power: Farm Women and Technology, 1913–1963* (Chapel Hill: University of North Carolina Press, 1993).

87. Jellison, "Women and Technology on the Great Plains," 148.

88. Ibid., 155. The author noted, however, that as late as 1940, electrified farm homes in several Great Plains states depended on their own home (isolated) plants rather than on central stations. Ibid., 148.

89. Simon Winchester, *The Men Who United the States: America's Explorers, Inventors, Eccentrics, and Mavericks, and the Creation of One Nation, Indivisible* (New York: Harper-Collins, 2012), 376.

90. Ibid., 376–77.

91. Ibid., 377.

92. Ibid., 378.

93. NRECA's website notes that it "is the national service organization for more than 900 not-for-profit rural electric cooperatives and public power districts providing retail electric service to more than 42 million consumers in 47 states." NRECA, "About Us," https://web.archive.org/web/20140827124934/http://www.nreca.coop/what-we-do/about-us, accessed 22 July 2021.

94. Doris Kearns Goodwin, *Leadership: In Turbulent Times* (New York: Simon and Schuster, 2018), digital edition. In the endnotes, Goodwin quoted from Leuchtenburg ("The lack of electric power divided the United States into two nations . . ."), citing a 2009 edition of *Franklin D. Roosevelt and the New Deal*.

95. Goodwin, *Leadership*, 90; my emphasis. Goodwin, by the way, overstated the work of government electrification agencies. Even in 1940, after some Hill Country residents started enjoying co-op-provided power, the REA and other public agencies had electrified 517,500 farms, and other public agencies energized 84,000 farms. These figures yield a total of about 600,000 farms nationwide—not the "millions" she described. In that year, private utilities sold power to 1,448,500 farms. US Department of Commerce, *Statistical Abstract of the United States, 1950* (Washington, DC: GPO, 1950), table 566, "Farm Electrification: 1930 to 1948," 476.

96. The popular history writer, Erik Larson, noted that when he wrote a book about the 1915 sinking of the *Lusitania*, he encountered a similar situation in which most accounts drew on the same—and often unreliable—sources. "[O]ne has to be very careful to sift and weigh the things that appear in books already published on the subject," he commented. "There are falsehoods and false facts, and these, once dropped into the scholarly stream, appear over and over again, with footnotes always leading back to the same culprits." Erik Larson, *Dead Wake: The Last Crossing of the Lusitania* (New York: Crown Publishers, 2015), 357.

97. Thomas P. Hughes, "The Electrification of America: The System Builders," *Technology and Culture* 20, no. 1 (January 1979): 153–60.

98. Leonard S. Hyman, *America's Electric Utilities: Past, Present and Future*, 4th ed. (Arlington, VA: Public Utilities Reports, 1992), 96.

99. The price of fifty power and light common stocks, known as the "*Electrical World* index," dropped from about 117 in September 1929 to 18.5 at the end of June 1932. "Utility Stocks and Bonds Lower," *Electrical World* 100 (2 July 1932): 5; and John L. Neufeld, *Selling Power: Economics, Policy, and Electric Utilities before 1940* (Chicago: University of Chicago Press, 2016), 115–18.

100. For a contemporaneous discussion of utilities' influence on educators, see CQ Press, "Public Utilities' Propaganda in the Schools," *Editorial Research Reports 1928*, vol. 2 (Washington, DC: CQ Press, 1928), http://library.cqpress.com/cqresearcher/cqresrre1928062800, accessed 13 August 2019. *Editorial Research Reports* was published starting in 1923 before being renamed *CQ Researcher* in 1991 (after being purchased by Congressional Quarterly in 1956). CQ Press, "About *CQ Researcher Online* and Related Resources," https://library.cqpress.com/cqresearcher/static.php?page=aboutcqr, accessed 23 September 2019.

101. Arthur R. Taylor, "Losses to the Public in the Insull Collapse: 1932–1946," *Business History Review* 36, no. 2 (Summer 1962): 188–204. Interestingly, Insull in 1913 had spoken about the possibilities for holding company abuses, such as the issuance of securities whose value exceeded the operating firms' capitalization. Samuel Insull, "Electrical Securities," address delivered 30 October 1913, in William E. Keily, ed., *Central-Station Electric Service: Its Commercial Development and Economic Significance as Set Forth in the Public Addresses (1897–1914) of Samuel Insull* (Chicago: privately printed, 1915), 440.

102. Hyman, *America's Electric Utilities*, 105. For a contemporaneous summary of the law, see Buel W. Patch, "Roosevelt Policies in Practice," *CQ Researcher*, 25 September 1936, http://library.cqpress.com/cqresearcher/cqresrre1936092500, accessed 12 July 2019.

103. Michel Foucault, "Society Must Be Defended," in Mauro Bertani and Alessandro Fontana, eds., trans. David Macey, *Lectures at the Collège de France, 1975–76* (New York: Picador, 2003), lecture of 28 January 1976. Foucault wrote that the "traditional function of history, from the first Roman annalists' until the late Middle Ages, and perhaps the seventeenth century or even later, was to speak the right of power and to intensify the luster of power." He also noted that history can be used, "[l]ike rituals, coronations, funerals, ceremonies, and legendary stories," as "an intensifier of power" (66).

104. Of course, not everyone felt so fondly about FDR. A trenchant critic since the 1930s, journalist John T. Flynn wrote *The Roosevelt Myth* (San Francisco: Fox and Wilkes, 1998), a scathing rebuttal of the traditional Depression-era narrative, originally in 1948, with a revision in 1956 and a reprint in 1998.

105. Several policy and law articles in recent years have been more critical of the REA than earlier works, though some still drew on Brown's history and works by Cooke when they discussed the early history of the organization. Historians such as Abby Spinak and Brent Cebul also have taken fresh and insightful looks at the formation and actions of the REA that diverge from traditional interpretations. See chapters 12 and 13.

CHAPTER TWO: **Unattractive Economics in the Rural Electricity Market**
Epigraph: "Iowa's Utility Service—A Handbook concerning the Electric Light and Power, the Electric Railway, and the Gas Industries," no date, likely 1927, in US Senate, 70th Congress, 1st sess., doc. 92, pt. 4, *Utility Corporations* (Washington, DC: GPO, 1929), exhibit 1453, 46.

1. "Report of Committee on Electricity in Rural Districts," NELA, *Proceedings* 40 (1911): 457.

2. "Report of the Committee on Electricity in Rural Districts: Central States," NELA, *Proceedings* 48 (1913): 225.

3. Mel Gorman, "Charles F. Brush and the First Public Electric Street Lighting System in America," *Ohio Historical Quarterly* 70, no. 2 (April 1961): 128–44; "New Lights in Broadway: Brush's System Again Tested Last Evening," *New York Times*, 20 December 1880, 1; and David E. Nye, *Electrifying America: Social Meanings of a New Technology, 1880–1940* (Cambridge, MA: MIT Press, 1990), 31.

4. Edison's technical and managerial feats are described in Thomas P. Hughes, "The Electrification of America: The System Builders," *Technology and Culture* 20, no. 1 (January 1979): 124–61.

5. Thomas P. Hughes, *Networks of Power: Electrification in Western Society, 1880–1930* (Baltimore, MD: Johns Hopkins University Press), 42, 137–39; and "Niagara's Power in Buffalo," *New York Times*, 17 November 1896, 9.

6. Electric Bond and Share Company, *The Commercial Development of the Electric Light and Power Industry* (New York: Electric Bond and Share Co., 1911), 8.

7. Insull effectively created a monopoly of Chicago's power companies as early as 1898. "Chicago's Electric Lights," *New York Times*, 18 May 1898, 7.

8. Samuel Insull, "Standardization, Cost System of Rates, and Public Control," in William E. Keily, ed., *Central-Station Electric Service: Its Commercial Development and Economic Significance as Set Forth in the Public Addresses (1897–1914) of Samuel Insull* (Chicago: privately printed, 1915), 34–47.

9. I explore the nature of incrementally improving generating technology in *Technology and Transformation in the American Electric Utility Industry* (New York: Cambridge University Press, 1989).

10. Samuel Insull, "A Quarter-Century Central-Station Anniversary Celebration in Chicago—1887–1912," in Keily, *Central-Station Electric Service*, 326–27. Much of this paragraph comes from my own books, *Technology and Transformation in the American Electric Utility Industry*, 19–25; and *Power Loss: The Origins of Deregulation and Restructuring in the American Electric Utility System* (Cambridge, MA: MIT Press, 1999), 23–31.

11. Commonwealth Edison, "Rate Schedule 'C'—Large Light and Power Service," in Illinois Public Utilities Commission, *Public Utilities Rates in the State of Illinois: Electric Rates* (Springfield, IL: Schnepp and Barnes, 1917), 94.

12. Ibid., 96–98.

13. Commonwealth Edison, "Rate Schedule 'A'—General Service," in ibid., 92–93.

14. In rate theory, this declining block structure was not as sound as schedules that included demand charges. But determining the peak consumption of thousands of small customers simply was not cost effective. G. P. Watkins, *Electrical Rates* (New York: D. Van Nostrand, 1921), 137.

15. For more on the logic of block-rate structures, see Herman H. Trachsel, *Public Utility Regulation* (Chicago: Richard D. Irwin, 1947), 254; and Hirsh, *Technology and Transformation*, 218–19.

16. Commonwealth Edison, "Rate Schedule 'A'—General Service," 93.

17. As demand for power continued to increase, electric companies needed to build new power plants. Happily, as I noted in *Technology and Transformation*, incremental innovation in power equipment enabled manufacturers to reduce the unit cost of producing electricity, at least through the 1960s. Hence, companies' promotion of electricity sales constituted part of a "grow and build" strategy that reduced the cost (and price) of electricity while allowing firms to earn healthy profits. Customers and suppliers both seemed to benefit from this strategy.

18. US Bureau of the Census, *Central Electric Light and Power Stations and Street and Electric Railways, 1912* (Washington, DC: GPO, 1915), 19; and Edward B. Meyer, *Underground Transmission and Distribution for Electric Light and Power* (New York: McGraw Hill, 1916).

19. Samuel Insull, "Centralization of Energy Supply," in Keily, *Central-Station Electric Service*, 448–50. I determined the length of the block illustrated by Insull by looking at

the same block today (bounded by Kenmore Ave., Ainslie Ave., Winthrop Ave., and Lawrence Ave.) on Google Maps, 8 September 2016.

20. New Hampshire Agricultural Experiment Station, *Progress of Agricultural Experiments—1925*, Bulletin no. 221 (Durham, NH: University of New Hampshire, 1926), 36.

21. F. W. Duffee and G. W. Palmer, *Turn On the Light Heat & Power*, Circular 163 (Madison: Extension Service of the College of Agriculture, University of Wisconsin, 1923), 28.

22. Frank D. Paine and Frank J. Zink, *Electric Service for the Iowa Farm*, Report no. 3, *Iowa State College Official Publication* 27, no. 8 (27 June 1928): 7.

23. "Rural Electric Service Committee," NELA, *Proceedings* 83 (1926): 150.

24. W. J. Greene, "Rural Electric Service Costs Analyzed," *Electrical World* 80, no. 13 (23 September 1922): 656.

25. Ibid.

26. M. R. Lewis, *Progress Report of the Idaho Committee on the Relation of Electricity to Agriculture* (Moscow, ID: Idaho CREA, 1926), 19, citing Luther R. Nash, *Economics of Public Utilities* (New York: McGraw Hill, 1925), 384.

27. Duffee and Palmer, *Turn On the Light*, 35.

28. E. A. Stewart, "Utilization of Electricity in Agriculture," *Agricultural Engineering* 5, no. 6 (June 1924): 126.

29. Paine and Zink, *Electric Service for the Iowa Farm*, 6.

30. Ibid., 6–7.

31. "Questions Asked by Farmers about Rural Electric Service, with Appropriate Answers," Appendix B to "Rural Electric Service Committee," NELA, *Proceedings* 81 (1924): 69.

32. Ibid., 70.

33. Walter R. Baker, "Thesis on the Economics of the Application of Electricity to Country Life" (BS thesis, Oregon Agricultural College, 1909), 2.

34. Keily, *Central-Station Electric Service*, 327.

35. David S. Weaver, "A Rural Electrification Survey," *Agricultural Engineering* 16, no. 9 (September 1935): 371. As late as February 1935, the Federal Power Commission reported that half of all farms connected to central stations drew fewer than 30 kWh per month. Federal Power Commission, *Rural Electric Service: Monthly Bills, Rural Line Construction Costs and Practices* (Washington, DC: GPO, 1936), v.

36. "Report of the Committee on Electricity in Rural Districts: Central States," 173.

37. Ibid., 149.

38. The use of electricity enabled the manufacturing sector to revolutionize production and increase overall efficiencies, which translated to increased economic activity and growing electricity consumption. Warren D. Devine Jr., "From Shafts to Wire: Historical Perspective on Electrification," *Journal of Economic History* 43, no. 2 (June 1983): 351.

39. The Department of Agriculture's inflation-adjusted index of commodity prices nationally went from 104 in July 1914 to 142 by December 1916. The index grew to 178 on average for 1917, 206 in 1918, and 217 in 1919. It peaked at 236 in May 1920. Data from USDA, National Agricultural Statistics Service, https://quickstats.nass.usda.gov/results /656D4D03-4B70-308B-AC34-D4CFCB2326A5, accessed 26 August 2016.

40. Frederick Strauss and Louis H. Bean, *Gross Farm Income and Indices of Farm Production and Prices in the United States, 1869–1937*, USDA Technical Bulletin no. 703 (Washington, DC: GPO, 1940), 16.

41. Economist Victor Zarnowitz defined the contraction of the period between January 1920 and July 1921 as a "major depression," though less severe than the Great Depression that began in August 1929. Zarnowitz, "Recent Work on Business Cycles in Historical Perspective: A Review of Theories and Evidence," *Journal of Economic Literature* 23, no. 2 (June 1985): 528.

42. National statistics from USDA, National Agricultural Statistics Service, https://quickstats.nass.usda.gov/results/23F7BD84-5A73-355E-B079-15753C198270, accessed 29 August 2016. Similar data come from Eaton G. Osman, ed., *The Price-Current-Grain Reporter Year Book for 1923* (Chicago: Price Current-Grain Reporter, 1923), 5; and Strauss and Bean, *Gross Farm Income*, 16.

43. Overall commodity prices dropped from a peak index of 236 in May 1920 to 112 in June 1921. The next time the index rose to 1920 levels was in 1946, when the average for the year reached 236 after spiking to 271 for one month (October 1946). Data from USDA, National Agricultural Statistics Service, https://quickstats.nass.usda.gov/results/656D4D03-4B70-308B-AC34-D4CFCB2326A5, accessed 26 August 2016.

44. Data from Strauss and Bean, *Gross Farm Income*, 16. A related metric, average net income on farms, showed a decline of more than half from 1920 to 1921, only to recover to about 79 percent of its 1920 level in 1929, but declining to 55 percent in 1930. Bruce L. Gardner, "Farm Income and Expenses: 1910–1999," in *Historical Statistics of the United States, Millennial Edition Online*, at http://hsus.cambridge.org/HSUSWeb/toc/showDownloadableTable.do?id=Da1266-1356, accessed 29 August 2016.

45. Royden Stewart, "Rural Electrification in the United States: Part II—National Development, 1924–1935," *Edison Electric Institute Bulletin* 9 (October 1941): 415; and Gene Smiley, "The U.S. Economy in the 1920s," EH.net, http://eh.net/encyclopedia/the-u-s-economy-in-the-1920s, accessed 10 August 2014.

46. L. C. Gray and O. E. Baker, *Land Utilization and the Farm Problem*, USDA Misc. Pub. no. 97 (Washington, DC: GPO, 1930), v.

47. Strauss and Bean, *Gross Farm Income*, 16.

48. David C. Wheelock, "Changing the Rules: State Mortgage Foreclosure Moratoria during the Great Depression," *Federal Reserve Bank of St. Louis Review* 90, no. 6 (2008): 573; Randal R. Rucker, "The Effects of State Farm Relief Legislation on Private Lenders and Borrowers: The Experience of the 1930s," *American Journal of Agricultural Economics* 72, no. 1 (February 1990): 25–34; and Lee J. Alston, "Farm Foreclosure Moratorium Legislation: A Lesson from the Past," *American Economic Review* 74, no. 3 (June 1984): 445–57.

49. Federal Reserve, *Banking and Monetary Statistics, 1914–1941* (Washington, DC: Federal Reserve, 1943, reprinted 1976), table 3, 19.

50. Lee J. Alston, "Farm Foreclosures in the United States during the Interwar Period," *Journal of Economic History* 43, no. 4 (1983): 888.

51. J. E. Stanford, East Texas Chamber of Commerce, "Minutes of Meeting of Texas Committee on Relation of Electricity to Agriculture Held at Dallas, Thursday, January 5, 1928," in US Senate, 70th Cong., 1st sess., doc. 92, pt. 74, *Utility Corporations* (Washington, DC: GPO, 1930), exhibit 2344, 709.

52. Verne Russell Hillman, "A Study of the Plans and Policies of Power Companies in Dealing with Rural Customers" (MS thesis, Virginia Polytechnic Institute, 1930), 46.

53. US Bureau of the Census, *Fifteenth Census of the United States: 1930, Agriculture*, vol. 4 (Washington, DC: GPO, 1932), 518.

54. Homer C. Price, "Farm Tenancy: A Problem in American Agriculture," *Popular Science Monthly* 72 (January 1908): 44.

55. Benjamin Horace Hibbard, "Farm Tenancy in the United States," *Annals of the American Academy of Political and Social Science* 40, no. 1 (March 1912): 31; Howard A. Turner, *A Graphic Summary of Farm Tenure*, USDA Misc. Pub. no. 261 (Washington, DC: GPO, 1936); and Benjamin Hodge Nichols, "Thesis on Rural Electric Rates and Rural Line Extension Policies of the United States" (MS thesis, Oregon State Agricultural College, 1932), 14–15.

56. Henry A. Wallace, "The Problem of Farm Tenancy," *Scientific Monthly* 41, no. 1 (July 1935): 52–53.

57. Ibid., 53.

58. Ibid., 53–54.

59. US Special Committee on Farm Tenancy, *Farm Tenancy: Report of the President's Committee* (Washington, DC: GPO, 1937), 35; Turner, *A Graphic Summary of Farm Tenure*, 1.

60. Bryant Putney, "Farm Tenancy in the United States," *Editorial Research Reports 1935* (Washington, DC: CQ Press, 1935), 201–19; US Special Committee on Farm Tenancy, *Farm Tenancy*, 35; and Mississippi State Planning Commission, *Farm Tenancy in Mississippi*, release no. 7 (Jackson, MS: Works Progress Administration, 1937), 7.

61. These states' farmers also benefited from being located near industrialized—and electrified—communities, which were often linked by transmission lines, thereby enabling easier extension of power lines from central stations. E. A. White, "Eleventh Annual Report to the Committee on the Relation of Electricity to Agriculture by the Director," presented at Chicago meeting, 27 July 1934, typescript, in AU Special Collections, Agricultural Engineering Records, record group 503, accession no. 79-001, box 45 of 64, 4.

62. The emergence and social significance of electric streetcars is described in Nye, *Electrifying America*, 85–137.

63. For a history of the Milwaukee company, see Forrest McDonald, *Let There Be Light: The Electric Utility Industry in Wisconsin, 1881–1955* (Madison, WI: American History Research Center, 1957), 47–59.

64. US Bureau of the Census, *Census of Electrical Industries 1922, Central Electric Light and Power Stations* (Washington, DC: GPO, 1925), xii–xiii.

65. Martin G. Glaeser, *Outlines of Public Utility Economics* (New York: Macmillan, 1927), 59.

66. For examples of such efficiencies, see Devine, "From Shafts to Wires," 347–72.

67. Hirsh, *Power Loss*, 81.

68. Edison Electric Institute, *Statistical Yearbook of the Electric Utility Industry / 1987* (Washington, DC: Edison Electric Institute, 1988), table 1.

69. Julie Cohn, *The Grid: Biography of an American Technology* (Cambridge, MA: MIT Press, 2017), 38–39.

70. W. S. Murray et al., *A Superpower System for the Region between Boston and Washington* (Washington, DC: GPO, 1921); William Spencer Murray, *Superpower: Its Genesis and Future* (New York: McGraw-Hill, 1925); and Leonard DeGraaf, "Corporate Liberalism

and Electric Power System Planning in the 1920s," *Business History Review* 64, no. 1 (Spring 1990): 1–31.

71. Glaeser, *Outlines of Public Utility Economics*, 66–67.

72. Lemont Kingsford Richardson, *Wisconsin REA: The Struggle to Extend Electricity to Rural Wisconsin, 1935–1955* (Madison: University of Wisconsin Experiment Station, 1960), 9.

73. "Customers Exceed 23,400,000," *Electrical World* 93, no. 1 (5 January 1929): 31; and "Domestic Use Increasing," *Electrical World* 93, no. 1 (5 January 1929): 29–30. To help maintain positive relations with their new residential customers and to help reduce public criticisms of utilities, several companies aggressively sold shares of utility securities to them. "Public Buys Utility Stocks in Unprecedented Volume," *Electrical World* 93, no. 1 (5 January 1929): 32; and "Customer Ownership of Public Utilities," *World's Work* 45, no. 4 (February 1923): 362–63.

74. Hillman, "A Study of Plans and Policies of Power Companies," 42.

75. Thomas K. McCraw, *Prophets of Regulation* (Cambridge, MA: Harvard University Press, 1984), 300–309.

76. Miller-Warren Energy Lifeline Act of 1975, AB 167, adding section 739 ("Gas and electric baseline . . .") to California's Public Utilities Code, Stats. 1975, c. 1010, sec. 1. Also see Daryl Lembke, "Utility Rate Break for Homes: PUC Policy Will Cut Cost of Household Gas, Power," *Los Angeles Times*, 17 September 1975, 1.

CHAPTER THREE: **Business Attitudes toward Farmers in the 1920s**

Epigraph: Business Men's Commission on Agriculture, *The Condition of Agriculture in the United States and Measures for Its Improvement* (New York: National Industrial Conference Board, 1927), 116.

1. Delos F. Wilcox, *Municipal Franchises: A Description of the Terms and Conditions upon Which Private Corporations Enjoy Special Privileges in the Streets of American Cities*, vol. 1 (Rochester, NY: Gervaise Press, 1910), 27–28.

2. David B. Danbom, "Rural Education Reform and the Country Life Movement, 1900–1920," *Agricultural History* 53, no. 2 (April 1979): 462–74.

3. Liberty Hyde Bailey, *Report of the Commission on Country Life* (New York: Sturgis and Walton, 1911), 45; "Roosevelt to the Farmers: Betterment of Rural Life One of the Nation's Great Needs," *New York Times*, 24 August 1910; 2; and David B. Danbom, *Born in the Country: A History of Rural America* (Baltimore, MD: Johns Hopkins University Press, 1995), 169. The president's speech can be found in Theodore Roosevelt, "Rural Life," *Outlook* 95 (27 August 1910): 919–22.

4. Bailey, *Report of the Commission*, 19–20.

5. Ibid., 18.

6. Danbom, *Born in the Country*, 173.

7. USDA, *Yearbook of the United States Department of Agriculture, 1913* (Washington, DC: GPO, 1914), 37.

8. Danbom, *Born in the Country*, 173.

9. Brandeis's use of "scientific management" principles to challenge railroad company policies is described in Oscar Kraines, "Brandeis and Scientific Management," *Publications of the American Jewish Historical Society* 41, no. 1 (September 1951): 41–60; and Melvin I. Urofsky, *Louis D. Brandeis: A Life* (New York: Pantheon, 2009).

10. "Gauge on Producer Urged by Vanderlip," *New York Times*, 28 April 1914, 1.

11. Forrest Crissey, "Speeding Up the Farm—Business Men at Wheel Make Good," *Banker-Farmer* 1, no. 6 (May 1914): 2.

12. D. A. Wallace, "Why the Business Man and the Farmer Must Get Together," *Banker-Farmer* 1, no. 10 (September 1914): 15.

13. US Department of Commerce, *Historical Statistics of the United States: Colonial Times to 1970* (Washington, DC: GPO, 1974), table, Series F 125–129, "Gross Domestic Product Originating in Private Farm and Nonfarm Sectors and Government," 232.

14. Edwin T. Layton Jr., *The Revolt of the Engineers: Social Responsibility and the American Engineering Profession* (Cleveland, OH: Press of Case Western Reserve University, 1971).

15. David J. Goldberg, *Discontented America: The United States in the 1920s* (Baltimore, MD: Johns Hopkins University Press, 1999), 10, 166–73.

16. Hoover gained a positive reputation by applying engineering principles to logistical problems, such as food distribution in the United States and Europe during and after the war. Described by Morris Cooke as "the engineering method personified," Hoover "combined technical excellence, professional dedication, and eminent public service" while standing for "professionalism, progressivism, and the engineering approach to social problems." Layton, *Revolt of the Engineers*, 179, 189, 191–93.

17. Kim McQuaid, "Young, Swope and General Electric's 'New Capitalism': A Study in Corporate Liberalism, 1920–33," *American Journal of Economics and Sociology* 36 (July 1977): 323–34.

18. Some historians have labeled these efforts in the business world as corporate liberalism, an approach that improved working conditions dramatically. But they did not arise because of managers' altruistic feelings. Rather, corporate leaders sought to placate workers as a means to co-opt potentially destructive forces that could have reshaped capitalism by introducing forms of socialism. Barbara G. Brents, "Capitalism, Corporate Liberalism and Social Policy: The Origins of the Social Security Act of 1935," *Mid-American Review of Sociology* 9, no. 1 (Spring 1984): 25; James Weinstein, *The Corporate Ideal in the Liberal State, 1900–1918* (Boston: Beacon Press, 1968); and Kim McQuaid, "Corporate Liberalism in the American Business Community, 1920–1940," *Business History Review* 52, no. 3 (Autumn 1978): 342–68.

19. Rakesh Khurana, *From Higher Aims to Hired Hands: The Social Transformation of American Business Schools and the Unfulfilled Promise of Management as a Profession* (Princeton, NJ: Princeton University Press, 2007).

20. David E. Nye, *America's Assembly Line* (Cambridge, MA: MIT Press, 2013).

21. Goldberg, *Discontented America*, 170.

22. This imagery is discussed in Tarla Rai Peterson, "Jefferson's Yeoman Farmer as Frontier Hero: A Self Defeating Mythic Structure," *Agriculture and Human Values* 7 (Winter 1990): 9–19; and David B. Danbom, "Romantic Agrarianism in Twentieth-Century America," *Agricultural History* 65, no. 4 (1991): 1–12.

23. Danbom, *Born in the Country*, 175.

24. Julius H. Barnes, "Dollar Wheat," *Nation's Business* (September 1923): 13–14.

25. John Philip Gleason, "The Attitude of the Business Community toward Agriculture during the McNary-Haugen Period," *Agricultural History* 32, no. 2 (April 1958): 127–38.

26. Julius H. Barnes, "Is There a 'National' Farm Problem?" *Nation's Business* (January 1927): 17.

27. In 1920, the Census Bureau reported 54.158 million Americans living in urban areas and 51.553 million people residing in rural districts. US Department of Commerce, *Historical Statistics*, table, Series A 57–72, "Population in Urban and Rural Territory," 11.

28. Alfred N. Goldsmith and Austin C. Lescarboura, *This Thing Called Broadcasting* (New York: Henry Holt, 1930), 249. At least one academic seemed to agree with the notion of a brain drain from the countryside. Carl C. Taylor, a sociology professor at North Carolina State College, acknowledged in 1924 that the "rural problem" meant, to some observers, a drift of the formerly agrarian population to cities, which robbed the farms of the "best minds and most ambitious citizens" such that the hinterland became "a decadent civilization." Taylor, "The Rise of the Rural Problem," *Journal of Social Forces* 2, no. 1 (November 1923): 29.

29. Goldsmith, *This Thing Called Broadcasting*, 258–59.

30. F. E. St. John, "Electricity on the Farm as a Practical Money Saver," *System on the Farm* 4, no. 3 (March 1919): 164.

31. W.C.I., "Is the Farmer behind the Times?" *Indiana Farmer's Guide*, 10 July 1920, 14.

32. E. R. McIntyre, "Hic Jacet—the Hick," *Independent*, 10 April 1926, 421. Many farmers spurned this characterization, feeling that they had already taken advantage of several modern technologies (see chapter 5). Historian Leah Glaser noted that 92 percent of farm women surveyed in 1923 "found pride in their work, generally enjoyed rural culture, and resented the Country Life Movement's suggestion that they lived like drudges." Glaser, *Electrifying the Rural American West: Stories of Power, People, and Place* (Lincoln: University of Nebraska Press, 2009), 22. Also see Emily Hoag Sawtelle, "The Advantages of Farm Life: A Study by Correspondence and Interviews with Eight Thousand Farm Women," unpublished manuscript, USDA, Bureau of Agricultural Economics, March 1924, 1, Internet Archive, https://archive.org/stream/CAT31046460#page/n1/mode/2up, accessed 16 September 2016.

33. E. P. Edwards, comments made after presentation of "Report of the Committee on Electricity on the Farm: Central States," NELA, *Proceedings* 48 (1913): 289.

34. Walter H. Johnson, "Address of President Walter H. Johnson," NELA, *Proceedings* 81 (1924): 5.

35. Owen D. Young, "Farm Electrification in New York State and How It Can Be Achieved," *Economic World* 29 (18 April 1925): 545.

36. Ibid.

37. While imploring ruralites to become more modern and willing to consume increasing amounts of electricity, utility executives appear to have become acutely aware of farmers' objections to the haughty attitudes of business folk. In a 1928 book for power company men speaking before public audiences, compiled by NELA, the author offered specific advice about the importance of tact: "Care should be taken by the speaker to refrain from appearing to tell the farmer how to run his farm. Farmers are just like others, in that they do not always like suggestions. In fact, they are sensitive on this point, and have a right to be, for their city cousins for years have been telling them how to run their farms, expounding their own theories of scientific agricultural operations. The man in the farming business resents this. Whether the advice given is right or wrong

has become immaterial. There has been so much of it that it quickly arouses his resentment." Public Speaking Committee of the NELA, *NELA Handbook* (New York: NELA, 1928), 69–70.

38. L. S. Wing, "Rural Electrification from an Economic and Engineering Standpoint," *Transactions of the American Society of Agricultural Engineers* 20 (June 1926): 42–43.

39. Ibid., 45–50. Wing pointed out that, even as the cost of living had grown 75 percent since 1883, the rate for lighting had dropped from 20 cents to 7.5 cents per kilowatt-hour (45).

40. Ibid., 43.

41. "Agricultural Engineers for Power Companies," *Electrical World* 85, no. 26 (27 June 1925): 1383, and reprinted with the same title in *Agricultural Engineering* 6, no. 12 (December 1925): 291. The word "hobby" is derived from the word "hobby-horse." Hence, if someone rides the "economic hobby" or "social hobby," he or she appears to have inclinations that focus on economic or social concerns. In a 1914 article dealing with why farmers left rural areas for cities, the author referred to a "reformer who had the economic hobby" and who explained the exodus from farms in exclusively financial terms. E. C. Lindemann, "Play Activities in Rural Life," *Michigan State Farmers' Institutes, Institute Bulletin* no. 20 (September 1914): 390.

42. William Spencer Murray, *Superpower: Its Genesis and Future* (New York: McGraw-Hill, 1925), 204–5.

43. "How Stands Rural Electrification?" *Electrical World* 99 (28 May 1932): 962, also quoted in Harry Slattery, *Rural America Lights Up* (Washington, DC: National Home Library Foundation, 1940), 25; emphasis in the original.

44. Letter from "Rural electrification committee of privately owned utilities" to Morris L. Cooke, administrator, REA, 24 July 1935, in US House of Representatives, *Rural Electrification Hearing before the Committee on Interstate and Foreign Commerce . . . on S. 3483*, 12–14 March 1936 (Washington, DC: GPO, 1936), 34, https://hdl.handle.net /2027/umn.31951d031367408.

45. Several authors used the term "decentralization" in the 1910s and 1920s to describe the movement of industries away from big cities. The term should not be confused with "distributed generation," a modern concept characterizing the production of electricity in small power plants rather than in the more typical, large-scale central stations.

46. Thomas Adams, "Regional and Town Planning," in *Proceedings of the Eleventh National Conference on City Planning* (Niagara Falls and Buffalo, NY, 26–28 May 1919), 77; and Harold L. Platt, "World War I and the Birth of American Regionalism," *Journal of Policy History* 5, no. 1 (1993): 139.

47. C. B. Purdom, "New Towns for Old," *Survey Graphic* 7 (May 1925): 169–72, reprinted in Carl Sussman, ed., *Planning the Fourth Migration: The Neglected Vision of the Regional Planning Association of America* (Cambridge, MA: MIT Press, 1976), 132. Garden cities would balance "commerce and civic structure—all encompassed within a spatial arrangement that restores a sense human scale to urban life." Mark Luccarelli, *Lewis Mumford and the Ecological Region: The Politics of Planning* (New York: Guilford Press, 1995), 77.

48. Lewis Mumford, "The Fourth Migration," *Survey Graphic* 7 (May 1925): 130–33, reprinted in Sussman, *Planning the Fourth Migration*, 55–64.

49. Luccarelli, *Lewis Mumford and the Ecological Region*, 80; and Sussman, *Planning the Fourth Migration*, 61–63.

50. Robert W. Bruère, "Giant Power—Region-Builder," *Survey Graphic* 7 (May 1925): 161–64, 188, reprinted in Sussman, *Planning the Fourth Migration*, 114.

51. Ibid., 116.

52. Howard P. Segal, "'Little Plants in the Country': Henry Ford's Village Industries and the Beginning of Decentralized Technology in Modern America," *Prospects* 13 (October 1988): 181–223; and Howard P. Segal, *Recasting the Machine Age: Henry Ford's Village Industries* (Amherst: University of Massachusetts Press, 2008).

53. "Henry Ford Dooms Our Great Cities," *Literary Digest* 83, no. 7 (15 November 1924): 13.

54. Guy E. Tripp, *Electric Development as an Aid to Agriculture* (New York: G. P. Putnam's Sons, 1926), 39.

55. Middle West Utilities Company, *America's New Frontier* (Chicago: Middle West Utilities Co., 1929), 29.

56. Ibid., 13. The original source of the quotation is Samuel Insull, "Twenty-Five Years of Electric Power," the text of a talk given by Insull on 17 November 1927 (Chicago: self-published, 1927), 6.

57. Middle West Utilities, *America's New Frontier*, 29, also emphasized in Insull, "Twenty-Five Years of Electric Power," 7. Secretary of Agriculture W. M. Jardine endorsed this approach (and is referred to in *America's New Frontier*), in "The Town Comes to the Farmer," *Saturday Evening Post*, 4 May 1929, 37, 217–18, 221–22. An executive in 1934 observed that the utility industry remained ready to pursue the goal of industrial decentralization "[i]f society wishes." P. H. Powers, "5,000 Kw.-Hr. per Rural Customer," *Electrical World* 103 (9 June 1934): 843.

58. The hopes of such a movement, fervently advocated by the RPAA and others, never materialized, and the trend toward greater urbanization and centralization occurred instead. Sussman, *Planning the Fourth Migration*; and Luccarelli, *Lewis Mumford and the Ecological Region*.

59. James H. Shideler, "'Flappers and Philosophers,' and Farmers: Rural-Urban Tensions of the Twenties," *Agricultural History* 47, no. 4 (October 1973): 289.

60. Despite the occasional success of socialists to win election to the US Congress and as mayors of major American cities in the first two decades of the twentieth century, socialism increasingly became viewed with contempt, especially after installation of a Communist regime in Russia during World War I. With fears of Communism spreading westward (as evidenced by the Red Scare of 1919 to 1920), the business community derided the approach in which government assumed many activities previously performed by corporations. In a 1924 speech, Secretary of Commerce Herbert Hoover observed that while "[s]ocialism may have a place with some of the nations of Europe . . . , [i]t has no place with us." "Socialism Unmasked," *Los Angeles Times*, 20 September 1924, 1. A good description of the threat of socialism, the Red Scare, and the business environment of the 1920s can be found in Goldberg, *Discontented America*, 40–65.

61. Slattery, *Rural America Lights Up*, 5–6.

62. Ibid., 6.

63. Ernest R. Abrams, *Power in Transition* (New York: Scribner's, 1940), 33.

CHAPTER FOUR: **The Lure and Lore of Rural Electrification**

Epigraph: Armin Karl Neubert, "The Application of Electricity to Agriculture" (BS thesis, University of Wisconsin, 1916), 1.

1. "The Farmer's Use for Electricity," *Isolated Plant* 2, no. 5 (May–June 1910): 27.

2. Historians have nicely documented the role of electricity in people's visions of the future. See David E. Nye, *Electrifying America: Social Meanings of a New Technology, 1880–1940* (Cambridge, MA: MIT Press, 1992); Carolyn Marvin, *When Old Technologies Were New: Thinking about Electric Communication in the Late Nineteenth Century* (Oxford: Oxford University Press, 1988); and Jennifer L. Lieberman and Ronald R. Kline, "Dream of an Unfettered Electrical Future," *Configurations* 25, no. 1 (Winter 2017): 1–27. In his study of American fictional and nonfictional works dealing with imaginary futures in the late nineteenth century, Kenneth M. Roemer noted that electricity stood out as one of three of the most mentioned technologies. The other two, aluminum and railroads, employed vast amounts of electricity, thus reinforcing the importance of the energy form in the popular mindset. Roemer, *The Obsolete Necessity: America in Utopian Writings, 1888–1900* (Kent, OH: Kent State University Press, 1976), 111.

3. "Edison's Victories with the Electric Light," *Omaha Daily Herald*, 25 December 1879, 4. The Menlo Park demonstration was widely reported in, for example, "The Electric Wizard," *Chicago Daily Tribune*, 22 December 1879, 8; "Edison's Electricity," *Washington Post*, 25 December 1879, 1; "Edison's Light; A Successful Exhibition at Menlo Park Last Night," *New York Herald*, 25 December 1879, 3; and "The Electric Light; The Great Inventor Claims That He Has Solved the Problem. Lamps Can Be Furnished at Cost of Twenty-Five Cents Each, and Will Neither Burn, Fuse, nor Melt," *Chicago Daily Tribune*, 28 December 1879, 3. The use of electric lights to inspire awe is explored in David E. Nye, *American Technological Sublime* (Cambridge, MA: MIT Press, 1994), 143–72.

4. Thomas P. Hughes, *Networks of Power: Electrification in Western Society, 1880–1930* (Baltimore, MD: Johns Hopkins University Press, 1983), 41.

5. Nye, *Electrifying America*, 191.

6. E. E. Whaley, "Electric Light Plants for Farm Use: How Science and Mechanics Have Solved One of the Farmer's Problems," *Farm Engineering* 1, no. 3 (January 1914): 12.

7. "Harvesting by Electricity," *Electrician* 1, no. 8 (August 1882): 153.

8. "Electricity on an Illinois Farm," *Electrical World* 55, no. 5 (3 February 1910): 282.

9. Frank Koester, *Electricity for the Farm and Home* (New York: Sturgis and Walton, 1913), 21.

10. Whaley, "Electric Light Plants for Farm Use," 12.

11. W. M. Hurst and L. M. Church, *Power and Machinery in Agriculture*, USDA Misc. Pub. no. 157 (Washington, DC: GPO, 1933), 57.

12. Carrie A. Meyer, "The Farm Debut of the Gasoline Engine," *Agricultural History* 87, no. 3 (2013): 287–313.

13. Studies of how tractors increased productivity and altered social and economic conditions on the farm include Robert E. Ankli and Alan L. Olmstead, "The Adoption of the Gasoline Tractor in California," *Agricultural History* 55, no. 3 (July 1981): 213–30; Allan G. Bogue, "Changes in Mechanical and Plant Technology: The Corn Belt, 1910–1940," *Journal of Economic History* 53, no. 1 (March 1983): 1–25; and Sally Clarke, "New Deal Regulation and the Revolution in American Farm Productivity: A Case Study of the

Diffusion of the Tractor in the Corn Belt, 1920–1940," *Journal of Economic History* 51, no. 1 (March 1991): 101–23.

14. C. D. Kinsman, "An Appraisal of Power Used on Farms in the United States," USDA Bulletin no. 1348 (Washington, DC: USDA, 1926): preface and 1.

15. Otto Doederlein, "The Electric Plow in Germany," *Scientific American Supplement*, no. 1035 (2 November 1895): 16542. An illustration in this article shows the plow attached to wires from a cart that carries a dynamo (an electric generator). The generator, in turn, is powered by a steam engine–driven tractor.

16. Franklin L. Pope, "Electricity," *Engineering* 3 (May 1892): 254–55.

17. Frank Koester, "Electric Power Applications on the Farm," *Electrical Review and Western Electrician* 60, no. 16 (30 April 1912): 745.

18. Valentin Confesor, "Electricity on the Farm" (BS thesis, University of Illinois, 1912), 12.

19. Koester, *Electricity for the Farm and Home*, 102–3.

20. "Bound to Come," *Farm Journal* 30, no. 3 (March 1906): 100.

21. H. H. Hosford, "Electricity on the Farm," *Ohio Farmer*, 17 December 1896, 458.

22. "Electricity on the Farm," *Southern Cultivator* 54, no. 6 (June 1896): 299.

23. C.H.G., "Electricity on the Farm," *Ohio Farmer*, 29 October 1903, 325.

24. "Farms Run by Electricity," *Current Literature* 51 (October 1911): 459.

25. "The Farmer's Use for Electricity," 24.

26. In almost all the literature I consulted, the term "farmer" referred to a male. Women on the farm were often described as "farm women" or "farmers' wives." In this book, I have generally maintained the description used in my sources, even though I am keenly aware that women on farms often did the same (or more) work than done by men.

27. Herman Russell, "Electricity on the Farm and the Influence of Irrigation from a Central-Station Standpoint," NELA, *Proceedings* 39 (1910): 198.

28. "The Farmer's Use for Electricity," 24.

29. M.D.H., "Electricity on the Farm," *Indiana Farmer's Guide*, 17 April 1920, 39.

30. Clarence D. Warner, "Electricity in Agriculture," Bulletin no. 16 of the *Hatch Experiment Station of the Massachusetts Agricultural College* (January 1892): 3.

31. Joseph Priestley, *The History and Present State of Electricity: With Original Experiments*, vol. 1 (London: Bathurst and Lowndes, 1775), 172.

32. Clark C. Spence, "Early Uses of Electricity in American Agriculture," *Technology and Culture* 3, no. 2 (Spring 1962): 142–43.

33. W. H. Weekes, "Observations on Electro-Culture, with Details of an Experiment Illustrating the Influence of Electric Currents on the Economy of Vegetation," *Electrical Magazine* 2 (July 1845): 113–14.

34. Observer, "Electro-Culture," *British Farmer's Magazine*, n.s., 9 (October 1845): 339; and "Agriculture in Scotland—No. 13," *American Agriculturalist* 4, no. 10 (October 1845): 338–39.

35. C. W. Siemens, "Electro-Horticulture," *Journal of the American Agricultural Association* 1, nos. 3 and 4 (1891): 195–99.

36. Elihu Thomson, "Future Electrical Development," *New England Magazine* 6 (1892): 633.

37. "A New Method of Electric Culture," *Scientific American Supplement* 36, no. 931 (4 November 1893): 14878.

38. A summary of some of these experiments can be found in "Electricity as a Stimulant to Plant Growth," *Scientific American* 92, no. 23 (10 June 1905): 458. Brief descriptions of Lemström's work are contained in obituaries published in *Electrician* 54, no. 1 (21 October 1904): 22; and *Nature* 71, no. 1832 (8 December 1904): 129.

39. Selim Lemström, "Experiments on the Influence of Electricity on Plants," *Report of the Sixty-Eighth Meeting of the British Association for the Advancement of Science Held at Bristol in September 1898* (London: John Murray, 1899): 808–9.

40. Koester, *Electricity for the Farm and Home*, 261–74.

41. Adolph David Bullerjahn, "Electro-Culture Considered in Connection with the Problem of Improving the Load Factor of an Electric Power Plant" (BS thesis, University of Wisconsin, 1913), 5.

42. F. S. Fletcher, "Electro-Culture as a Central Station Possibility," *Electrical Review and Western Electrician* 70, no. 17 (28 April 1917): 705. Electroculture did not ultimately become a significant element of modern agricultural techniques. The USDA sponsored studies, with a report in 1926 concluding that its researchers could not find any increase in crop production using electrical current. Lyman J. Briggs, A. B. Campbell, R. H. Heald, and L. H. Flint, "Electroculture," USDA Bulletin no. 1379 (Washington, DC: GPO, 1926).

43. Summaries of research include "Electroculture of Plants: An Account of Some Recent Experiments," *Scientific American Supplement* 67, no. 1730 (27 February 1909): 135; "Electro-Culture of the Soil," *Scientific American Supplement* 79, no. 2044 (6 March 1915): 151; "Electro-Culture: A Résumé of the Literature and Summary of Facts from Scattered Sources," *Scientific America Supplement* 79, no. 2051 (24 April 1915): 258–59; "Electro-Culture of Crops: A Review of Recent Developments," *Scientific American Supplement* 84, no. 2186 (24 November 1917); and Ingvar Jorgensen and Walter Stiles, "The Electroculture of Crops," *Scientific American Supplement* 85, no. 2214 (8 June 1918): 366–68.

44. Putnam A. Bates, "Electricity on the Farm," *Transactions of the American Institute of Electrical Engineers* 19, no. 2 (1912): 1985.

45. Ibid., 1988.

46. Ibid., 1995. Ensilage, also known as silage, is a form of animal food (fodder) produced by shredding the entire green plant of grass crops and often storing the resulting material in a silo, where anaerobic fermentation takes place and helps preserve the food while also boosting its nutritional value. J. M. Wilkinson, K. K. Bolsen, and C. J. Lin, "History of Silage," in Dwayne R. Buxton, Richard E. Muck, and Joseph H. Harrison, eds. *Silage Science and Technology*, Agronomy Monograph no. 42 (Madison, WI: American Society of Agronomy, 2003): 1–30.

47. Bates, "Electricity on the Farm," 1998–99, 2003.

48. Putnam A. Bates, "Light and Power on the Farm," *Scientific American* 104, no. 15 (15 April 1911): 380, 392–93; T. Commerford Martin and Putnam A. Bates, "Central Station Current for Farms," *Scientific American* 105, no. 13 (23 September 1911): 273–74; Putnam A. Bates, "Farm Electric Lighting by Wind Power," *Scientific American* 107, no. 13 (28 September 1912): 262–63; Putnam A. Bates, "Harvesting Ice by Electric Power," *Scientific American* 107, no. 18 (12 October 1912): 299–300; Putnam A. Bates, "Electricity and Spray Irrigation,"

Scientific American 107, no. 18 (2 November 1912): 364; Putnam A. Bates, "How Electricity Makes the Dairy Cleaner," *Scientific American* 107, no. 23 (7 December 1912): 482, 498–99; and Putnam A. Bates, "Agriculture, Electricity and Irrigation," *Scientific American* 108, no. 21 (24 May 1913): 472, 479.

49. Koester, *Electricity for the Farm and Home*, 3–5. Biographical information about Koester comes from *Encyclopedia Americana* (1920), Wikisource, http://en.wikisource .org/wiki/The_Encyclopedia_Americana_%281920%29/Koester,_Frank, accessed 25 February 2013.

50. Nye, *Electrifying America*, 295–96.

51. E. P. Edwards, "The Neglect of the Power Problem," *Transactions of the American Society of Agricultural Engineers* 4 (December 1910): 86–97.

52. General Electric Company, *Electricity on the Farm*, bulletin A4115 (Schenectady, NY: General Electric, 1913), 11–12. This issue superseded another GE publication of the same name, bulletin B-4038, from 1911, as noted on the issue's cover.

53. General Electric Company, *Electricity on the Farm*, 25, 33.

54. Ibid., 34.

55. Ibid.

56. Savings accrued from about 22 percent on a 40-acre farm to about 37 percent on a 160-acre farm. Ibid., 44–45.

57. John Liston, "General Electric Lecture Service," *General Electric Review* 19, no. 3 (March 1916): 236–40. The movie *Back to the Farm* can be seen at Internet Archive, https://archive.org/details/back_to_the_farm, accessed 30 January 2019.

58. Ibid., at about 24:47.

59. *Farm Light and Power Year Book: Dealers' Catalog and Service* (New York: Farm Light and Power Publishing Co., 1922).

60. I could not locate a first edition of this publication, though it is referred to in early 1926 advertisements and in newspaper articles, such as "Booklet Tells Function of Electricity on Farm," *Wakefield* (Michigan) *News*, 7 February 1925, 5. The *G-E Farm Book*, 2nd ed. (Schenectady, NY: General Electric, 1926), Internet Archive, https://archive.org /details/TheG-eFarmBook, accessed 6 July 2018.

61. *G-E Farm Book*, 14. Continuing its publicity campaign, the company issued other publications, such as the late 1920s' *Electric Light on the Farm and in the Rural Districts*, which offered advice on the best uses of GE lamps in the farm home and outbuildings. Not only do electric lights reduce fires often caused by kerosene lamps or exposed flames; they also enhance social gatherings, enable easy nighttime reading, illuminate work done in the barn and stables, and encourage hens to lay more eggs. A. L. Powell and A. D. Bell, *Electric Light on the Farm and in the Rural Districts*, Lighting Data Bulletin LD 153A (Harrison, NJ: General Electric, no date, but likely 1928), 5–6, 8, 11, 16, 26.

62. Kenneth Lipartito, "Picturephone and the Information Age: The Social Meaning of Failure," *Technology and Culture* 44, no. 1 (January 2003): 77.

63. Ibid., 77–78.

64. Sheila Jasanoff and Sang-Hyun Kim, "Containing the Atom: Sociotechnical Imaginaries and Nuclear Power in the United States and South Korea," *Minerva* 47, no. 2 (June 2009): 120.

65. Simone M. Müller and Heidi J. S. Tworek, "Imagined Use as a Category of Analysis: New Approaches to the History of Technology," *History and Technology* 32, no. 2 (2016): 106.

CHAPTER FIVE: **Farmers on Their Own**

Epigraph: "What Electricity Is Doing," *Farm Mechanics* 6, no. 4 (February 1922): 164.

1. Today, we often use the terms "self-generation" or "distributed generation" for power systems not connected to the grid of transmission lines connected to large-scale power plants. More discussion of such systems is contained in this chapter's conclusion.

2. Donald C. Shafer, "The New Hired Man: Harnessing the Home Stream," *Saturday Evening Post* 182, no. 49 (4 June 1910): 17.

3. David R. Cooper, *Water Power for the Farm and Country Home* (Albany: New York State Water Supply Commission, 1911), 3–4.

4. Frederick Irving Anderson, *Electricity for the Farm: Light, Heat, and Power* (New York: Macmillan, 1915), xvii, xix, xxi, xxiii.

5. "Electricity on the Farm," *Electrical World* 48, no. 15 (13 October 1906): 698.

6. T. Commerford Martin and Putnam A. Bates, "Convert Abandoned Mills into Electric Power Plants: Electricity on the Farm Made Possible by Co-operation," *Scientific American* 105, no. 9 (26 August 1911): 184, 196.

7. "The Story of an Electric Farm," *Popular Electricity* 4, no. 4 (August 1911): 289–97. The farmer's experience is also described in "Farms Run by Electricity," *Current Literature* 51, no. 4 (October 1911): 459–60.

8. "Rancher Has His Own Power Plant at Slight Expense," *Home Demonstration News*, typed newsletter from State College of Washington, Cooperative Extension Work in Agriculture and Home Economics, Pullman, WA, vol. 2, no. 14 (February, March, and April 1920), in WSU Special Collections. After being founded in 1890 as Washington Agricultural College, the institution was renamed the State College of Washington in 1905 (also known as Washington State College). It became Washington State University in 1959. Washington State University, "Inspiring Washington: 125 Years, and Counting," https://timeline.wsu.edu, accessed 29 July 2021.

9. A. B. Crane, *Let the Creek Light Your Home*, Extension Service Bulletin no. 124 (Pullman: State College of Washington, 1924), 2.

10. A. M. Daniels, C. E. Seitz, and J. C. Glenn, *Power for the Farm from Small Streams*, Farmers' Bulletin no. 1430 (Washington, DC: GPO, 1925).

11. Calculation from data in Wesley Clair Mitchell, ed., *Income in the United States, Its Amount and Distribution, 1909–1919*, vol. 2 (New York: National Bureau of Economic Research, 1922), 313, https://www.nber.org/chapters/c9420.pdf, at http://papers.nber.org/books/mitc22-1, accessed 22 June 2017. This source noted that 2.2 percent of farmers in 1913 had incomes exceeding $2,000.

12. USDA, 1920 Census Publications, "Farms and Farm Property," 24, USDA Census of Agriculture Historical Archive, http://agcensus.mannlib.cornell.edu/AgCensus/censusParts.do?year=1920, accessed 22 June 2017.

13. Carol Lee, "Wired Help for the Farm: Individual Electric Generating Sets for Farms, 1880–1930" (PhD diss., Pennsylvania State University, 1989), 103.

14. David E. Nye, *Consuming Power: A Social History of American Energies* (Cambridge, MA: MIT Press, 1998), 110–11. The typical windmill pump has its origin in the work of Daniel Halladay, who patented a self-governing and low-maintenance machine in 1854. T. Lindsay Baker, *A Field Guide to American Windmills* (Norman: University of Oklahoma Press, 1985), 5–15. A description of an early machine can be found in "Improved Governor for Wind Mills," *Scientific American* 10, no. 4 (7 October 1854): 25. More than a million water pumping machines populated the American Midwest and West, according to P. W. Carlin, A. S. Laxon, and E. B. Muljadi, "The History and State of the Art of Variable-Speed Wind Turbine Technology," *Wind Energy* 6 (2003): 129.

15. Putnam Bates, "Farm Electric Lighting by Wind Power," *Scientific American* 107, no. 13 (28 September 1912): 262.

16. *Farm Light and Power Year Book: Dealers' Catalog and Service* (New York: Farm Light and Power Publishing, 1922), 41.

17. Ibid. For details on making improvised wind-powered generators, see J. Leo Ahart, "Don't Waste All the Wind!," *Farm Journal* 59, no. 3 (March 1935): 4; and Ahart, "Wind-Made Power," *Farm Journal* 59, no. 4 (April 1935): 15.

18. Advertisement in *Purdue Agriculturist* 16, no. 8 (May 1922): 128. A description of (and praise for) the machine is also included in "Electricity: A City Luxury for Every Farm," *Farm Light and Power Year Book*, 43.

19. "Electricity: A City Luxury for Every Farm," 47. Carol Lee noted that the Aeroelectric wind generator constituted "the first commercial wind electric plant for sale to farmers." Lee, "Wired Help for the Farm," 115.

20. More details on the device can be found in "Generating Electricity on the Farm," *Commercial America* 18, no. 9 (July 1921): 49. Also see Ronald R. Kline, *Consumers in the Country: Technology and Social Change in Rural America* (Baltimore, MD: Johns Hopkins University Press, 2000), 102–3.

21. *Farm Light and Power Year Book*, 47–48.

22. "Generating Electricity from the Wind," *Current Opinion* 71, no. 1 (July 1921): 119. Besides the Perkins machine, a 1922 catalog listed turbines offered by fifty-four other manufacturers. *Farm Light and Power Year Book*, 322–23.

23. E. G. McKibben and J. Brownlee Davidson, *Wind Electric Plants*, Bulletin no. 297 (Ames: Agricultural Experiment Station, Iowa State College of Agriculture and Mechanic Arts, 1933), 267–68, in ISU Special Collections, Davidson papers, "Writings—Wind Electric Plants 1933," RS 9/2/11, folder 11/56.

24. Ibid., 271.

25. Ibid., 274. A 1923 report from the Wisconsin Extension Service came to a similar conclusion: It noted that the wind electric plants' "cost of operation . . . is of course very low, but the initial investment is higher and interest will be higher." F. W. Duffee and G. W. Palmer, *Turn On the Light*, Circular 163 (Madison: Extension Service of the College of Agriculture, University of Wisconsin, July 1923): 14–15.

26. Letter from M. L. Nichols, agricultural engineer, Alabama Polytechnic Institute, to R. E. Lambert, Darlington, AL, 7 December 1928, in AU Special Collections, Agricultural Engineering Records, record group 503, accession no. 79-001, box 24 of 64. In a response to a similar inquiry, Nichols noted that a Hebco wind electric plant had a good

reputation for service in the Midwest, but it could only be counted on when "little electric load is all that is demanded." Letter from M. L. Nichols, to C. C. Lowder of Mobile, AL, 24 November 1928, in ibid.

27. The story of the establishment of the Wincharger company can be found in Arthur Van Vlissingen, "Riches from the Wind," *Popular Science Monthly* 132, no. 5 (May 1938): 55. More on the company's history can be read at Wincharger.com, https://www.wincharger.com/index.php/history/wincharger-history, accessed 29 May 2013. Also see the Wincharger Corporation website, Internet Archive, web.archive.org/web/20150519014158/http://www.windcharger.org/Wind_Charger/Wincharger.html, accessed 11 August 2015. This site noted that the Zenith Radio Corporation ordered fifty thousand units and bought 51 percent of the firm. A recent description of the chargers consists of Leslie C. Mcmanus, "Charged by the Wind," *Farm Collector* magazine, https://www.farmcollector.com/equipment/charged-by-the-wind, accessed 14 July 2021.

28. "Electric Plants Offer Power to Everyone," *Popular Mechanics* 69, no. 6 (June 1938): 850–52.

29. "50c a Year for Farm Electric Power," advertisement in *Farm Journal* 60, no. 12 (December 1936): 62.

30. "Rural Electrification Is Here!," advertisement in *Popular Mechanics* 67, no. 5 (May 1937): 151A. Also see "I've Got Rural Electrification without Waiting for the High Line!," advertisement in *Popular Mechanics* 67, no. 6 (June 1937): 159A.

31. W. R. Bullock, "The Farmer's Use for Electricity," *Isolated Plant* 2, no. 5 (May–June 1910): 24–27. The *Isolated Plant* appeared to be aimed at engineers and others operating equipment for generating electricity in large buildings and factories that did not draw on central station power. This issue, though, focused on farm use of isolated generation units.

32. Editorial, "Let There Be Light," *Isolated Plant* 2, no. 5 (May–June 1910): 1. For more on advantages of the tungsten lights, see James Dixon, "The Small Isolated Plant," *Isolated Plant* 2, no. 5 (May–June 1910): 3.

33. E. E. Whaley, "Electric Light Plants for Farm Use," *Farm Engineering* 1, no. 3 (January 1914): 12. While the low voltage may have made the system safer than using central station–provided 110-volt electricity, it also decreased overall efficiency. "Lighting Country Homes by Private Electric Plants," *Isolated Plant* 2, no. 5 (May–June 1910): 15.

34. "Light Your Place by Electricity," advertisement in *Farm Journal* 32, no. 1 (January 1908): 29.

35. "A Safe Light Wherever You Need It," advertisement, *Farm Journal* 35, no. 11 (November 1911): 581.

36. George W. Kable and R. B. Gray, *Report on C.W.A. National Survey of Rural Electrification* (Washington, DC: USDA, 1934), 55.

37. *Home-Farm Power and Lighting: A Book of Helpful Instruction* (Cincinnati, OH: American Automobile Digest, 1920), 3.

38. R. F. Bucknam, *An Economic Study of Farm Electrification in New York*, Bulletin 496 (Ithaca, NY: Cornell University Agricultural Experiment Station, 1929), 19.

39. Morris Llewellyn Cooke, "National Plan for the Advancement of Rural Electrification under Federal Leadership and Control with State and Local Cooperation and as a

Wholly Public Enterprise," typescript document, undated but likely 1934, from Franklin D. Roosevelt Library, Hyde Park, NY, 2. The term "Delco plants" was used generically in an oral history interview of Roy Hintgen, whose family owned an electric appliance shop in a small Minnesota town. D. Jerome Tweton, *The New Deal at the Grass Roots: Programs for the People in Otter Tail County, Minnesota* (St. Paul: Minnesota Historical Society Press, 1988), 136.

40. Apparently, Kettering was motivated to adapt the small cash register motors for use in automobiles after learning that someone had died while hand-cranking an engine. Arthur Pound, *The Turning Wheel: The Story of General Motors through Twenty-Five Years, 1908–1933* (Garden City, NY: Doubleday, 1934), 272. Also see "Business: All Change!," *Time* 21, no. 2 (9 January 1933), 65–69. Kettering was featured on the cover of this issue.

41. "Distributing Organization History of Delco Light and Frigidaire," typewritten text, in Scharchburg Archives, folder 79-10.3-1. Also see Pound, *The Turning Wheel*, 273–74.

42. Delco-Light, *The Delco-Light Story* (Dayton, OH: Delco-Light, 1922), 7.

43. T. Commerford Martin and Stephen Leidy Coles, *The Story of Electricity*, vol. 2 (New York: M. M. Marcy, 1919), 250–51. A similar story was told in Pound, *The Turning Wheel*, 273. Kettering's sensitivity to farmers' needs was also reported by a colleague at Delco, Louis Ruthenburg, in "Ten Great Years with 'Boss" Kettering," *Ward's AutoWorld* 5 (April 1969), pt. 2, 50.

44. The book by Frederick I. Anderson, *Electricity for the Farm: Light, Heat, and Power by Inexpensive Methods from the Water Wheel or Farm Engine* (New York: Macmillan, 1915), written before the introduction of Delco-Light sets, contained two chapters on using gasoline engines for making electricity and for storing it in batteries.

45. The advantage of the direct drive feature is described in "Farm Lighting Outfits," *Automotive Industries*, 19 September 1918, in clippings file at Scharchburg Archives.

46. Charles F. Kettering and William A. Chryst, Delco-Light, "System of Electrical Generation," patent 1,341,327, filed 27 October 1915, granted 25 May 1920. The benefits of air cooling were described in L. S. Keilholtz, "Development of Farm Electric Light and Power Plants," *Agricultural Engineering* 2, no. 5 (May 1921): 109–10. Of course, the machine did not remain maintenance-free; a 1918 service manual noted that "[e]verything in the nature of machinery requires attention of some kind from the user" and that ignoring service requirements "results in a less efficient system." Domestic Engineering Co., *Information for the User of Delco-Light*, 2nd ed. (Dayton, OH: Domestic Engineering Co., 1918), 2, in Scharchburg Archives, folder 2008.16.04.

47. *The History of Frigidaire*, mimeographed undated document, probably written around 1949, in Scharchburg Archives, folder 79-10.1-37, pt. 2, 2. Also, a typewritten document, "Distributing Organization History of Delco Light and Frigidaire," in Scharchburg Archives, folder 79-10.3-1, listed the incorporation date of the Domestic Engineering Company as 11 February 1916. "First Delco Light Shipment Made by Local Plant," *Dayton Sunday News*, 16 April 1916, in clippings file at Scharchburg Archives.

48. "Delco-Light Brings City Conveniences to the Country," advertisement in *Historic Missouri*, Sixteenth Annual Publication of the American Federation of Catholic Societies (Kansas City, MO: Tingle-Titus Printing Co., 1917), 50.

49. Ibid.

50. "Delco-Light—Electricity for Every Farm," advertisement in *Prairie-Farmer's Reliable Directory of Farmers and Breeders*, Montgomery County, Illinois (Chicago: Prairie Farmer Publishing Co., 1918), 4.

51. Delco-Light, *The Delco-Light Story*, various pages.

52. "Why the Delco-Light Washer Washes Cleaner," in ibid., 77.

53. Martin and Coles, *The Story of Electricity*, 255.

54. Pound, *The Turning Wheel*, 463. Also, *The History of Frigidaire*, pt. 2, 13; and "General Motors Absorbs Delco," *New York Times*, 27 September 1919, 22.

55. *The History of Frigidaire*, pt. 3, 1. A General Motors letter to stockholders in 1925 noted that "Delco-Light Company . . . , a subsidiary of General Motors, is now the world's largest maker of electric refrigerators. That product is Frigidaire, and it leads in its field of *electric* refrigeration just as General Motors cars lead the field of transportation." "Frigidaire, the Electric Refrigerator," in John J. Raskob papers, Hagley Digital Archives, https://digital .hagley.org/m473_20100721_120?solr_nav%5Bid%5D=686461e23c3de3a3d794&solr _nav%5Bpage%5D=0&solr_nav%5Boffset%5D=1#page/1/mode/2up, accessed 26 September 2019.

56. "Delco-Light Company," *Sweet's Architectural Catalogue*, sec. C (New York: Sweet's Catalogue Service, 1927), C2878.

57. Letters from Frank Genske of Wauwatosa, WI, 25 April 1925, and Joseph Holzem of North Milwaukee, WI, 1 May 1925, in Scharchburg Archives, folder 87-11.17-105. The phrases in italics were underlined in the originals.

58. "Marine Hardware and Accessories at the Show," *Rudder* 33, no. 3 (March 1917): 28.

59. "Delco-Light Company," *Sweet's Architectural Catalogue*, C2878.

60. Delco-Light, *The Delco-Light Story*, 16.

61. *Chicago Defender*, 20 October 1917, 210.

62. "Kilmarnock, VA," *Afro-American*, 5 December 1925, 14.

63. In an application that offered portable electric power, the US Forest Service used a Delco-Light unit, secured to a truck, to provide power to an electric auger for drilling holes in dead trees, in which explosives were placed. John F. Krafsic, "Felling Snags with Explosives," *DuPont Magazine* 22, no. 6 (June 1928): 19–20, Hagley Digital Archives, https://digital .hagley.org/1928_22_06#page/1/mode/2up, accessed 26 September 2019.

64. *The History of Frigidaire*, pt. 2, 3 and 5.

65. "Delco Light Salesmen in a Convention," *Dayton Sunday News*, 19 March 1916, in clippings file at Scharchburg Archives.

66. "Program and Directory of the Second Annual Delco-Light Convention," Dayton, OH, 6–9 March 1918, in Scharchburg Archives, folder 79-10.3-4.

67. *Better Delco-Light Selling*, internal publication, vol. 3, no. 2 (1 January 1919), 1, and vol. 3, no. 29 (27 June 1919), frontispiece, in Scharchburg Archives, folder 1979.010.3 -003.

68. *The History of Frigidaire*, pt. 2, 15.

69. "Delco-Light Increases Farm Efficiency," advertisement in *Motor World*, 11 September 1918, 105, in clippings file in Scharchburg Archives.

70. "Delco-Light: Electricity for Anyone Anywhere," advertisement in *Indiana Farmer's Guide*, 4 June 1921, 9.

71. "1/2 kilowatt Delco-Light Plant," advertisement in *Indiana Farmer's Guide*, 19 June 1920, 11.

72. "Back to 1917 Prices," advertisement in *Indiana Farmer's Guide*, 21 October 1922, 1099.

73. GMAC apparently started the financing program in 1920. "Acceptance Plan for Delco-Light," in *Automobile Topics*, trade magazine, 13 March 1920, in clippings file in Scharchburg Archives.

74. These terms applied for land-owning farmers. For tenant farmers, the total price remained the same, but GMAC expected a larger down payment—of $140.50. "Delco-Light Time Payment Plans," 1923, in Scharchburg Archives, folder 79-10.3-2.

75. "Delco-Light Business Larger," *Wall Street Journal*, 7 August 1926, 3.

76. "Delco Light Co. Planning $100,000,000 Expansion Program," *Christian Science Monitor*, 17 June 1926, 10; and "Make Your Farm Pay Bigger Profits with DELCO-LIGHT," advertisement in *Farm Journal* 52, no. 6 (June 1928): 25. This number of isolated plants on farms seems consistent with information provided by the Committee on the Relation of Electricity to Agriculture and conveyed by the American Farm Bureau Federation (AFBF) in 1927. In an annual report, the AFBF reported a huge increase in the number of farms that received central station electricity—up from 1924 by 86 percent to between 300,000 and 350,000 farms. "It is probable," the document noted, "that a number nearly as great are receiving service from individual lighting plants" (though not only from Delco-Light sets). AFBF, "The American Farm Bureau Federation in 1927: Summary of Year's Work as Presented by the Executive Secretary," 22. The exact number of isolated electric light plants, however, is difficult to determine. One needs to be careful about giving too much credence to Delco-Light's proclamations of more than three hundred thousand sold. Some of those plants may have been purchased by nonfarmers or for use outside the United States. But the overall estimate of about three hundred thousand used on farms (from all manufacturers) in the late 1920s is a number seen in several sources. Among them is the figure of 270,303 light plants in use by farmers as of 1 April 1930. W. M. Hurst and L. M. Church, *Power and Machinery in Agriculture*, USDA Misc. Pub. no. 157 (Washington, DC: GPO, 1933), 17.

77. "Mrs. O'Leary's Cow," advertisement in *National Service* 9, no. 1 (January 1921): 41. Of course, at the time of the Chicago fire, no electric lights for home or barn use existed.

78. "Increases Farm Efficiency," advertisement in *Indiana Farmer's Guide*, 2 November 1918, 19.

79. Delco-Light, *The Delco-Light Story*, 44. Similar claims about the economic value of the lighting system are made in the promotional booklet, "Delco-Light 'Pays for Itself,'" undated, but probably around 1919 based on testimonial letters printed in the booklet, in Scharchburg Archives, folder 79-10.3-2. A journalist in 1921 calculated that farm use of electrified machinery yielded savings of between $5 and $12 per week. The financial savings (based on an assumed labor cost of 50 cents per hour) came on top of the "greater comfort and enjoyment, of physical benefits and mental uplift" that electricity made possible. F. J. St. John, "Hire a Helper—2 Cents an Hour," *Farm Mechanics* 5, no. 3 (July 1921): 19.

80. "Shorter Hours, Bigger Profits," advertisement in *Farm Journal* 52, no. 3 (March 1928): 67.

81. "Increase Your Farm Profits with Delco-Light," advertisement in University of Minnesota student yearbook, *Agrarian*, 1935, 104, University of Minnesota Libraries, https://umedia.lib.umn.edu/item/p16022coll338:2474, accessed 14 May 2019.

82. Delco-Light, *The Delco-Light Story*, 31.

83. Ibid., 42. The company explicitly marketed to farm women, as is evidenced by offering a prize in 1919 for the best essay, written by a wife of a salesman, on "what Delco-Light Means to the Woman on the Farm." *Better Delco-Light Selling*, internal publication, vol. 3, no. 2 (16 October 1919), 2, in Scharchburg Archives, folder 1979.010.3-003.

84. Delco-Light, *The Delco-Light Story*, 43.

85. "A Brighter Christmas on the Farm with DELCO-LIGHT," advertisement in *Indiana Farmer's Guide*, 9 December 1922, 1273.

86. Kline, *Consumers in the Country*; Deborah Fitzgerald, *Every Farm a Factory: The Industrial Ideal in American Agriculture* (New Haven, CT: Yale University Press, 2003); and Joshua T. Brinkman and Richard F. Hirsh, "Welcoming Wind Turbines and the PIMBY ('Please in My Backyard') Phenomenon," *Technology and Culture* 58, no. 2 (April 2017): 335–67.

87. Delco-Light, *The Delco-Light Story*, 25, 36, 46, 67, 68.

88. Letter from A. C. Blaser, Mosely, MI, to Delco Light Co., 8 January 1921, in ibid., 45.

89. Letter from J. H. Osurim, Vicksburgh [*sic*], Mississippi, 16 October 1929, in Scharchburg Archives, folder 79-10.2-17.

90. *Better Delco-Light Selling*, internal publication, vol. 3, no. 2 (1 January 1919), 2, in Scharchburg Archives, folder 1979.010.3-003, emphasis in the original.

91. "Delco-Light on the Farm," advertisement in *Greensboro Daily News* 21, no. 90 (14 October 1919), 1.

92. "Electricity for the Farm," *Western Farmer*, 15 June 1917, 3, in clippings file in Scharchburg Archives.

93. "Electricity for Every Farm," advertisement in *Western Farmer*, 15 June 1917, 3, in clippings file in Scharchburg Archives; my emphasis.

94. "Delco-Light Brings City Conveniences to the Country," advertisement in *Historic Missouri*, Sixteenth Annual Publication of the American Federation of Catholic Churches (Kansas City, MO: Tingle-Titus Printing Co., 1917), 50; also see "Delco-Light: Keeps the Young Folks on the Farm," advertisement in *Indiana Farmer's Guide*, 22 June 1918, 21.

95. "Delco-Light Is Carrying the Comforts and Conveniences of the City into Farm Homes," *Indiana Farmer's Guide*, 21 February 1920, 15.

96. The article highlighted the Ohio farm of O. E. Bradfute, who made electricity from an isolated plant. At the time of the article's publication, Mr. Bradfute was vice president of the American Farm Bureau Federation and president of the Ohio Farm Bureau Federation. F. J. St. John, "All Modern Comforts—and More," *Farm Mechanics* 6, no. 4 (February 1922): 137.

97. The number of companies producing isolated generating sets seems to have contracted after Delco-Light entered the market. In 1922, just after the recession ended, the *Farm Light and Power Year Book* listed fifty-nine firms that sold internal combustion–based

power plants. *Farm Light and Power Year Book* (New York: Farm Light and Publishing Co., 1922), 4, 35, 39.

98. Lawrence L. Koontz, "Delco Light Investigation" (MS thesis, Virginia Polytechnic Institute, 1929), 3–6.

99. Ibid., 16.

100. On average, household customers of power from central station plants paid 7.1 cents per kWh in 1927 and 6.9 cents per kWh in 1928, down from 7.7 cents per kWh in 1920. US Department of Commerce, *Statistical Abstract of the United States, 1933* (Washington, DC: GPO, 1933), table 363, "Average Retail Prices of Electricity for Household Use," 325.

101. Koontz, "Delco Light Investigation," 17. A similar logic applied to the fixed costs of a central stations; more consumption meant lower cost per unit of electricity sold.

102. Ibid., 21.

103. E. W. Lehmann and F. C. Kingsley, *Electric Power for the Farm*, Bulletin no. 332 (Urbana: University of Illinois Agricultural Experiment Station, 1929), 472.

104. General Electric Company, *Electricity on the Farm* (Schenectady, NY: General Electric, 1913), Publication no. A4115, 62, Internet Archive, https://archive.org/details /ElectricityOnTheFarm, accessed 26 May 2016.

105. Duffee and Palmer, *Turn On the Light*, 10–12.

106. Ibid.

107. E. E. Brackett and E. B. Lewis, "Unit Electric Plants for Nebraska Farms: A Survey of Present Conditions and a Study of Types of Plants," *Bulletin of the Nebraska Agricultural Experiment Station*, no. 235 (Lincoln: University of Nebraska, College of Agriculture, Experiment Station, May 1929): 16–17.

108. Duffee and Palmer, *Turn On the Light*, 21.

109. Ibid., 63.

110. Brackett and Lewis, "Unit Electric Plants," 28.

111. Lehmann and Kingsley, "Electric Power for the Farm," 473.

112. J. P. Schaenzer, *Rural Electrification* (New York: Bruce Publishing, 1935), 105.

113. Frank J. G. Duck, "When Will the High-Line Reach You . . . If Ever?," *Breeder's Gazette* 103, no. 3 (March 1938): 16–19.

114. NELA reported 249,342 isolated "farm plants" (likely excluding a much smaller, but undetermined, number of water- and wind-generator systems) as of 1 April 1930; 596,014 farms (9.5 percent of the total number of farms) obtained high-line electric service. NELA, *Progress in Rural and Farm Electrification for the 10 Year Period 1921–1931* (New York: NELA, 1932), 8.

115. Emily Hoag Sawtelle, "The Advantages of Farm Life: A Study by Correspondence and Interviews with Eight Thousand Farm Women," unpublished manuscript, March 1924, 10, Internet Archive, https://archive.org/stream/CAT31046460#page/n1/mode /2up, accessed 19 September 2016.

116. Duffee and Palmer, *Turn On the Light*, 23, 26.

117. "Co-operation Needed in Rural Service," *Electrical World* 80, no. 23 (2 December 1922): 1216. In 1933, the former chairman of the New York Public Service Commission reiterated this view, observing that isolated farm plants constituted introductions to the benefits of electricity rather than competitors to high-line service. William A. Prendergast, *Public Utilities and the People* (New York: D. Appleton-Century, 1933), 250.

118. J. C. Martin, "Rural Line Development," *NELA Bulletin* 10, no. 10 (October 1923): 609.

119. "Notes on Progress of Farm Electrification," *NELA Bulletin* 11, no. 1 (January 1924): 24; and R. H. Timmons, "'Dirt Farmers' of Kansas Display Keen Interest in Farm Electrification," *NELA Bulletin* 11, no. 3 (March 1924): 136.

120. G. C. Neff, "Electric Power and the Farmer," *NELA Bulletin* 10, no. 4 (April 1923): 196; and "Committee on Relation of Electricity to Agriculture Organized," NELA *Bulletin* 10, no. 10 (October 1923): 629.

121. Authors of articles on isolated generation plants sometimes advised readers to spend a bit more to wire their homes in anticipation of obtaining high-line power at a later date. Reginald Trautschold, "Wiring the Farm for Electricity," *Farm Mechanics* 6, no. 6 (June 1921): 41.

122. T. W. Norcross, "A New Deal in Rural Electrification: A National Plan," report prepared by the National Power Policy Committee, handwritten date of 6 May 1935, 14–15, at USDA, National Agricultural Library Digital Collections, http://handle.nal .usda.gov/10113/CAT10930569, accessed 8 February 2017.

123. In modern parlance, the grid constitutes the usually interconnected "collection of generators, transmission lines, substations, and related infrastructure" operated to provide electricity across the country. See Julie Cohn, "When the Grid Was the Grid: The History of North America's Brief Coast-to-Coast Interconnected Machine," *Proceedings of the IEEE* 107, no. 1 (January 2019): 232.

124. Data in nominal dollars from USDA, Economic Research Service, Farm and Wealth Statistics, https://data.ers.usda.gov/reports.aspx?ID=17831, accessed 15 October 2019.

125. A list of American Civil War "alternate histories" can be found at Wikipedia, https://en.wikipedia.org/wiki/American_Civil_War_alternate_histories, accessed 15 October 2019.

126. John K. Brown, "Not the Eads Bridge: An Exploration of Counterfactual History of Technology," *Technology and Culture* 55, no. 3 (July 2014): 524.

127. Discussions of contingency, indeterminate moments in history, and counterfactuals can be found in Brown, "Not the Eads Bridge," 521–59, and essays by Eric Schatzberg, "Counterfactual History and the History of Technology," and Lee Vinsel, "The Value of Counterfactual Analysis: Investigating Social and Technological Structure," on the "Technology's Stories" website of the Society for the History of Technology, http://www.technologystories.org/category/counterfactuals, accessed 13 December 2016.

128. Hannah Ritchie and Max Roser, "Access to Energy," November 2019, OurWorldInData.org, https://ourworldindata.org/energy-access, accessed 24 January 2021. Also see Rockefeller Foundation, "One Billion People Don't Have Access to Electricity and This Map Shows You Who," Mashable, https://mashable.com/2017/09/15/one-billion-people -dont-have-access-to-electricity/#kfVeEq_EsOqt, accessed 22 August 2018.

129. Ute Hasenöhrl, "Rural Electrification in the British Empire," *History of Retailing and Consumption* 4, no. 1 (2018): 10–27. Also see Moses Chikowero, "Subalternating Currents: Electrification and Power Politics in Bulawayo, Colonial Zimbabwe, 1894–1939," *Journal of Southern African Studies* 33, no. 2 (June 2007): 287–306, which emphasizes the means by which a local government in colonial Zimbabwe employed electricity to privilege white settlers and to maintain "racial separation" from the majority Black population.

130. The potential benefits of DG are described in G. Pepermans, J. Driesen, D. Hae-seldonckx, R. Belmans, and W. D'haeseleer, "Distributed Generation: Definition, Benefits and Issues," *Energy Policy* 33, no. 6 (April 2005): 787–98.

131. See, for example, Nathaniel J. Williams, Paulina Jaramillo, Jay Taneja, and Taha Selim Ustun, "Enabling Private Sector Investment in Microgrid-Based Rural Electrification in Developing Countries: A Review," *Renewable and Sustainable Energy Reviews* 52 (2015): 1268–81. Many policy problems remain in using the described DG approach, of course, as noted in Subhes C. Bhattacharyya and Debajit Palit, "Mini-Grid Based Off-Grid Electrification to Enhance Electricity Access in Developing Countries: What Policies May Be Required?," *Energy Policy* 94 (July 2016): 166–78.

132. Stakeholders use the terms "microgrids" and "minigrids" without commonly accepted definitions. One definition suggests that a microgrid acts "as a single controllable entity with respect to the grid," but that it can either be connected or remain independent of the grid. Minigrids have been described as small networks independent of a grid, and "potentially coupled with [an] energy storage system." Peter Asmus and Adam Wilson, "Microgrids, Mini-Grids, and Nanogrids: An Emerging Energy Access Solution Ecosystem," Energy Access Practitioner Network, 31 July 2017, http://energyaccess.org/news/recent-news/microgrids-mini-grids-and-nanogrids-an-emerging-energy-access-solution-ecosystem, accessed 5 September 2018.

133. "In Africa, Microgrids Are Changing People's Lives," *Beam* magazine, https://medium.com/thebeammagazine/microgrids-are-building-a-better-future-for-populations-in-remote-areas-46d06b0c9966, accessed 30 August 2018. Also see Marianne Zeyringer, Shonali Pachauri, Erwin Schmid, Johannes Schmidt, Ernst Worrell, and Ulrich B. Morawetz, "Analyzing Grid Extension and Stand-Alone Photovoltaic Systems for Cost-Effective Electrification of Kenya," *Energy for Sustainable Development* 25 (2015): 75–86; and "Africa Microgrids," Microgrids Projects, http://microgridprojects.com/africa-microgrids, accessed 30 August 2018.

134. SOLA Future Energy, "Microgrids in Africa: Rethinking the Centralised Electricity Grid," 9 May 2018, Sola Group, https://www.solafuture.co.za/news/microgrids-africa-rethinking-centralised-electricity-grid, accessed 30 August 2018.

135. Sachiko Graber, Oladiran Adesua, Chibuikem Agbaegbu, Ifeoma Malo, and James Sherwood, "Electrifying the Underserved: Collaborative Business Models for Developing Minigrids under the Grid," Rocky Mountain Institute publication, 2019, https://rmi.org/insight/undergrid-business-models, accessed 3 June 2020; and "Nigeria: How N146m Solar Grid Powers Torankawa Village for 1 Year," *Daily Trust*, 7 January 2020, AllAfrica, https://allafrica.com/stories/202001070010.html, accessed 20 July 2020. The Torankawa plant can operate as part of a connected grid or independent of it. Also see Sachi Graber and Ebun Ayandele, "Reliable and Affordable Electricity for Nigeria: Growing the Minigrid Market," 28 August 2018, Rocky Mountain Institute, https://www.rmi.org/reliable-and-affordable-electricity-for-nigeria-growing-the-minigrid-market, accessed 30 August 2018; and Nigerian Economic Summit Group, *Minigrid Investment Report: Scaling the Nigerian Market* (Abuja: Nigerian Renewable Energy Roundtable, 2018), Rocky Mountain Institute, https://www.rmi.org/wp-content/uploads/2018/08/RMI_Nigeria_Minigrid_Investment_Report_2018.pdf, accessed 30 August 2018.

136. Todd Levin and Valerie M. Thomas, "Can Developing Countries Leapfrog the Centralized Electrification Paradigm?," *Energy for Sustainable Development* 31 (2016): 98. A 2017 study of the twenty "high-impact" countries in sub-Saharan Africa and Asia indicated that only about 1 percent of financing commitments—from 2013–14, a total of about $20 billion—went for decentralized energy; about 50 percent went for grid-connected renewables, 18 percent for transmission and distribution, 20 percent for grid-connected fossil fuel power, and 8 percent for market support. Sustainable Energy for All, *Energizing Finance: Scaling and Refining Finance in Countries with Large Energy Access Gaps* (Washington, DC: Sustainable Energy for All, 2017), 23, https://www.seforall.org/sites/default/files/2017_SEforALL_FR4_PolicyPaper.pdf, accessed 30 August 2018. Also see Sustainable Energy for All, *Missing the Mark: Gaps and Lags in Disbursement of Development Finance for Energy Access* (Washington, DC: Sustainable Energy for All, 2017), https://www.seforall.org/sites/default/files/2017_SEforALL_FR1.pdf, accessed 30 August 2018.

137. These systems are not necessarily cheap, though, requiring support from government, "impact investors," foundations, or nongovernmental organizations. While grid-served customers in Nigeria pay about $0.10 per kWh, minigrids offer power from about $0.25 to $0.70 per kWh. Diesel self-generation, though, costs between $0.50 and $1.50 per kWh. Sachi Graber and James Sherwood, "Using Undergrid Minigrids to Drive Development in Thousands of Sub-Saharan African Communities," 24 August 2018, Rocky Mountain Institute, https://www.rmi.org/using-undergrid-minigrids-to-drive-development-in-thousands-of-sub-saharan-african-communities, accessed 30 August 2018.

CHAPTER SIX: **Utility Interest in Rural Electrification Awakens**

Epigraph: Wigginton E. Creed, "Some Economic and Social Aspects of Interconnection," in NELA, *Proceedings* 83 (1926): 142.

1. Gavin Wright, *The Political Economy of the Cotton South* (New York: W. W. Norton, 1978).

2. US Census figures show that in 1910, about 54 percent of Americans lived in rural areas (but not necessarily on farms). That percentage dropped to about 49 percent in 1920. US Department of Commerce, *Statistical Abstract of the United States, 1930* (Washington, DC: GPO, 1930), table 39, "Urban and Rural Population, by States," 46.

3. An Act to Establish a Department of Agriculture, 12 Stat. 387, chap. 72, 15 May 1862, 7 USC 2201.

4. An Act Donating Public Lands to the Several States and Territories Which May Provide Colleges for the Benefit of Agriculture and Mechanic Arts, 12 Stat. 503, chap. 130, 2 July 1862, 7 USC 301.

5. An Act to Establish Agricultural Experiment Stations in Connection with the Colleges Established in the Several States . . . , 24 Stat. 440, chap. 314, 2 March 1887, 7 USC 361a; and An Act to Provide for Cooperative Agricultural Extension Work . . . , 38 Stat. 372, chap. 79, 8 May 1914, 7 USC 341.

6. "Electricity in Agriculture," *Pacific Rural Press* 43, no. 19 (7 May 1892): 434.

7. "Report Amending and Favoring S. 1170, to Establish Electrical Experiment Station for Farming," 20 March 1894 (Senate Rep. 271, 2nd sess., vol. 1), as noted in US

Superintendent of Documents, *Catalogue of the Public Documents of the Fifty-Third Congress and of All Departments of the Government of the United States* (Washington, DC: GPO, 1896), 440. US Senate, "Report to Accompany S. 1170," 53rd Cong., 2nd sess., Report no. 271, 20 March 1894. For more on Senator Peffer's efforts, see Clarke C. Spence, "Early Uses of Electricity in American Agriculture," *Technology and Culture* 2, no. 2 (Spring 1962): 145–46.

8. Liberty Hyde Bailey, *Report of the Commission on Country Life* (New York: Sturgis and Walton, 1911), 68–69.

9. Ibid., 135.

10. "A Bureau of Farm Power," *NELA Bulletin* 6, no. 3 (November 1912): 126; and "Proposed United States Bureau of Farm Power," *Electrical World* 60, no. 10 (7 September 1912): 488. The bill's text was published, with commentary and a list of supporting organizations, in "Bureau of Farm Power," *Gas Engine* 14, no. 9 (September 1912): 461, 478.

11. "Farmers' Gas-Engines: Questions Answered by H. R. Brate," *Farm Journal* 37, no. 2 (February 1913): 99.

12. "The Convention of the American Society of Agricultural Engineers," *Science* 37, n.s., no. 945 (7 February 1913): 236.

13. "A Bureau of Farm Power," *Scientific American* 107, no. 6 (10 August 1912): 110.

14. T. C. Martin, "Report of the Committee on Progress," NELA, *Proceedings* 46 (1913): 105.

15. "U.S. Government to Solve Farm Power Problems," *Flaming Sword* 27, no. 1 (January 1913): 30–32.

16. The bill is listed as HR 25782 in 1912 (48 Cong. Rec. 9161, 1912), HR 3989 in 1913 (50 Cong. Rec. 285, 1913), HR 13766 (51 Cong. Rec. 121, 1914), and HR 646 in 1916 (53 Cong. Rec. 122, 1916).

17. USDA, *Domestic Needs of Farm Women*, Report no. 104 (Washington, DC: GPO, 1915), 4, 30, 31–32.

18. David E. Nye, *Electrifying America: Social Meanings of a New Technology, 1880–1940* (Cambridge, MA: MIT Press, 1992), 118.

19. F. Houghton, "Electricity on the Farm," *Ohio Farmer*, 4 October 1894, 261.

20. "Farm and Garden: Electricity on the Farm," *Watchman*, 11 February 1897, 30.

21. "Topics in Season," *Farm Journal* 31, no. 3 (March 1907): 147.

22. "The Railroad and the Farmer," *Air Line News* 1 (August 1907): 8–9, cited in H. Roger Grant, *Electric Interurbans and the American People* (Bloomington: Indiana University Press, 2016), 62. The "Air Line" was a proposed high-speed railway line between Chicago and New York.

23. In 1914, Herman Russell was the assistant general manager of the power company. According to Jack Sears, *Club Men of Rochester in Caricature* (Rochester, NY: Roycrofters, 1914), 136, he had also won the city's tennis championship for three years. "Men in the Industry," *Gas Age* 43 (1 April 1919): 374.

24. Herman Russell, "Electricity on the Farm and the Influence of Irrigation from a Central-Station Standpoint," NELA, *Proceedings* 39 (1910): 196–204; and Herman Russell, "Central-Station Power for Farming and Irrigation," *Electrical World* 55, no. 22 (2 June 1910): 1465.

25. "The Farm of Edison Light and Power," *Electrical Review and Western Electrician* 61, no. 5 (3 August 1912): 206–8; and Leavitt Edgar, "Commercial Methods at Boston," *Electrical World* 62, no. 18 (1 November 1913): 903.

26. Leavitt L. Edgar, "Modern Merchandizing," Association of Edison Illuminating Companies, *Minutes of the Twenty-Ninth Annual Meeting (Thirty-Fourth Convention) of the Association of Edison Illuminating Companies, 9–12 September, 1913* (New York, AEIC, 1913), 149–54.

27. Ibid., 150; and C. H. Miles, "The Development and Application of Electricity to Agriculture," *Proceedings of the New England Section of the NELA* 4 (1912): 100–102.

28. Association of Edison Illuminating Companies, *Minutes*, 151–53.

29. Miles, "The Development and Application of Electricity to Agriculture," 100.

30. L. L. Elden, "Load Factor and Power Factor—How to Improve Them," *Proceedings of the New England Section of the NELA* 3 (1911): 151.

31. Forrest McDonald, *Insull: The Rise and Fall of a Billionaire Utility Tycoon* (Chicago: University of Chicago Press, 1962), 137–42. Also see Samuel Insull, "The Production and Distribution of Energy," in William E. Keily, ed., *Central-Station Electric Service: Its Commercial Development and Economic Significance as Set Forth in the Public Addresses (1897–1914) of Samuel Insull* (Chicago: privately printed, 1915), 358–77.

32. Forrest McDonald, *Let There Be Light: The Electric Utility Industry in Wisconsin, 1881–1955* (Madison, WI: American History Research Center, 1957), 14.

33. Ibid., 146. The utilities were managed by Insull's holding company, North West Utilities Company, incorporated in 1918, which was in turn controlled by Middle West Utilities Company. John Moody, *Moody's Analyses of Investments, Part III: Public Utility Investments* (New York: Moody's Investors Service, 1920), 944.

34. McDonald, *Let There Be Light*, 147. Neff is listed (albeit incorrectly as Grover W. Neff—incorrect middle initial) as the manager of the Southern Wisconsin Power Company in a 1914 business directory. R. L. Polk & Co., *Wisconsin State Gazetteer and Business Directory, 1913–1914* (Chicago: R. L. Polk & Co., 1914), 419. In 1916, he is described as the general superintendent of the Wisconsin River Power Company and Southern Wisconsin Power Company in Prairie du Sac. *Proceedings of the American Institute of Electrical Engineers* 35, no. 6 (June 1916): 154. Early biographical information about his family comes from L. W. Royse, ed., *A Standard History of Kosciusko County, Indiana* (Chicago: Lewis Publishing, 1919), 470–71.

35. McDonald, *Let There Be Light*, 147–48, 148n37.

36. Philip J. Funigiello, *Toward a National Power Policy* (Pittsburgh, PA: University of Pittsburgh Press, 1973), 33n5.

37. "290,000 Tons of Coal Saved in Single Year; 2 Plants in Operation," *Wisconsin State Journal*, 16 May 1920, news clipping from Wisconsin Historical Society online, https://www.wisconsinhistory.org/Records/Newspaper/BA4134, accessed 12 July 2020.

38. "Middle West Utilities," *United States Investor* 33, no. 1 (24 June 1922): 1287.

39. Don Sterns, "Consolidations Increase in Prairie States," *Electrical World* 91, no. 21 (26 May 1928): 1104.

40. REA's third administrator, Harry Slattery, described Neff as a one of a few "[p]rogressive utility leaders" and one who warned his utility colleagues in 1923 to stop neglecting farm

electrification. Harry Slattery, *Rural America Lights Up* (Washington, DC: National Home Library Foundation, 1940), 9. In a 1948 retrospective piece, the REA's first administrator, Morris Cooke, described Neff as among the "more progressive executives" of the utility industry. Morris Llewellyn Cooke, "The Early Days of the Rural Electrification Idea: 1914–1936," *American Political Science Review* 42, no. 3 (June 1948): 439.

41. The California network in 1910 consisted of 1,920 miles of transmission lines of which 150 miles operated at 100,000 volts. Lines supplying power at 11,000 volts or less were considered distribution lines. Paul M. Downing, "The Developed High-Tension Net-Work of a General Power System," *Journal of Electricity, Power and Gas* 24, no. 19 (7 May 1910): 429. Also see John Martin, "How Old Yuba Came to Be Built," *Pacific Service Magazine* 9, no. 1 (June 1917): 45–46.

42. A description of electrification in Visalia, viewed as "typical of a modern intensive farming district," can be found in Putnam A. Bates, "Electricity on the Farm," *Transactions of the American Institute of Electrical Engineers* 19, no. 2 (1912): 1987–89. An excellent review of early rural electrification in the Golden State consists of James C. Williams, *Energy and the Making of Modern California* (Akron, OH: University of Akron Press, 1997), 218–36.

43. David E. Nye, "Electrifying the West, 1880–1940," in Rob Kroes, ed., *The American West as Seen by Europeans and Americans* (Amsterdam: Free University Press, 1989), 193.

44. W. W. Wheeler, "On the Cost of Irrigation by Electrically Driven Pumps from Transmission Services," *Electrical Review* 47, no. 13 (23 September 1905): 465–66; H. Day Hanford, "The Use of Electric Power in Pumping Water for Irrigation," *Journal of Electricity, Power, and Gas* 34, no. 20 (15 May 1915): 392; and Bernard A. Etcheverry, *Irrigation Practice and Engineering*, vol. 1, *Use of Irrigation Water and Irrigation Practice* (New York: McGraw Hill, 1915).

45. The 1908 study was reported in O. L. Waller, *Cost of Pumping for Irrigation*, Bulletin no. 103 (Pullman: State College of Washington Extension Service, 1923), 13–14.

46. "Electric Irrigation Plants," *Journal of Electricity, Power and Gas* 28, no. 9 (2 March 1912): 209.

47. "40 Acres of Irrigation," General Electric advertisement in *Pacific Service Magazine* 9, no. 1 (June 1917): i. On the following page, the Allis-Chalmers Manufacturing Company advertised its high-efficiency centrifugal pumps for irrigation.

48. O. L. Waller, "Cost of Pumping for Irrigation," *Journal of Electricity, Power and Gas* 37, no. 18 (28 October 1916): 340; S. B. Shaw, "Rainfall and Agricultural Power Use," *Journal of Electricity, Power and Gas* 37, no. 13 (23 September 1916): 242–43; and "Building the Agricultural Load," *Journal of Electricity* 45, no. 10 (15 November 1920): 459.

49. James C. Williams, "Otherwise a Mere Clod: California Rural Electrification," *IEEE Technology and Society Magazine* 7 (December 1988): 13–16.

50. G. R. Kenny, "Unusual Features of an Irrigation Pumping Load," *Journal of Electricity* 38, no. 11 (1 June 1917): 448–49.

51. The railroad car, which was received "with a bang amidst red fire and rockets," is described in E. A. Weymouth, "Our Travelling Electrical Exhibit," *Pacific Gas and Electric Magazine* 3, no. 12 (May 1912): 443–46. Also see Frederick S. Myrtle, "How 'Pacific Service' Blazed the Trail," *Pacific Service Magazine* 16, no. 6 (October 1925): 174–93; and "Building the Agricultural Load," 458.

52. Weymouth, "Our Traveling Electrical Exhibit," 445.

53. Myrtle, "How 'Pacific Service' Blazed the Trail," 175.

54. "Electric Irrigation Plants," 209.

55. "Save Labor on Your Farm This Winter," advertisement in *Pacific Rural Press* 89 (27 February 1915): 272.

56. "Building the Agricultural Load," 458, 460.

57. Williams, "Otherwise a Mere Clod," 14.

58. "Electrifying the West's Most Important Industry," *Journal of Electricity* 45, no. 10 (15 November 1920): 462–63.

59. NELA, *Progress in Rural and Farm Electrification for the 10 Year Period 1921–1931* (New York: NELA, 1932), 5.

60. US Department of Commerce, Bureau of the Census, *Central Electric Light and Power Stations and Street and Electric Railways* (Washington, DC: GPO, 1915), 155.

61. B. D. Moses, *Electrical Statistics for California Farms*, Circular 316 (Berkeley: University of California, College of Agriculture, Agricultural Experiment Station, 1929), 7, Internet Archive, https://archive.org/details/electricalstatis316mose, accessed 10 July 2021.

62. "Statistics of the Electric Light and Power Industry, 1929," NELA, *Proceedings* 87 (1930): 1428.

63. NELA, *Progress in Rural and Farm Electrification*, 5.

64. B. D. Moses, "Rural Electrification in California," no. 437 "of a series of talks dealing with practical applications of electricity in farming, presented under the auspices of the Rural Electrification Section of the General Electric Company, and broadcast by WGY (790 kc.) and W2XAF (9530 kc), Schenectady, N.Y." in UCD Special Collections, B. D. Moses Collection, box 1 of 1, call number D-181, scrapbook #2, folder 3.

65. Edward J. Crosby, *The Story of the Washington Water Power Company and Its Part in the History of Electric Service in the Inland Empire, 1889–1930 Inclusive* (no publication data listed), 20, in WSU Special Collections, folder 19, cage 207, Washington Water Power collection.

66. James E. Davidson, vice president of Pacific Power and Light Company, in "The Tireless Farmer: Results from Ruralizing Electricity," Northwest Electric Light and Power Association, *Fifth Annual Convention, Papers, Reports and Discussions* (1913): 188.

67. Comment by M. C. Osborn, Washington Water Power Company, in ibid., 189.

68. L. J. Smith, "A Report on the State Wide Survey of Rural Electric Lines in Representative Parts of the State of Washington," no page number, submitted October 1925 to the Washington CREA, in WSU Special Collections, cage 607, Washington Farm Electrification Committee Records.

69. NELA, *Progress in Rural and Farm Electrification*, 5.

70. Ibid.

71. Ibid., 5, 7.

72. Williams, "Otherwise a Mere Clod," 18.

73. Another exception existed in some New England and mid-Atlantic states, where a large percentage of farms obtained electrical service by the early 1930s. (Rhode Island, Connecticut, and Massachusetts saw more than 50 percent of their farms electrified, for example.) The high proportion of these electrified farms resulted from their proximity to

urban (and industrial) communities, such that power companies could readily connect them with wires coming from existing transmission and intercity distribution networks. Even so, these eastern farms did not use as much power as their brethren in California: Massachusetts farms drew an average of 1,012 kWh per year (in 1930), more than those in any nearby state. But that amount still paled in comparison to the 14,677 kWh consumed by the typical California farm in the same year. US Federal Power Commission, *National Power Survey*, interim report no. 1 (Washington, DC: GPO, 1935), 53; and George W. Kable and R. B. Gray, *Report on C.W.A. National Survey of Rural Electrification* (Washington, DC: USDA, 1934), 52–53.

74. Farms in irrigation states consumed 1.513 billion kWh in 1931; those in non-irrigation states used 365 million kWh. NELA, *Progress in Rural and Farm Electrification*, 7.

75. Federal Power Commission, *Rural Electric Service*, Rate Series no. 8 (Washington, DC: GPO, 1936), 17.

CHAPTER SEVEN: **The Unexpected Public Relations Value of Rural Electrification**

Epigraph: M. T. D. Crocker, in discussion printed following "Report of the Rural Lines Committee," in NELA, *Proceedings* 77 (1922): 129.

1. A. Cupler, "Rural Electrification in the United States," *Editorial Research Reports*, vol. 3 (Washington, DC: CQ Press, 1926), http://library.cqpress.com/cqresearcher/cqres rre19260834000, accessed 17 July 2017.

2. Abby Spinak, "'Not Quite So Freely as Air': Electrical Statecraft in North America," *Technology and Culture* 61, no. 1 (January 2020): 71–108.

3. A list of NELA members is contained in NELA, *Proceedings* 1 (1885): 242.

4. "Report of the Committee on Revision of Constitution," NELA, *Proceedings* 8 (1890): 282.

5. "Constitution of the National Electric Light Association," adopted 1 June 1921, NELA, *Proceedings* 76 (1921): 1320–26.

6. In 1907, one Ohio manager reported that his firm had started experimenting with rural service, "and the results are very encouraging." F. H. Plaice, "Establishing Day Circuits in Towns of 10,000 Population and Under," NELA, *Proceedings* 31 (1907): 180.

7. The committee's origin is described in Royden Stewart, "Rural Electrification in the United States: Part I—The Pioneer Period, 1906–1923," *Edison Electric Institute Bulletin* (September 1941): 382; and John G. Learned, "Report of Committee on Electricity in Rural Districts," NELA, *Proceedings* 40 (1911): 448. The early work of the committee is described in "Promoting the Use of Electricity in Farming Districts," *Electrical World* 57, no. 14 (1911): 820–21.

8. Learned, "Report of Committee," 448, 520.

9. Ibid., 457.

10. Ibid., 519.

11. Ibid., 521.

12. Comment by J. C. Parker, chairman of the Eastern States report, "Third Commercial Session," NELA, *Proceedings* 48 (1913): 136.

13. J. C. Parker, "Report of the Committee on Electricity on the Farm: Eastern States," in ibid., 145.

14. C. W. PenDell, "Report of the Committee on Electricity on the Farm: Central States," in ibid., 226.

15. Ibid., 188.

16. Ibid., 164.

17. "Handling Rural Business in Ohio," *Electrical World* 76, no. 1 (3 July 1920): 29–30.

18. "Demand for Rural Service Based on Economic Reasons," *Electrical World* 78, no. 17 (23 October 1920): 817–19.

19. "Rural Service a New Central-Station Problem," *Electrical World* 76, no. 17 (23 October 1920): 815.

20. M. O. DellPlain, following presentation of C. W. PenDell, "Report of the Committee on Electricity on the Farm: Central States," 290.

21. "Farm Line Construction," NELA, *Proceedings* 76 (1921): 732.

22. Between 1910 and 1930, the urban population in the United States grew about 64 percent while the rural population increased 8 percent. Data calculated from US Census Bureau, "Table 1. Urban and Rural Population: 1900 to 1990," released October 1995, https://www.census.gov/population/censusdata/urpop0090.txt, accessed 29 July 2019; and James H. Shideler, "*Flappers and Philosophers*, and Farmers: Rural-Urban Tensions of the Twenties," *Agricultural History* 47, no. 4 (October 1973): 283–99.

23. "Farm Line Construction," 732.

24. G. C. Neff, in discussion following "Report of Overhead Systems Committee," NELA, *Proceedings* 76 (1921): 851.

25. Ibid., 851–52.

26. "Business Revival Dominates the N.E.L.A. Convention: Fourth General and Executive Session," *Electrical World* 79, no. 20 (20 May 1922): 999.

27. "Electric Service and Farm Productiveness," *Electrical World* 78, no. 22 (26 November 1921): 1060.

28. "Work Being Done by the Rural Lines Committee of the N.E.L.A.," *NELA Bulletin* 9, no. 4 (April 1922): 224.

29. G. C. Neff, "The Prospect of Business from the Farm," *Electrical World* 72, no. 1 (7 January 1922): 27.

30. Ibid.

31. Ibid.

32. G. C. Neff, in discussion following "Report of the Rural Lines Committee," NELA, *Proceedings* 77 (1922): 128.

33. Ibid.

34. Ibid.

35. J. C. Martin, in discussion following "Report of the Rural Lines Committee," 128.

36. Comments by M. T. D. Crocker, in ibid., 129.

37. For a primer on railroad regulation, see Thomas K. McCraw, *Prophets of Regulation* (Cambridge, MA: Harvard University Press, 1984), 1–79.

38. Richard F. Hirsh, *Power Loss: The Origins of Deregulation and Restructuring in the American Electric Utility System* (Cambridge, MA: MIT Press, 1999), 14–31.

39. M. N. Baker, ed., *The Municipal Year Book* (New York: Engineering News Publishing Co., 1902), xxxv, 165, 201, 215. This source notes that Chicago had both public- and company-provided electricity, with the city offering street lighting services only.

40. Leonard S. Hyman, *America's Electric Utilities: Past, Present, and Future* (Arlington, VA: Public Utilities Reports, 1983), 71.

41. David W. Wilma and Walt Crowley, *Power for the People: A History of Seattle City Light* (Seattle: University of Washington Press, 2010). The California city's entrance into the power business is described in Vincent Ostrom, *Water and Politics: A Study of Water Policies and Administration in the Development of Los Angeles* (Los Angeles: Haynes Foundation, 1953), 59–61. Also see US Bureau of the Census, *Census of Electrical Industries 1922, Central Electric Light and Power Stations* (Washington, DC: GPO, 1925), 8–11.

42. "Study Fuel Economy in New York State," *Electrical World* 70, no. 26 (29 December 1917): 1260, which discusses the use of interconnection of companies in New York State to gain efficiencies. Also see "Electrical Interconnection to Conserve Fuel," *Electrical World* 71, no. 1 (5 January 1918): 12–15. For a recent history of transmission and interconnection, see Julie A. Cohn, *The Grid: Biography of an American Technology* (Cambridge, MA: MIT Press, 2017).

43. Thomas P. Hughes, *Networks of Power: Electrification in Western Society, 1880–1930* (Baltimore, MD: Johns Hopkins University Press, 1983), 393–403, described the origin and value of holding companies. A less scholarly and more polemical account of the Electric Bond and Share Company can be found in J. F. Christy, *The Power Trust vs. Municipal Ownership* (Chicago: Public Ownership League of America, 1929).

44. Hirsh, *Power Loss*, 43–46, described the declining expertise of regulatory bodies and their "capture" by utility companies.

45. US Bureau of the Census, *Census of Electrical Industries 1922*, 8–11. In 1927, the number of municipal power organizations dropped to 2,198; five years later, the number totaled 1,802. US Bureau of the Census, *Census of Electrical Industries, 1932: Central Electric Light and Power Stations* (Washington, DC: GPO, 1934), 4.

46. The public power advocates promoted their views in newspaper editorials, pamphlets, books, and conferences. See, for example, *Annals of the American Academy of Political and Social Science* 57, no. 1 (January 1915), which contained presentations made at the 1914 Conference of American Mayors. Topics included the value of public and private ownership of utilities and the benefits of state regulation and holding companies. Also see Carl D. Thompson, *Municipal Electric Light and Power Plants in the United States and Canada* (Chicago: Public Ownership League of America, 1922); and Delos. F. Wilcox, *The Indeterminate Permit in Relation to Home Rule and Public Ownership: A Report* (Chicago: Public Ownership League of America, 1926).

47. Passage of the Federal Power Act of 1920 gave authority to a newly created Federal Power Commission to oversee hydroelectric power development on navigable waterways. Congressional Research Services, "The Federal Power Act (FPA) and Electricity Markets," CRS Report no. R44783, 10 March 2017, https://www.everycrsreport.com/reports/R44783 .html, accessed 6 August 2021. Debate over government ownership of facilities continued thereafter. During the 1924 presidential campaign, for example, Robert La Follette, senator from Wisconsin, tried to insert a provision into the Republican platform calling for such ownership. "La Follette Looms over Republicans' Hopes as Specter," *Washington Post*, 9 June 1924, 3; and George T. Odell, "Public Ownership of Utilities Is New to Wisconsin Platform," *Christian Science Monitor*, 9 June 1924, 4. Commerce Secretary Herbert Hoover discredited the proposal as a form of European-style "socialism," which would "increase our cost

of service, decrease our national efficiency, undermine our democracy, [and] destroy the fundamentals upon which our nation has become great." "Socialism Unmasked," *Los Angeles Times*, 20 September 1924, 1. The secretary had earlier spoken out against government domination of "production and distribution of commodities and services" in Herbert Hoover, *American Individualism* (Garden City, NY: Doubleday, Page & Co., 1922), 54–55. For a thorough political history of the public power fight, see Jay L. Brigham, *Empowering the West: Electrical Politics before FDR* (Lawrence: University Press of Kansas, 1998).

48. Richard Lowitt, *George W. Norris: The Persistence of a Progressive, 1913–1933* (Urbana: University of Illinois Press, 1971), 22–30. For more on the controversy, see Robert W. Righter, *The Battle over Hetch Hetchy: America's Most Controversial Dam and the Birth of Modern Environmentalism* (New York: Oxford University Press, 2005), 131–32; and George W. Norris, *Fighting Liberal: The Autobiography of George W. Norris* (New York: Macmillan, 1945), 162.

49. "Ford Shoals Bid Dies in Senate—until December," *Chicago Daily Tribune*, 5 June 1924, 18; "Ford's Shoals Bid Is Withdrawn," *Washington Post*, 14 October 1924, 1; "Chronology of Muscle Shoals 1824–1929," *Congressional Digest* 9, no. 5 (May 1930): 130–32; and Preston J. Hubbard, *Origins of the TVA: The Muscle Shoals Controversy, 1920–1932* (Nashville, TN: Vanderbilt University Press, 1961).

50. Norman Wengert, "Antecedents of TVA: The Legislative History of Muscle Shoals," *Agricultural History* 26, no. 4 (October 1952): 141–47.

51. G. Cullom Davis, "The Transformation of the Federal Trade Commission, 1914–1929," *Mississippi Valley Historical Review* 49, no. 3 (1962): 443.

52. "The 'Power Trust,'" *New York Times*, 11 February 1925, 20; "Shoals Deadlock Hits Supply Bills . . . Norris Describes 'Trust,'" *New York Times*, 20 December 1924, 3; and Arthur Sears Henning, "'Power Trust' Perils Nation, Norris Warns," *Chicago Daily Tribune*, 1 March 1931, 1. Also see Ronald C. Tobey, *Technology as Freedom: The New Deal and the Electrical Modernization of the American Home* (Berkeley: University of California Press, 1996), 40–48.

53. For a discussion of the origin of the FTC investigation, see Lowitt, *George W. Norris*, 360–64.

54. Ibid., 31; and Samuel P. Hays, *Conservation and the Gospel of Efficiency: The Progressive Conservation Movement, 1890–1920* (Pittsburgh, PA: University of Pittsburgh Press, 1999), 180–82.

55. Lowitt, *George W. Norris*, 55.

56. Gifford Pinchot, *The Power Monopoly: Its Make-Up and Its Menace* (Milford, PA: n.p., 1928).

57. Thomas Parke Hughes, "Technology and Public Policy: The Failure of Giant Power," *Proceedings of the IEEE* 64, no. 9 (September 1976): 1361–71.

58. An early biography of Cooke is Kenneth E. Trombley, *The Life and Times of a Happy Liberal: A Biography of Morris Llewellyn Cooke* (New York: Harper, 1954). Among several biographies of Pinchot is Char Miller, *Gifford Pinchot and the Making of Modern Environmentalism* (Washington, DC: Island Press, 2001).

59. "Copy of letter from former governor Gifford Pinchot, of Pennsylvania, dated March 15, to Governors of All States," 15 March 1927, in US Senate, 70th Cong., 1st sess., doc. 92, pt. 4, *Utility Corporations* (Washington, DC: GPO, 1930), exhibit 1957, 417.

60. "Governor's Message of Transmittal," in Morris L. Cooke and Judson C. Dicker-man, *Report of the Giant Power Survey Board to the General Assembly of the Commonwealth of Pennsylvania* (Harrisburg, PA: Telegraph Printing Co., 1925), xiii.

61. Of the ten-point summary of Giant Power, "Electric service for the rural popula-tion" stood at number seven. Ibid., 14.

62. The report notes that private utilities did not attack the rural market because they were busy "meeting the very great and urgent demands for power in the large industrial centers" and because the high rates charged to farmers (if they could obtain central station service) meant ruralites could not afford to consume much electricity. Ibid., 37.

63. Harold Evans, "The World's Experience with Rural Electrification," *Annals of the American Academy of Political and Social Science* 118 (March 1925): 42.

64. US Bureau of the Census, *Central Electric Light and Power Stations and Street and Elec-tric Railways with Summary of Electrical Industries, 1912* (Washington, DC: GPO, 1915), 155.

65. The Petersburg farmers appear to have created an organization that simply made ad hoc arrangements with a local power company rather than incorporating into a formal entity, such as a co-op, according to the US census report description. Ibid., 154–55.

66. Udo Rall, US Federal Emergency Relief Administration, "A Study of Cooperative Consumer Associations for Rural Electrification," typescript report R.E. 34845, 16 May 1935, 6–7, HathiTrust Digital Library, https://hdl.handle.net/2027/uiug.30112069450564, accessed 11 July 2021.

67. "Cooperative Associations for Supply of Electric Current," *Monthly Labor Review* 46, no. 1 (January 1938): 110.

68. Rall, "A Study of Cooperative Consumer Associations for Rural Electrification," 5.

69. Ibid., 6. Examples of failures of co-ops are listed in ibid., 5–6.

70. Ibid., 11.

71. Carl D. Thompson, *Confessions of the Power Trust* (New York: E. P. Dutton, 1932), 623; and Hughes, "Technology and Public Policy," 1370–71.

72. Giant Power planners claimed not to advocate public ownership, despite Evans's statement, except as a last resort. Gifford Pinchot, "Giant Power," NELA, *Proceedings* 82 (1924): 46–50.

73. "Pinchot Takes Radical Stand," *Electrical World* 85, no. 8 (21 February 1925): 421.

74. "Pinchot Swallows the Wooden Minnow of Public Ownership," in US Senate, 70th Cong., 1st sess., doc. 92, pts. 5 and 6, *Utility Corporations* (Washington, DC: GPO, 1930), exhibit 2982, 367–73.

75. "Superpower versus Giant Power," address of J. J. Buell, delivered 12 Novem-ber 1925, in ibid., exhibit 1957, 80–85. Other defenses of the utility industry (and attacks of Giant Power) can be seen in "A Disappointing Report," *Electrical World* 85, no. 9 (28 February 1925): 443–44; and "What 'Giant Power' Has of Good Will Survive the Bill's Defeat," *Electrical World* 87, no. 8 (20 February 1926): 389.

76. See, for example, Robert W. Bruère, "Giant Power—Region-Builder," *Survey Graphic* 7 (May 1925): 161–64, 188, reprinted in Carl Sussman, ed., *Planning the Fourth Migration: The Neglected Vision of the Regional Planning Association of America* (Cam-bridge, MA: MIT Press, 1976), 111–20.

77. Keith Fleming, *Power at Cost: Ontario Hydro and Rural Electrification, 1911–1958* (Montreal: McGill-Queen's University Press, 1992), 27–30.

78. George W. Kable and R. B. Gray, *Report on C.W.A. National Survey of Rural Electrification* (Washington, DC: USDA, 1934), 59. For a more modern treatment of Ontario's experiences with rural electrification, see Fleming, *Power at Cost*. The Ontario Hydro efforts likely contributed to the American public power movement, but I could not find evidence that it directly pushed NELA into creating the Rural Lines Committee. When searching for stories in major American newspapers about Ontario's efforts in the years after 1920, I found few. And while Ontario Hydro clearly sought to increase the number of farmers who received electricity, its efforts often were criticized by Canadian newspapers and politicians. Of course, the perception of Ontario's successes (rather than the reality) may have influenced NELA leaders to pursue rural electrification more emphatically.

79. US Senate, 70th Cong., 1st sess., doc. 92, pt. 1, *Utility Corporations* (Washington, DC: GPO, 1930), exhibit 38, 365. Also see William S. Murray, *Electric Utilities in Canada and the United States: Government Owned and Controlled Compared with Privately Owned and Regulated* (New York: NELA, 1922), 2. Elsewhere, Murray opined that government ownership of utilities "would multiply the scandals connected with the distribution of patronage similar to the illogical development of useless waterways and of ill-considered irrigation projects. Political expediency would dictate policies," he continued, "and local and bloc demands would greatly enlarge the dimensions of the notorious Congressional pork barrel." William Spencer Murray, *Superpower: Its Genesis and Future* (New York: McGraw-Hill, 1925), 205.

80. NELA, *Political Ownership and the Electric Light and Power Industry* (New York: NELA, 1925), 8–10. Among other sources, the document cites a highly publicized document, produced under the auspices of the Smithsonian Institution but written by a utility consultant, which also challenged the claims made by supporters of the Ontario Hydro system. Samuel S. Wyer, *Niagara Falls: Its Power Possibilities and Preservation*, Pub. 2820 (Washington, DC: Smithsonian Institution, 1925).

81. Some of the billed items and payments are listed in US Senate, 70th Cong., 1st sess., doc. 92, pt. 4, *Utility Corporations* (Washington, DC: GPO, 1930), exhibit 1749, 232 and exhibit 1753, 237. Stewart's work was described as part of the utility industry's propaganda efforts in a 1928 report, CQ Press, "Public Utilities' Propaganda in the Schools," *Editorial Research Reports 1928*, vol. 2 (Washington, DC: CQ Press, 1928), http://library.cqpress.com/cqresearcher/cqresrre1928062800; and US Senate, 70th Cong., 1st sess., doc. 92, pt. 71A, *Utility Corporations* (Washington, DC: GPO, 1934), 115–16, 167, 258, 353–55, 359, 428, 431.

82. E. A. Stewart, "Electricity in Rural Districts Served by the Hydroelectric Power Commission of the Province of Ontario, Canada, 1926, "in US Senate, 70th Congress, 1st sess., doc. 92, pt. 1 *Utility Corporations* (Washington, DC: GPO, 1929), exhibit 39, 365–66.

83. E. A. Stewart, *Electricity in Rural Districts Serviced by the Hydro-Electric Power Commission of the Province of Ontario, Canada* (N.p.: n.p., 1926), 5. Stewart's report was reviewed by Ontario Hydro-Electric engineers, who advised the author about several inaccuracies. But Stewart apparently made no changes in the report, despite the request to do so by the HEPC's chairman. In response to a telegram sent from Toronto in June 1928 asking if Professor Stewart represented the University of Minnesota in publishing the study without making corrections, W. C. Coffey, the Department of Agriculture dean,

indicated that Stewart did the work on his own time. Telegram from Rogers Star, To-ronto, Canada, to University of Minnesota president L. D. Coffman, 21 June 1928; letter from W. C. Coffey, dean and director, University of Minnesota Department of Agricul-ture, to Professor William Boss, 21 June 1928; and telegram from W. C. Coffey to Rogers Star, 21 June 1928, in UM Archives, UARC, Ag. Eng. 339, box 4 / ALD 4.1. Stewart re-signed from the University of Minnesota on 1 November 1928 after being offered the position of president of the Northwestern Public Utilities Company. Memo from William Boss, chief of the University of Minnesota Division of Agricultural Engineering, to W. C. Coffey, dean, University of Minnesota Department of Agriculture, 11 September 1928, in ibid.; and letter from William Boss to Dr. E. A. White, director of the national CREA, 9 November 1928, in UM Archives, UARC, Ag. Eng. 339, box 6 / ALD4.1, folder "Commit-tee on Relation of Electricity to Agriculture, 1928–1929."

84. Lowitt, *George W. Norris*, 262.

85. F.G.R. Gordon, "Here Is the Truth about Ontario Hydroelectric," *Boston Herald*, 4 December 1927, in US Senate, 70th Cong., 1st sess., doc. 92, pt. 3, *Utility Corporations* (Washington, DC: GPO, 1930), exhibit 842, 316.

86. Other studies of the Ontario system are summarized in Thompson, *Confessions of the Power Trust*, 507–12. A recent and compelling study of the way in which the utility indus-try discredited public power is Naomi Oreskes, "The Fact of Uncertainty, the Uncertainty of Facts and the Cultural Resonance of Doubt," *Philosophical Transactions of the Royal Society A* 373 (2015), https://doi.org/10.1098/rsta.2014.0455, accessed 20 December 2020.

87. Wigginton E. Creed, "Electricity on California Farms," radio address over KGO, 21 October 1926, in *Pacific Service Magazine* 16, no. 11 (January 1927): 361.

88. "Report of W. H. Ude, Chairman of Executive Committee, Public Relations Sec-tion, Northwest Electric Light and Power Association," in US Senate, 70th Cong., 1st sess., doc. 92, pts. 10–16, *Utility Corporations* (Washington, DC: GPO, 1930), exhibit 4249, 668. The proposal's defeat was the "result of the very thorough program of educa-tion . . . to inform the electors." Ibid., 669.

89. The Shelton (WA) *Mason County Journal* editorial was cited in "Washington Is Aware of the Bone Bill Dangers," *Journal of Electricity* 53, no. 2 (15 July 1924): 42.

90. Part of the editorial in the San Leandro (CA) *Reporter* is cited in "The Farmer and the Bone Free Power Bill," *Journal of Electricity* 53, no. 3 (1 August 1924): 81.

91. George F. Oxley, director of publicity, NELA, to Horace M. Davis, director, Nebraska Commission on Public Utility Information, 2 February 1924, in US Senate, 70th Cong., 1st sess., doc. 92, pts. 10–26, *Utility Corporations* (Washington, DC: GPO, 1930), exhibit 4149, 314.

92. F. S. Dewey, Kansas City Power & Light, Kansas City, Missouri, to Horace M. Da-vis, secretary, Middle West division, NELA, and director of the Nebraska Committee on Public Utility Information, 16 December 1925, in US Senate, 70th Cong., 1st sess., doc. 92, pts. 5 and 6, *Utility Corporations* (Washington, DC: GPO, 1930), exhibit 2988, 570.

93. Oxley letter to Davis, 314.

94. Minutes of Public Relations National Section, Electric Light Association, execu-tive committee, 25 July 1927, in US Senate, 70th Cong., 1st sess., doc. 92, pts. 10–26, *Utility Corporations* (Washington, DC: GPO, 1930), exhibit 4134, 215.

95. Minutes of 19 January 1927 meeting, in US Senate, 70th Cong., 1st sess., doc. 92, pts. 10–26, *Utility Corporations* (Washington, DC: GPO, 1930), exhibit no. 4134, 212.

96. Letter from Horace M. Davis, director of the Nebraska Utilities Information Bureau, to Mrs. Chatie Coleman Westinius, editor of the *Headlight*, Stromsburg, NE, 4 January 1926, exhibit 4372, in ibid., 831.

97. Bill Luckin, *Questions of Power: Electricity and Environment in Inter-War Britain* (Manchester: Manchester University Press, 1990), 73–90.

98. Karl Ditt, "The Electrification of the Countryside: The Interests of Electrical Enterprises and the Rural Population in England, 1888–1939," in Paul Brassley, Jeremy Burchardt, and Karen Sayer, eds., *Transforming the Countryside: The Electrification of Rural Britain* (London: Routledge, 2017), 26.

99. R. Borlase Matthews, *Electro-Farming: Or the Application of Electricity to Agriculture* (London: Ernest Benn, 1928), 2. Also see R. Borlase Matthews, "Electro-Farming, or the Applications of Electricity to Agriculture," *Journal of the Institution of Electrical Engineers* 60, no. 311 (July 1922): 725–41. Matthews became a director of a power company in 1930 and died tragically in an unsuccessful attempt to rescue his son in a swimming accident in 1943. "Mr. R. Borlase Matthews," obituary in *Nature* 152, no. 3857 (2 October 1943): 379.

100. Critics of rural electrification challenged Matthews's optimistic visions of farm electrification. See B. M. Jenkin, "Discussion on "Electro-Farming, or the Applications of Electricity to Agriculture," *Journal of the Institution of Electrical Engineers* 60, no. 311 (July 1922): 743; and Sir Daniel Hall, "Applications of Electricity to Agriculture: Discussion," *Journal of the Institution of Electrical Engineers* 64, no. 3561 (August 1922): 813.

101. C. Damier Whetham, "Research Work by the Society in 1924: I—Electric Power in Agriculture," *Journal of the Royal Agricultural Society of England* 85 (1924): 246–51.

102. Exceptions existed, such as when farms were located near town generating plants, and distribution lines could be extended without needing to raise voltages with transformers. Ibid., 263–64.

103. Ibid., 248, 259, 267–70.

104. Paul Brassley, "Electrifying Farms in England," in Brassley, Burchardt, and Sayer, eds., *Transforming the Countryside*, 91. Only after nationalization of the British electrical industry in 1948 did rural customers receive greater attention. Almost total electrification occurred by end of the 1960s. Ibid., 94. Also see Leslie Hannah, *Electricity before Nationalisation: A Study of the Development of the Electricity Supply Industry in Britain to 1948* (Baltimore, MD: Johns Hopkins University Press, 1979), 189–92.

105. Reflecting idealistic principles as nations recovered from the Great War, the meeting's leaders sought to demonstrate how participants could exploit the "latest achievements in science and engineering" to advance "true human progress" and yield a genuine "international morality." *The Transactions of the First World Power Conference*, vol. 1, *Power Resources of the World* (London: Percy Lund Humphries, 1924), vii–viii. Supported by British government agencies and industry trade groups, the conference attracted representatives from the United States, France, Italy, Japan, "and other great nations of the world." In what appeared as a symbolic and substantive measure, conference organizers also invited Germany to participate, "which she did in a manner to seal the spirit of complete goodwill in

which the Conference assembled." Ibid., ix. *Science* magazine observed that the event constituted "from many points of view, the most notable gathering of its kind ever convened" to review "the power problems of the world with a completeness that has never before been attempted." "The World Power Conference," *Science* 60, no. 1550 (12 September 1924): 237–38. The United States' contingent consisted of 147 engineers, professors, and power company officials, including Samuel Insull, who served as his country's honorary vice chairman. *The Transactions of the First World Power Conference,* vol. 1, 1501–06; and "World Power Conference," in *Sixth Annual Report of the Federal Power Commission* (Washington, DC: GPO, 1926), 13.

106. F. H. Krebs, "Technical Development and Financial Organisation, Including Co-operative Schemes for Electricity Supply in Agriculture in Denmark," in *The Transactions of the First World Power Conference,* vol. 4 (London: Percy Lund Humphries, 1924), 250.

107. Ibid., 244. Also see V. Faaborg-Andersen, "Design of Low Tension Lines and Installations for Rural Electrification," in *The Transactions of the First World Power Conference,* vol. 4 (London: Percy Lund Humphries, 1924), 235–41.

108. NVE (Norges vassdrags- og energidirecktorat—Norwegian water resources and energy directorate), *Overview of Norway's Electricity History* (Oslo, Norway: NVE, 2017), 2–4, publikasjoner.nve.no/rapport/2017/rapport2017_15.pdf, accessed 2 April 2018.

109. Ragnar Nilsen, "Rural Modernisation as National Development: The Norwegian Case 1900–1950," *Norsk Geografisk Tidsskrift—Norwegian Journal of Geography* 68, no 1 (2014): 50–58. Specific and directly comparable data on rural electrification rates in many countries are hard to come by. The NVE report notes that in 1936 "the national electricity access rate was 74%," but that number includes rural and urban areas. In the northernmost (and slowest-to-be-electrified) part of the country, the "household electrification rate"—not necessarily the farm electrification rate—was about 30 percent in the late 1930s. NVE, *Overview of Norway's Electricity History,* 5.

110. O. Ganguillet, "L'électricité dans les menages et dans l'agriculture," *The Transactions of the First World Power Conference,* vol. 4, 583–88.

111. Kable and Gray, "Report on C.W.A. National Survey of Rural Electrification," 59–62.

112. V. C. Lagendijk, *Electrifying Europe: The Power of Europe in the Construction of Electricity Networks* (Eindhoven: Technische Universiteit Eindhoven, 2008), 23–24.

113. Even with these reported gains, only a small percentage of the country's 84 million peasants received electric service. Jonathan Coopersmith, *The Electrification of Russia, 1880–1926* (Ithaca, NY: Cornell University Press, 1992), 67, 138, 236–37, 242, 243.

114. Evans, "The World's Experience with Rural Electrification," 33.

115. Ibid., 31.

116. Ibid., 31–33.

117. Federal Emergency Administration of Public Works (Morris Cooke, chairman), *Report of the Mississippi Valley Committee of the Public Works Administration* (Washington, DC: GPO, 1934), 51.

118. "Administrator Calls United States 'Backward' in Rural Electrification," *Rural Electrification News* 1, no. 1 (September 1935): 19. As Abby Spinak notes, Ontario Hydro

already subsidized half the cost of rural line construction (the result of a 1921 law), and in 1930, it provided loans for purchases of electrical appliances. Facing a power surplus, the agency also offered rural customers free power for various appliances. Spinak, "'Not Quite So Freely as Air,'" 94.

119. REA, *The REA Guide: An Outline for Rural Highlines* (Washington, DC: REA, 1936), 17–18.

120. Robert T. Beall, "Rural Electrification," in USDA, *Yearbook of Agriculture, 1940* (Washington, DC: GPO, 1940), 790–91.

121. Historian David Nye observes that Americans generally viewed electricity as a commodity—a product made by corporations for sale like other commodities. In Europe, by contrast, people often viewed electricity as a social service that governments should control. David E. Nye, *Electrifying America: Social Meanings of a New Technology, 1880–1940* (Cambridge, MA: MIT Press, 1990), 140–41.

122. "U.S. behind Japan in Rural Electricity," *Washington Post*, 11 August 1935, 2.

123. F. A. Gaby, listed as chief engineer of the HEPC, "Electrical Service for Rural Districts," part of P. T. Davies, "Utilization of Power in Canada," in *The Transactions of the First World Power Conference*, vol. 4, 28.

124. H. S. Bennion, "United States Leads in Rural Electrification," *Edison Electric Institute Bulletin* 3, no. 11 (November 1935): 409–13.

125. Ronald R. Kline, *Consumers in the Country: Technology and Social Change in Rural America* (Baltimore, MD: Johns Hopkins University Press, 2000), 133. Kline argued that "the NELA first took up the matter of rural lines in 1911—the year Ontario Hydro began to plan a system for rural service" even though NELA created its Committee on Electricity in Rural Districts in November 1910. "Report of Committee on Electricity in Rural Districts," NELA, *Proceedings* 40 (1911): 448. It still remains likely, however, that NELA members were motivated by Ontario Hydro's renewed emphasis on rural electrification that began in 1921.

CHAPTER EIGHT: **The Industry Organizes the CREA**

Epigraph: "Of What Use Is Electricity on the Farm?," *Electrical World* 80, no. 6 (5 August 1922): 261.

1. F. W. Duffee and G. W. Palmer, *Turn On the Light*, Circular 163 (Madison: Extension Service of the College of Agriculture, University of Wisconsin, July 1923), 29.

2. Orville M. Kile, *The Farm Bureau through Three Decades* (Baltimore, MD: Waverly Press, 1948); "Huge League of Farm Bureaus Launched Here," *Chicago Daily Tribune*, 4 March 1920, 22; Frank Ridgway, "Efficiency Men to Study Needs of Our Farmers," *Chicago Daily Tribune*, 5 March 1920, 16; "Farm Bureaus Are Federated," *Los Angeles Times*, 4 March 1920, 15; and Orville Merton Kile, *The Farm Bureau Movement* (New York: Macmillan, 1921).

3. James H. Shideler, *"Flappers and Philosophers*, and Farmers: Rural-Urban Tensions of the Twenties," *Agricultural History* 47, no. 4 (October 1973): 284.

4. American Farm Bureau Federation, *Report of the Executive Secretary: The Federation's Third Year* (Chicago: AFBF, 1922), 8.

5. G. C. Neff, "Rural Lines Committee," NELA, *Proceedings* 79 (1923): 45.

6. Ibid., 46. The name of the organization was listed incorrectly in NELA's *Proceedings* as "the Committee on Electricity as Related to Agriculture." Ibid.

7. Royden Stewart, "Bringing Power to the Farm: Article II—National Development 1924–35: The CREA," *Public Utilities Fortnightly* 27, no. 11 (22 May 1941): 652; "Conference on the Relation of Electricity to Agriculture," *Agricultural Engineering* 4, no. 4 (April 1923): 62–63; and "Committee on Relation of Electricity to Agriculture Organized," *Agricultural Engineering* 4, no. 10 (October 1923): 166–67.

8. "Electricity and Its Relation to Agriculture," in American Farm Bureau Federation, *Annual Report of the Secretary for the Year 1923* (Chicago: AFBF, 1923), 15. The AFBF's involvement in the creation of the national CREA is described in Kile, *The Farm Bureau through Three Decades*, 78–80.

9. "Our New President," *Agricultural Engineering* 2, no. 1 (January 1921): 6.

10. A brief biography of White and a description of the founding of the ASAE can be found in Ronald B. Furry, *A Pioneering Department: Evolution from Rural Engineering to Biological and Environmental Engineering at Cornell University, 1907–2007* (Ithaca, NY: Internet-First University Press, 2007), 7, 23, https://ecommons.cornell.edu/handle/1813 /7642, accessed 30 July 2015.

11. Edwin T. Layton Jr., *The Revolt of the Engineers: Social Responsibility and the American Engineering Profession* (Cleveland, OH: Press of Case Western Reserve University, 1971).

12. Robert E. Stewart, *Seven Decades That Changed America: A History of the American Society of Agricultural Engineers, 1907–1977* (St. Joseph, MI: American Society of Agricultural Engineers, 1979), 2.

13. *Twenty-First Biennial Report of the Iowa State College of Agriculture and Mechanic Arts Made to the Governor of Iowa for the Biennial Period July 1, 1903 to June 20, 1905* (Des Moines, IA: B. Murphy State Printer, 1906), 6–7. This report noted that "the Iowa State College is, we believe, the first institution to take up this work in an extensive manner." Also see Iowa State University, "Department of Agricultural and Biosystems Engineering 125th Points of Pride," https://www.cals.iastate.edu/content/department-agricultural -and-biosystems-engineering-125th-points-pride, accessed 28 July 2021.

14. Furry, *A Pioneering Department*, 11.

15. *Bulletin of the Virginia Polytechnic Institute, Catalogue, April 1913* (Roanoke, VA: Stone Printing and Manufacturing Co., 1913), 32.

16. E. A. White, "The President's Annual Address," *Agricultural Engineering* 3, no. 1 (January 1922): 12; and "President White Blows Whistle for 1921 Kick-Off," *Agricultural Engineering* 2, no. 1 (January 1921): 6.

17. Deborah Fitzgerald, *Every Farm a Factory: The Industrial Ideal in American Agriculture* (New Haven, CT: Yale University Press, 2003), 22–23.

18. "Program of the Seventeenth Annual Meeting of the American Society Agricultural Engineers [*sic*—no "of"], in VT Special Collections, Papers of Julian Ashby Burruss (1919–45), RG2/8, box 7, folder 469, "Burruss, C. E. Seitz," with letter from Seitz to Burruss, 3 November 1923. A published summary explained, "An outstanding feature of this program was the very evident fact that the rural electrification idea is rapidly gaining momentum." "ASAE Meeting Breaks All Records," *Agricultural Engineering* 4, no. 12 (December 1923): 199.

19. "Rural Electrification," *Agricultural Engineering* 4, no. 10 (October 1923): 158.

20. The Hatch Act of 1887 provided funds for establishing agricultural experiment stations, while the 1914 Smith-Lever Act extended money to land-grant schools to fund extension services.

21. *Electrical World* observed that the "significant thing is that the character of thinking exhibited by [the professor] is going on in the ranks of men who are the trusted advisers of the farmer." "Agricultural Thinking on Electric Service Problem," *Electrical World* 82, no. 2 (14 July 1923): 62. Of course, as noted in chapter 3, some farmers did not trust experts—even those working at land-grant colleges.

22. E. P. Edwards, electrical engineer for General Electric, "The Neglect of the Power Problem," *Transactions of the American Society of Agricultural Engineers* 4 (December 1910): 86.

23. Frank Ridgway, "Electrification of Rural America Is Still Experimental: Experts Advise against Haste in Connection with Project," *Chicago Sunday Tribune*, 13 September 1925, pt. 2, 14.

24. Neff, "Rural Lines Committee," 46.

25. Historian Audra Wolfe noted that agricultural engineers—and especially extension agents—encouraged farmers to adopt electrification as a social good, whereas utility managers sought to help ruralites only when doing so helped their companies as well. In statements made by the CREA director and others, it appears that agricultural engineers toned down their concerns for social modernization and acceded to the industry, the only entity able to help farmers obtain electric service. By working to achieve the companies' goals, however, the engineers knew that ruralites would benefit as well. Audra J. Wolfe, "'How Not to Electrocute the Farmer': Assessing Attitudes towards Electrification on American Farms, 1920–1940," *Agricultural History* 74, no. 2 (Spring 2000): 515–29.

26. "Electricity and Its Relation to Agriculture," 16.

27. E. A. White, "The Investigation Program," *CREA Bulletin* 1, no. 1 (10 September 1924): 4.

28. White established a subset of problems for investigation, involving the possible use of electricity for water supply, field work, refrigeration, and plant growth. Ibid., 4–5. "Foundation for the Start," *CREA Bulletin* 1, no. 1 (10 September 1924): 3.

29. Minnesota CREA, *The Red Wing Line*, brochure likely published in 1924, inside cover and pages 2–3, though the booklet has no page numbers, found in WSU Special Collections, TK 4018 M56 1924, and in UM Archives, 621.393 M66. Also see E. A. Stewart, J. M. Larson, and J. Romness, *The Red Wing Project on Utilization of Electricity in Agriculture* (N.p. [St. Paul?]: University of Minnesota Agricultural Experiment Station, no date, but likely 1927 or 1928), 2, University of Minnesota Libraries Digital Conservancy, http://conservancy.umn.edu/handle/11299/48685, accessed 15 September 2014.

30. "Preliminary Announcement: Iowa State Organization on the Relation of Electricity to Agriculture and Iowa's Rural Community for Electrical Development," typescript found in in WSU Special Collections, Leslie John Smith Papers (rural electrification, 1923–46) folder, cage 5034, no date, but likely 1923 or 1924; F. D. Paine and J. B. Davidson, "Farm Electricity to Be Studied," *Iowa Engineer* 25, no. 2 (November 1924): 9; and "Iowa's Rural Community for Electrical Development," *Iowa State College of Agriculture and Mechanic Arts Official Publication* 24, no. 1 (3 June 1925): 5. About Eloise Davison, see Ronald R.

Kline, "Agents of Modernity: Home Economists and Rural Electrification, 1925–1950," in Sarah Stage and Virginia B. Vincenti, eds., *Rethinking Home Economics: Women and the History of a Profession* (Ithaca, NY: Cornell University Press, 1997), 241; Ronald R. Kline, *Consumers in the Country: Technology and Social Change in Rural America* (Baltimore, MD: Johns Hopkins University Press, 2000), 136; and Amy S. Bix, *Girls Coming to Tech! A History of American Engineering Education for Women* (Cambridge, MA: MIT Press, 2013), 51.

31. The "Rural Electrification Directory," in *CREA Bulletin* 3, no. 3 (15 June 1927): 11–24, listed twenty-three committees in the following states: Alabama, California, Idaho, Illinois, Indiana, Iowa, Kansas, Michigan, Minnesota, Missouri, Nebraska, New Hampshire, New York, Ohio, Oklahoma, Oregon, Pennsylvania, South Carolina, South Dakota, Texas, Virginia, Washington, and Wisconsin. In the *CREA Bulletin* 4, no. 1 (30 January 1928): 5, New Jersey was added to this list. Louisiana and Maryland were included in a map of states that had committees in "Report on Farm Electrification Research," *CREA Bulletin* 6, no. 1 (June 1931): 4, bringing the number to twenty-six.

32. "Relation between State and National Committees," *CREA Bulletin* 1, no. 1 (10 September 1924): 5. NELA's Rural Electric Service Committee reiterated the value of the agricultural engineers' contributions. In a 1924 report, its chairman, Grover Neff, argued that ASAE's membership, consisting of "practically every agricultural engineer from the colleges and universities of this country . . . should logically become a clearing house . . . between electrical and agricultural interests on all information relating to rural electrification." G. C. Neff, "Rural Electric Service Committee," NELA, *Proceedings* 81 (1924): 65.

33. Alabama Farm Bureau Federation, *Electrifying Agriculture: Progress Report of Alabama Farm Bureau Committee on the Relation of Electricity to Agriculture* (Montgomery: Alabama Farm Bureau, [1925?]), 6.

34. "Project Statement," dated 18 November 1923, showed a budget for one University of Minnesota project at $5,000 for three years. WSU Special Collections, Leslie John Smith Papers, (Rural electrification, 1923–1946) folder, cage 5034 (1 of 2); "Estimate of Budget for Studies and Investigations on the Relation of Electricity to Agriculture for the State of Washington (For One Year Beginning January 1, 1926)," in WSU Special Collections, College of Agriculture, box 103, folder 4423, "Washington Committee on the Relation of Electricity to Agriculture—Minutes, 1925–1945." Accounting figures for the Minnesota CREA are listed in US Senate, 70th Cong., 1st sess., doc. 92, pt. 4, *Utility Corporations* (Washington, DC: GPO, 1930), exhibit 1749, 230–32. A summary of the first two years of the CREA's work can be found in E. A. White, "Features of Rural Service," *Electrical World* 85, no. 25 (20 June 1925): 1307–8; and S. H. McCrory, "Problems Involved and Methods Used in Promoting Rural Electrification," *Journal of Farm Economics* 12 (April 1930): 321–22.

35. "Payments to the Committee on Relation of Electricity to Agriculture," in US Senate, 70th Cong., 1st sess., doc. 92, pt. 61, *Utility Corporations* (Washington, DC: GPO, 1934), exhibit 5584, 175.

36. Harry Slattery, *Rural America Lights Up* (Washington, DC: National Home Library Foundation, 1940), 21, 22.

37. Ibid., 22.

38. Ibid., 22–23. A similar estimate (perhaps based on Slattery's) of a total of $2.4 million spent on the CREA-associated work is included in Royden Stewart, "Rural Elec-

trification in the United States: Part II—National Development, 1924–1935," *Edison Electric Institute Bulletin* 9 (October 1941): 411.

39. Stewart, Larson, and Romness, *The Red Wing Project*, 2.

40. Stewart taught in high schools and, from 1917 to 1920, as an assistant professor of physics at Kansas State Agricultural College, having earned a bachelor's degree from the University of Chicago in 1915. He taught general physics and agricultural physics in the Farm Engineering program and applied for membership in ASAE in 1921. Biographical information assembled from "KSAC Loses Another Man," *Evening Kansan-Republican* (Newton, KS), 30 August 1920, 1; University of Minnesota, *Bulletin of the University of Minnesota* 24, no. 9 (27 April 1921): 56–57; "Applicants for Membership," *Agricultural Engineering* 2, no. 9 (September 1921): 199; University of Minnesota, *Bulletin of the University of Minnesota* 24, no. 47 (27 March 1922): 31; and University of Minnesota, *Bulletin of the University of Minnesota* 26, no. 26 (5 July 1923): 60.

41. "Project Statement," dated 18 November 1923, 1.

42. Ibid., 2.

43. Stewart, Larson, and Romness, *The Red Wing Project*, 2.

44. E. A. Stewart, "Problems on the Utilization of Electricity in Agriculture That Were Proposed at the Equipment Conference Held at Red Wing, Minn., October 8 and 9, 1924," *CREA Bulletin* 1, no. 3 (15 November 1924): 10–12; and E. A. Stewart, "The Minnesota Project on the Utilization of Electricity in Agriculture," *CREA Bulletin* 1, no. 4 (15 December 1924): 1–7.

45. Stewart, Larson, and Romness, *The Red Wing Project*, 26.

46. Ibid., 27.

47. Ibid., 128.

48. Charles F. Stuart, "Will Electricity Pay Its Own Way on the Farm?" *Forbes* (19 July 1924): 486.

49. Ibid., 487.

50. Ibid.

51. Ibid. 492.

52. "Problem of Farm Service Being Solved," *Public Service Magazine* 37, no. 4 (October 1924): 107.

53. "Farm Service Exhibit at Minnesota State Fair," *Electrical World* 84, no. 15 (11 October 1924): 810.

54. "Electricity to End Farm Drudgery," *Popular Mechanics* 44, no. 2 (August 1925): 260.

55. Ibid., 253.

56. "Shows Electricity as Useful on Farm," *New York Times*, 7 October 1928, N16.

57. C. F. Stuart, in discussion following "Rural Electric Service Committee, NELA, *Proceedings* 85 (1928): 148.

58. Stewart, Larson, and Romness, *The Red Wing Project*, 27.

59. Stewart, "The Minnesota Project," 7.

60. Minnesota CREA, *The Red Wing Experimental Electric Power Line*. The movie is available (with parts on display) at the Bakken Museum, Minneapolis, MN. No date listed on the film, but probably 1928 or 1929.

61. Ibid.; emphasis in the original.

62. The industry's assets totaled nearly $11 billion. "Statistics of the Electric Light and Power Industry, 1929," NELA, *Proceedings* 87 (1930): 1425; and William J. Hagenah, "Performance of Utilities under State Regulation," NELA, *Proceedings* 87 (1930): 64.

63. "Statistics," 1424, 1429.

64. US Congress, Temporary National Economic Committee, "Investigation of Concentration of Economic Power," Monograph no. 26, *Economic Power and Political Pressures* (Washington, DC: GPO, 1941), 152. The $1 million figure seems to represent efforts to influence parties in ways that would be viewed today as unethical—namely, the dissemination of articles and editorials published in newspapers after payment by NELA and without attribution. For all forms of advertising, such as promotions about the declining cost of electric service, the utility industry spent between $25 million and $30 million in 1923 alone. Ibid., 153.

CHAPTER NINE: **State Committees Work to Resolve Uncertainties**

Epigraph: Robert Stewart, "Electricity in the Service of Agriculture," *Bankers' Magazine* 116, no. 2 (February 1928): 195.

1. National CREA director E. A. White wrote to Arthur Capper, senator from Kansas, in 1925 indicating that "the different agricultural conditions in different parts of the country" merited individual research efforts performed by various state committees. Letter from E. A. White to Senator Arthur Capper, 3 January 1925, 68th Cong., 2nd sess., 66 Cong. Rec. 1533 (1925).

2. F. W. Duffee and G. W. Palmer, *Turn On the Light*, Circular 163 (Madison: Extension Service of the College of Agriculture, University of Wisconsin, July 1923), 31. The authors are listed as agricultural engineers who also were part of the experiment station staff. University of Wisconsin Agricultural Experiment Station, *Science Serves Wisconsin Farms: Annual Report of the Director, 1921–1922* (Madison: University of Wisconsin, 1923), unnumbered page after 122.

3. "Wisconsin Utility Associations Unite," *Electrical World* 79, no. 13 (1 April 1922): 647.

4. E. R. Meacham, "The Wisconsin Project," *CREA Bulletin* 1, no. 5 (26 January 1925): 1.

5. F. W. Duffee, "What the Wisconsin Project on the Application of Electricity to Agriculture Has Revealed, and Future Plans," typescript of talk given at annual meeting of the state CREA, 27 June 1927, 9, in UW Archives, Agricultural Engineering, Wisconsin Commission [*sic*] for the Application of Electricity to Agriculture (WCREA), series #9/12/5-1, box 1. The early days and organization of the Wisconsin project are described in Meacham, "The Wisconsin Project," 1–5. Forrest McDonald describes the creation of the Wisconsin CREA and the experimental line extended to farms—the second such line, after the one used by the Minnesota CREA participants—in *Let There Be Light: The Electric Utility Industry in Wisconsin, 1881–1955* (Madison, WI: American History Research Center, 1957), 287–89. The results of early experiments conducted by the Wisconsin CREA are described in "Report on Ripon Project to the Wisconsin Committee on the Application of Electricity to Agriculture," typescript in folder on "Report of Meeting Held May 8, 1925," in UW Archives, WCREA, series # 9/12/5-1, box 1. Weekly reports on the Ripon experiments in 1925 can be found in the folder titled "WCREA Reports of Experiments on Ripon Line," in UW Archives, WCREA, series # 9/12/5-1, box 2. A more formal

report of the early activities of the Wisconsin CREA can be found in Meacham, "The Wisconsin Project," 1–5.

6. Meacham, "The Wisconsin Project," 2–3.

7. Many results from early work of the Wisconsin CREA are contained (along with descriptions of other state committees' research) in "Correlation Outline of Investigations on the Application of Electricity to Agriculture," *CREA Bulletin* 3, no. 2 (4 January 1927): 3–24.

8. "Report on Ripon Project to the Wisconsin Committee on the Application of Electricity to Agriculture," typescript paper included in "Committee on the Application of Electricity to Agriculture," report of meeting held 8 May 1925, no page, in UW Archives, WCREA, series # 9/12/5-1, box 1.

9. Meacham, "The Wisconsin Project," 4.

10. The company's president, Thomas Martin, apparently was a strong rural electrification advocate whose interest in the subject is described in Leah Rawls Atkins, *Developed for the Service of Alabama: The Centennial History of the Alabama Power Company, 1906–2006* (Birmingham: Alabama Power Co., 2006), 125–29.

11. "Annual Report of the Department of Agricultural Engineering," Alabama Polytechnic Institute, typescript, 46, in AU Special Collections, Agricultural Engineering Records, record group 503, accession no. 79-001, box 8 (of 64) in folder marked "Report 1924."

12. For using just lighting and other small appliances, farm customers paid 9 cents per kWh. M. L. Nichols and E. C. Easter, "The Alabama Project," *CREA Bulletin* 1, no. 3 (15 November 1924): 4. Though the rate was perhaps lower than what might be justified (since it did not account for all the costs incurred from extending lightly used distribution lines), it allowed for a reasonable way to assess how much (and for what uses) farmers would consume electricity. "Such knowledge is necessary," a 1925 report noted, "before a rate can be fixed to fit both the company and the farmer." Alabama Farm Bureau Committee on the Relation of Electricity to Agriculture, *Electrifying Agriculture: Progress Report of Alabama Farm Bureau Committee on the Relation of Electricity to Agriculture* (Montgomery: Alabama Farm Bureau, [1925?]), 9, in AU Special Collections, call number HD 9688 A4.

13. "Rural Electric Development in Alabama," *CREA Bulletin* 2, no. 5 (6 April 1926): 8–11.

14. Ibid., 2–3.

15. Ibid., 11.

16. "Accomplishments in Rural Electric Development by Alabama Power Company during 1932," *Edison Electric Institute Bulletin* 2, no. 1 (January 1934): 9. In 1924, an editor of an Alabama Polytechnic Institute report observed, "It is obvious that it will be necessary for farmers to use electricity for other purposes than for light. To succeed it must be a paying proposition for all concerned[,] and volume use will be necessary to make it pay." Farmers would simply need to increase their use of electricity, perhaps by altering their farming techniques, such as by abandoning a "one crop system" of cotton, which apparently limited farmers' income substantially. Alabama Power Co., *Commemorating the Era of Rural Electrification in Alabama* (Tuscaloosa: Alabama Power Co., 1951), 9, quoting from P. O. Davis, "Shall We Electrify Our Farms?" *Progressive Farmer* (14 January 1924).

Put differently in a contemporaneous publication, "The company expects a reasonable return on its investment and the farmer expects a reasonable rate for service." Alabama Farm Bureau Committee, *Electrifying Agriculture*, 7.

17. Letter from Charles E. Seitz, Dept. of Agricultural Engineering, to Dr. Julian A. Burruss, president of VPI, 14 April 1924, describing the "meeting of representatives of farmers [sic] organizations, Electric Power Companies, the State Dept. of Agriculture, and the V.P.I." for creating the state committee, in VT Special Collections, Papers of Julian Ashby Burruss (1919–45), RG2/8, box 7, folder 545–1925, "Chas. E. Seitz, Burruss." A second conference in April 1924 established a goal of obtaining $12,500 from power companies so VPI's Agricultural Engineering Department could pursue two years of studies. Letter from Charles A. Seitz, professor of agricultural engineering, to Julian A. Burruss, president of VPI, 14 April 1924, in VT Special Collections, Papers of Julian Ashby Burruss (1919–45), RG2/8, box 7, folder 545–1924, "Chas. E. Seitz, Burruss;" Letter from Charles E. Seitz to members of the Committee on the Relation of Electricity to Agriculture, 20 February 1925; typescript, in VT Special Collections, Department of Agricultural Engineering papers, RG13/4, "Rural Electrification—General—1920s–30"; "Progress Report of the Committee on the Relation of Electricity to Agriculture, April 1925," typescript, in VT Special Collections, Department of Agricultural Engineering papers, RG13/4, "Rural Electrification—General—1920s–30"; and Charles A. Seitz, "The Utilization of Electricity in Agriculture," report attached to letter from Seitz to Burruss, 14 April 1924, in VT Special Collections, Papers of Julian Ashby Burruss (1919–45), RG2/8, box 7, folder 545–1924, "Chas. E. Seitz, Burruss."

18. Both examples are given in Charles E. Seitz, "Virginia's Progress in Rural Electrification," *Electrical South* 9 (March 1929): 60–63. The hen experiment is also described in J. A. Waller Jr., "A Report on the Present Status of Rural Electrification in Virginia by the Virginia Committee on the Relation of Electricity to Agriculture," *CREA Bulletin* 2, no. 6 (12 May 1926): 10. Researchers at other CREA affiliates performed almost identical experiments, with comparable results. The fact that various groups performed similar studies suggests that the engineers felt the need to demonstrate the validity of their results under different local circumstances. See "Poultry House Lighting," *CREA Bulletin* 7, no. 1 (November 1931): 199–207.

19. Hobart Beresford, "Progress Report of the Idaho Committee on the Relation of Electricity to Agriculture on Electrification of the Caldwell Substation of the College of Agriculture, University of Idaho," August 1927, 7, in UI Special Collections, Day-NW TK4018.I3; "Minutes of Meeting of the Executive Committee on the Idaho State Committee on the Relation of Electricity to Agriculture," typed minutes in bound volume, "Idaho Committee on the Relation of Electricity to Agriculture (Minutes of Meetings) 1925–1932," from University of Idaho Library stacks, call number Day-NW TK 4018 I3a 1925–32; and "The Use of Electricity on the Farm for 1931, University of Idaho Sub-Experiment Station, Caldwell, Idaho," in UI Special Collections, Agricultural Experiment Station, Department of Agricultural Engineering, "Rural Electrification Progress Report No. 8, Idaho Committee on the Relation of Electricity to Agriculture, Hobart Beresford, Secretary and Project Director, April 1932," typescript, vol. 8, which noted (on page 5) that while the three-year study ended in 1930, the interest in the project "has been sufficient to warrant the continuation of the study during 1931."

20. Hobart Beresford, *Farm Electrification in Idaho*, Bulletin no. 237 (Moscow: University of Idaho, Agricultural Experiment Station, 1940), 18.

21. Ibid., 5; and Idaho CREA, "Progress Report No. 5," issued May 1929, bound typescripts, in UI Special Collections, call number TK 4018 I3.

22. Frank D. Paine, "What Is It All About," in "Iowa's Rural Community for Electrical Development, Report No. 1," *Iowa State College of Agriculture and Mechanic Arts Official Publication* 24, no. 1 (3 June 1925): 8.

23. Eloise Davison, "Electrical Equipment in the Farm Home," in "Iowa's Rural Community," 10–11. Davison became the director of home economics of NELA and the representative of the American Home Economic Association of the national CREA, as was noted in the 1928 report of the Iowa CREA. Frank D. Paine and Frank J. Zink, "Electric Service for the Farm, Report No. 3," *Iowa State College of Agriculture and Mechanic Arts Official Publication* 27, no. 8 (27 June 1928): 9.

24. Foreword by Dean Genevieve Fisher of the Home Economics Division, in Harriett C. Brigham, Frank J. Zink, and Frank D. Paine, "Electric Service for the Farm, Report No. 6," *Iowa State College of Agriculture and Mechanic Arts Official Publication* 27, no. 12 (18 July 1928): 9.

25. Ibid., 5–6.

26. Ibid., 32.

27. Ibid., 10–11.

28. Ibid., 12.

29. Ibid., 19.

30. "Minutes of the First Meeting of the Washington State Advisory Committee on the Relation of Electricity to Agriculture," taken by CREA secretary (and Washington State College professor) L. J. Smith, 19 January 1925, in WSU Special Collections, College of Agriculture, box 103, in folder 4423, "Washington Committee on the Relation of Electricity to Agriculture, 1925–1945."

31. Harry L. Garver, *Electric Hay Hoists*, Popular Bulletin no. 139 (Pullman: State College of Washington Agricultural Experiment Station, 1928), HathiTrust Digital Library, https://hdl.handle.net/2027/uiug.30112019918207.

32. Edw. C. Johnson, "Highlights of the Washington Committee on the Relation of Electricity to Agriculture," typed manuscript of talk given "at the 21st annual meeting of the Washington Committee on the Relation of Electricity to Agriculture," April 1946, in WSU Special Collections, Leslie John Smith Papers, Rural Electrification (rural electrification, 1923–46) folder, cage 5034. The surveys are first described in L. J. Smith, "A Report on the State Wide Survey of Rural Electric Lines in Representative Parts of the State of Washington," typed manuscript submitted October 1925 to Washington CREA, in WSU Special Collections, College of Agriculture, box 103, in folder 4423, "Washington Committee on the Relation of Electricity to Agriculture, 1925–1945."

33. Ben D. Moses, "The Development of Farm Electrification in California," paper presented at Rural Electric Conference, 30 January 1950, 3–4, in UCD Special Collections, Walker Papers, call number D-111, box 18, folder 5.

34. H. B. Walker, "Research Reveals Opportunities and Responsibilities in Rural Load Building," paper written in 1940 for presentation at an undisclosed conference, 4–5, in UCD Special Collections, Walker Papers, call number D-111, box 12, folder 6.

35. Ibid., 4–8.

36. H. B. Walker, "The Function of the College of Agriculture and CREA in Rural Electrical Development," typescript, 5, in UCD Special Collections, Walker Papers, call number D-111, box 12, folder 6; and Walker, "Research Reveals Opportunities," 9–10.

37. "Rural Electric Service Committee" NELA, *Proceedings* 85 (1928): 140–41; and "Rural Electrification Project on National Scale," *Electrical World* 91, no. 17 (28 April 1928): 878. The project had a five-year lifespan, ending in June 1933. "National Rural Project Issues Final Report," *Electrical World* 102 (26 August 1933): 258. Also see *The National Rural Electric Project: What It Is—Its Purpose, Activities Cooperation and Personnel*, report no. 1 (College Park, MD: National Rural Electric Project, October 1931); and *Final Report of the National Rural Electric Project*, report no. M-16 (College Park, MD: National Rural Electric Project, July 1933). The purpose and funding of the project is described in US Senate, 70th Cong., 1st sess., doc. 92, pt. 61, *Utility Corporations* (Washington, DC: GPO, 1934), exhibit 5594, 193.

38. *Electricity on the Farm and in Rural Communities*, CREA Bulletin 4, no. 1 (30 January 1928): cover.

39. "Index," *CREA Bulletin* 4, no. 1 (30 January 1928): 135–36.

40. "Report of the Fifth Annual Meeting of the Rural Electric Project Leaders (June 1928)," typescript, 9, in WSU Special Collections, Washington Farm Electrification Committee Records, cage 607, Annual and Executive Minutes 1925–41, box 1, folder 1. In the first issue of the *CREA News Letter*, a less formal sister publication of the *CREA Bulletin*, the editor noted that the 1928 *Electricity on the Farm and in Rural Communities* had been purchased by "a fair percentage" of power companies that served rural customers. Distributed to farmers, often by meter readers, the publication, which includes "201 real farm pictures of such local personal interest," will engender "good will and better understanding." "New Books," *CREA News Letter* 1 (16 April 1928): 1–2.

41. "A Survey of Research," *CREA Bulletin* 6, no. 1 (June 1931): 3. This issue of the *CREA Bulletin* was titled *Report on Farm Electrification Research*.

42. Ibid., 5.

43. "Alarms," *CREA Bulletin* 6, no. 1 (June 1931): 12.

44. "Cooking with Electricity," *CREA Bulletin* 7, no. 1 (November 1931): 11. The study showed that an electric range required about 25 minutes per week to maintain versus about 181 minutes to clean a coal stove, replenish its fuel supply, kindle the fire, and dispose of ashes.

45. "Fire Protection," "Milking Machines," "Dairy Refrigeration," and "Ensilage Cutters," *CREA Bulletin* 7, no. 1 (November 1931): 89, 93–102, 102–6, 131–40, respectively.

46. "Poultry House Lighting," *CREA Bulletin* 7, no. 1 (November 1931): 199–201.

47. "An Indirect Aid to the Work of Farm Demonstrators," *Electrical World* 99 (20 February 1932): 351.

48. "Preface," *CREA Bulletin* 7, no. 1 (November 1931): 1.

49. John G. Learned, "Report of Committee on Electricity in Rural Districts," NELA, *Proceedings* 40 (1911): 465; and Franz Koester, "The Application of Electric Motors to Agricultural Operations," *Engineering Magazine* 31, no. 5 (August 1906): 657–66.

50. Frank Koester, "Electric Power Applications on the Farm," *Electrical Review and Western Electrician* 60, no. 16 (20 April 1912): 743.

51. Frank Koester, *Electricity for the Farm and Home* (New York: Sturgis & Walton, 1913), 161.

52. Ibid., 159. Born in 1876 (with the first name of Franz) and educated in Germany, Koester came to the United States in 1902. He published several books on electric power systems and city planning. Biographical entry in the 1920 edition of *The Encyclopedia Americana*, Wikisource, https://en.wikisource.org/wiki/The_Encyclopedia_Americana_(1920)/Koester,_Frank, accessed 5 February 2018.

53. H. Winfield Secor, "Electric Farm de Luxe," *Science and Invention* 10, no. 2 (June 1922): 119.

54. R. W. Trullinger, "Some Research Features of the Application of Electricity to Agriculture," *CREA Bulletin* 1, no. 2 (15 October 1924): 4–5.

55. Robert C. Williams, *Fordson, Farmall, and Poppin' Johnny: A History of the Farm Tractor and Its Impact on America* (Champaign: University of Illinois Press, 1987). Pricing information from Larry Gay, "History of the Ford Tractor," Brassworks, https://www.thebrassworks.net/history-of-the-ford-tractor.html, accessed 5 February 2018. Steam-powered tractors enjoyed a period of modest success, such that at their peak usage in 1910, they supplied about 17 percent of tractive power employed on farms. Williams, *Fordson, Farmall, and Poppin' Johnny*, 10–12; and Reynold M. Wick, *Steam Power on the American Farm* (Philadelphia: University of Pennsylvania Press, 1953).

56. "Electricity Used for Plowing Farm," *Reading (PA) Eagle*, 12 August 1927, 18.

57. *Roe Wireless Electric Plow*, brochure, no date, probably 1927, from LeRoy (NY) Historical Society.

58. "Electricity Used in Soil Cultivating," *Caledonia (NY) Advertiser*, 14 July 1927, 1. The same picture, with a description of the Roe plow appeared in the *Alma (KS) Signal*, 15 September 1927, 8. Reprinted from the *Fort Worth Star-Telegram*, a long description of the plow appeared in the *Honey Grove (TX) Signal*, 9 September 1927, 7.

59. "Electricity Used for Plowing Farm," 18.

60. "Electricity Speeds Up Crops: Electro-Cultivation Raises Plants in Record Time," *Science and Invention* 15, no. 7 (November 1927): 610.

61. "Electric Plow Kills All Pests," *Popular Science Monthly* 111, no. 5 (November 1927): 41.

62. "Science: Electric Plow," *Time* 10, no. 5 (1 August 1927), 23. Roe won a patent on the machine in 1929. US Patent 1,737,866 for "Method of and Apparatus for the Practice of Agriculture," filed 10 October 1923, granted 3 December 1929.

63. "Plow Sends Power into Soil It Tills: Demonstration of 'Electro-Culture' near Rochester Draws Many Farmers and Experts," *New York Times*, 21 August 1927, E1.

64. "Memorandum on National Electric Rural Project," letter dated 22 October 1928 from Paul S. Clapp, managing director of the NELA, in US Senate, 70th Cong., 1st sess., doc. 92, pts. 10–16, *Utility Corporations* (Washington, DC: GPO, 1930), exhibit 4132, 191. CREA affiliates had demonstrated their interest in the Roe plow earlier, by publishing a brief article about it in the first issue of the *CREA News Letter* in April 1928. "Promising New Uses Deserving Consideration," *CREA Bulletin* 1 (16 April 1928): 3. The direct current generator, the newsletter reported, was powered by a take-off shaft on the tractor to produce a 100,000-volt spark discharge, which flowed from the plow's coulter to the two in-ground plow blades. Ibid.

65. Letter from M. L. Nichols to R. B. Cray, USDA, Toledo, OH, 13 January 1928, in AU Special Collections, Agricultural Engineering Records, record group 503, accession no. 79-001, box 24 (of 64).

66. "Electric Plow," *CREA Bulletin* 6, no. 1 (June 1931): 25.

67. Ibid. Winterberg's thesis was called "Influence of Cultivation with an Electric Plow on Soil and Crop Response," as noted in University of Maryland, *Biennial Report of the University of Maryland and State Board of Agriculture* 27, no. 12 (December 1930): 52. Also see *Final Report of the National Rural Electric Project*, report no. M-16 (College Park, MD: National Rural Electric Project, July 1933), which noted (on page 11) that "the electric discharge did not increase or decrease yield, bacterial activity, available plant food, or the larval population of the soil."

68. "Electro-Culture," *CREA Bulletin* 6, no. 1 (June 1931): 26.

69. Robert Stewart, "Electricity in the Service of Agriculture," *Bankers Magazine* 116, no. 2 (February 1928): 195.

70. "Electric Plow," 26.

71. "Research Activities of Different Agencies," *CREA Bulletin* 6, no. 1 (June 1931): 5.

72. F. S. Dewey, "Threshing with Electricity in Iowa," *Electrical World* 64, no. 3 (22 August 1914): 378–79.

73. E. A. Stewart, J. M. Larson, and J. Romness, *The Red Wing Project on Utilization of Electricity in Agriculture*, publication of the University of Minnesota Agricultural Experiment Station, no date, but listed on website as 1927, 1, University of Minnesota Libraries Digital Conservancy, http://conservancy.umn.edu/handle/11299/48685, accessed 15 September 2014, 98.

74. Ibid., 114.

75. Ibid., 99.

76. Ibid., 114.

77. Ibid., 114, 118. A similar account appeared in "Threshing," *CREA Bulletin* 7, no. 1 (November 1931): 284–88. Though seeming moderately encouraging, these reports carried several caveats, as noted, to suggest that in-field threshing with electric motors was not practical. Moreover, other techniques for supplying portable power became more popular during the 1920s, as discussed in the next paragraph.

78. Stewart, Larson, and Romness, *The Red Wing Project*, 118.

79. Williams, *Fordson, Farmall, and Poppin' Johnny*, 61–63. Also see the advertisement from United States Rubber Company, "When You Buy a Fordson Buy a Little Giant Belt," with a picture of a Fordson tractor powering a threshing machine, *Power Farming* 29, no. 8 (August 1920): 42; and Goodyear advertisement for "Klingtite" farm belts, "Profitable Threshing and Goodyear Belts," *Power Farming* 29, no. 8 (August 1920): 5. An article in February 1920 noted, "The tractor has made practical the owning and operating of small threshers fitted with all labor-saving attachments by farmers individually or in small groups." H. L. Thomson, "Why Home Threshing Is Better," *Power Farming* 29, no. 2 (February 1920): 9, 20.

80. Stewart, Larson, and Romness, *The Red Wing Project*, 120. This conclusion seemed to verify the impression, based on surveys, held by Kansas CREA researchers as early as 1925. "The tractor is the competitor of the electric motor for farm belt work, particularly

for heavy loads, such as threshing, silo filling, and in some cases grinding and shredding." H. B. Walker, "Present Status of Rural Electrification in Kansas: Central Station Service," *Kansas State Agricultural College Bulletin* 9, no. 7 (1 July 1925): 19.

81. J. Romness, "The Trend in the Use of Electricity on the Farm," *Agricultural Engineering News Letter*, no. 33, Agricultural Extension Division, University Farm, University of Minnesota, St. Paul, 15 December 1934, 1.

82. Ibid.

83. J. Romness, "Common Uses of Electricity on the Farm," *Agricultural Engineering News Letter*, no. 47, Agricultural Extension Division, University Farm, University of Minnesota, St. Paul, 15 February 1936, 1.

84. M. L. Nichols, "Experiments with Solar Heating," 1926 or 1927, and M. C. Easter and M. L. Nichols, "Electric Range on the Farm," typed reports in AU Special Collections, Agricultural Engineering Records, record group 503, accession no. 79-001, box 8 (of 64), in folder marked "1923–24." Often, I found that documents were placed in incorrectly marked folders, as seems the case here, since work on CREA projects only began in 1924.

85. "Artificial Sunshine," *CREA Bulletin* 7, no. 1 (November 1931): 53–54.

86. "Ultra-violet Light for Poultry," *CREA Bulletin* 7, no. 1 (November 1931): 187–92. CREA director White admitted in 1938 that the results from similar experiments (which included livestock as well as poultry) "have not been conclusive." E. A. White, "Fifteenth Annual Report to the Committee on the Relation of Electricity to Agriculture by the Director," 20, typescript of presentation given at 15th annual meeting of CREA, Chicago, 12 October 1938, in ISU Special Collections, James R. Howard collection, box 7, folder 10.

87. "Insect Control by Radio Waves," in *The National Rural Electric Project: What It Is— Its Purpose, Activities Cooperation and Personnel*, 13, suggested the promise of using radio waves, though the failure was noted in the *Final Report of the National Rural Electric Project*, report no. M-16 (College Park, MD: National Rural Electric Project, July 1933), 11.

88. Walker, "Research Reveals Opportunities," 13. Walker refers to W. F. Gericke and J. R. Tavernetti, "Heating of Liquid Culture Media for Tomato Production," *Agricultural Engineering* 17, no. 4 (April 1936): 141–42.

89. Walker, "Research Reveals Opportunities," 5.

90. "Report on Ripon Project to the Wisconsin Committee on the Application of Electricity to Agriculture," typescript included in "Committee on the Application of Electricity to Agriculture," report of meeting held 8 May 1925, in UW Archives, WCREA, series #9/12/5-1, box 1.

91. Ben D. Moses, "Rural Electric Investigation in California," *Agricultural Engineering* 6, no. 7 (July 1925): 156. More advice on building and operating brooders appeared in J. E. Dougherty and B. D. Moses, *Construction and Operation of Electric Brooders*, Circular 325 (Berkeley: University of California College of Agriculture, Agricultural Experiment Station, 1931).

92. Oscar E. Anderson Jr., *Refrigeration in America* (Princeton, NJ: Princeton University Press, 1953), 196.

93. Jonathan Rees, *Refrigeration Nation: A History of Ice, Appliances, and Enterprise in America* (Baltimore, MD: Johns Hopkins University Press, 2013), 140–47.

94. Ibid., 158; and Anderson, *Refrigeration in America,* 197.

95. "Electric Domestic Refrigeration," NELA, *Proceedings* 81 (1924): 458. Such concerns likely explain why a survey of 1,786 electrified farms performed in New York State showed just 12 having refrigerators in 1926. R. F. Bucknam, *An Economic Study of Farm Electrification in New York,* Bulletin 496 (Ithaca, NY: Cornell University Agricultural Experiment Station, 1929), 17.

96. Brigham, Zink, and Paine, "Electric Service for the Farm, Report No. 6," 24. For other discussions of use of refrigerators (and their costs), see Shelley Nickles, "'Preserving Women': Refrigerator Design as Social Process in the 1930s," *Technology and Culture* 43, no. 4 (October 2002): 696; and David E. Nye, *Electrifying America: Social Meanings of a New Technology, 1880–1940* (Cambridge, MA: MIT Press, 1990), 275. A description of the market for farm use of refrigerators can be found in W. T. Ackerman, "Rural Refrigeration," *Refrigerating Engineering* 17, no. 1 (January 1929): 1–4. The modest use (and rejection) of gas-powered refrigerators is discussed in Ruth Schwartz Cowan, *More Work for Mother: The Ironies of Household Technology from the Open Hearth to the Microwave* (New York: Basic Books, 1983), 128–42.

97. Carroll Gantz, *Refrigeration: A History* (Jefferson, NC: McFarland & Co., 2015), 114. In mid-1929, *Electrical World* reported that refrigerators could be found in only 6.5 percent of wired (urban and rural) homes. "This Domestic Business," *Electrical World* 93, no. 21 (25 May 1929): 1039.

98. Ronald C. Tobey, *Technology as Freedom: The New Deal and the Electrical Modernization of the American Home* (Berkeley: University of California Press, 1996), 122–23; and "Model T Appliances," *Business Week,* 16 June 1934, 11.

99. Rees, *Refrigeration Nation,* 160–61.

100. "Domestic Refrigerators," *CREA Bulletin* 7, no. 1 (November 1931): 33–41.

101. Ronald R. Kline, "Resisting Development, Reinventing Modernity: Rural Electrification in the United States before World War II," *Environmental Values* 11, no. 3 (August 2002): 336.

102. M. S. Winder, "Saving the Farm for the Family," NELA, *Proceedings* 87 (1930): 62.

103. "Report of the Committee on Electricity in Rural Districts: Central States," NELA, *Proceedings* 48 (1913): 173; and "Rural Uses of Electricity in 1946," typescript in WSU Special Collections, Leslie John Smith Papers (rural electrification, 1923–46) folder, cage 5034.

104. Buel W. Patch, "Rural Electrification and Power Rates," *Editorial Research Reports 1934,* vol. 2 (Washington, DC: CQ Press, 1934): 381–96, in *CQResearcher,* http://library.cqpress.com/cqresearcher/cqresrre1934120400, accessed 13 April 2015.

105. Morris L. Cooke, "The New Viewpoint," *Rural Electrification News* 1, no. 2 (October 1935): 2.

106. John M. Carmody, "Rural Electrification: Progress and Future Prospects," *Journal of Farm Economics* 20, no. 1 (February 1938): 365.

107. REA, *Report of Rural Electrification Administration, 1938* (Washington, DC: GPO, 1939), 113.

108. Harry Slattery, *Rural America Lights Up* (Washington, DC: National Home Library Foundation, 1940), 21.

109. Ibid.

110. D. Clayton Brown, *Electricity for Rural America: The Fight for the REA* (Westport, CT: Greenwood Press, 1980), 3.

CHAPTER TEN: **Regulation and the Extension of Lines to Rural Areas**

Epigraph: Wisconsin Public Service Commission, 1936, 14 P.U.R. (N.S.) 25 in "Re Extensions of Rural Lines of Electric Utilities," in Francis X. Welch, ed., *Cases on Public Utility Regulation, with Supplemental Notes* (Washington, DC: Public Utilities Reports, 1938), cumulative supplement no. 2, 138.

1. In the early days of public utilities, city leaders "hailed with enthusiasm" the seekers of franchises, who would bring new, modern services that would help metropolises thrive economically. Delos F. Wilcox, *Municipal Franchises: A Description of the Terms and Conditions upon which Private Corporations Enjoy Special Privileges in the Streets of American Cities*, vol. 1 (Rochester, NY: Gervaise Press, 1910), 2–3.

2. New York State granted companies the right to do business, while political subdivisions—cities—gave companies the authority to obtain access to streets and other public resources. Leonora Arent, *Electric Franchises in New York City* (New York: Columbia University, 1919), 18.

3. Milo R. Maltbie, *Franchises of Electrical Corporations in Greater New York, a Report Submitted to the Public Service Commission for the First District* (New York: Public Service Commission for the First District, State of New York, 1911), 16.

4. Ibid., 8–9.

5. Delos Wilcox, a PhD graduate from Columbia University and New York City's chief of the Bureau of Franchises of the Public Service Commission, wrote that the city alderman, who worked with others to grant franchises, viewed "himself as a lord of privileges" but was too often "ignorant and incapable of constructive civic statesmanship." Nevertheless, he was easily "[f]lattered by the attentions of the courtly agents of corporate wealth" and "too glad to grant what franchise seekers have asked." Wilcox, *Municipal Franchises*, 8. A similar account is presented in W. H. Downey, "Regulation of Urban Utilities in Iowa," in Benjamin F. Shambaugh, ed., *Applied History* 1 (1912): 131.

6. Wilcox, *Municipal Franchises*, 29.

7. Richard F. Hirsh, *Power Loss: Deregulation and Restructuring in the American Electric Utility System* (Cambridge, MA: MIT Press, 1999), 20–23.

8. I. Leo Sharfman, "Commission Regulation of Public Utilities: A Survey of Legislation," *Annals of the American Academy of Political and Social Science* 53 (May 1914): 3–5; and George J. Stigler and Claire Friedland, "What Can Regulators Regulate? The Case of Electricity," *Journal of Law and Economics* 5 (October 1962): 4.

9. Though some commissions in their early years were led by acknowledged experts in utility practice, such as New York's Milo Maltbie, many bodies became populated by political cronies of governors (who appointed them) or others who had little interest in regulating effectively. They also came under the influence of the power companies they supposedly oversaw in a process known as "regulatory capture." See discussion in Hirsh, *Power Loss*, 41–46.

10. A discussion of the obligation of publicly regulated entities to serve is included in Jim Rossi, "The Common Law 'Duty to Serve' and Protection of Consumers in an Age of Competitive Retail Public Utility Restructuring," *Vanderbilt Law Review* 51 (1998): 1233–320;

Jim Rossi, "Universal Service in Competitive Retail Electric Power Markets: Whither the Duty to Serve?," *Energy Law Journal* 21, no. 1 (2000): 27–49; and Floyd L. Norton IV and Mark R. Spivak, "The Wholesale Service Obligation of Electric Utilities," *Energy Law Journal* 6, no. 2 (1985): 179–208.

11. Forrest McDonald, *Let There Be Light: The Electric Utility Industry in Wisconsin, 1881–1955* (Madison, WI: American History Research Center, 1957), 279n8.

12. The situation in which undeveloped territories lay between companies' specified monopoly zones is discussed in US Senate, 70th Cong., 1st sess., doc. 46, *Electric-Power Industry: Supply of Electrical Equipment and Competitive Conditions* (Washington, DC: GPO, 1928), 208–13.

13. Utilities providing gas, water, and electric services could sometimes refuse to extend infrastructure to customers in rural areas even if the companies had earned franchises or charters, as was made clear in a 1924 annotation to an order published by the Oregon regulatory body. It noted, "The refusal of utility companies to place undue burdens upon present customers by extending service into sparsely settled territory has been sustained by the Commissions in some instances." "Oregon Public Service Commission, Re: Establishment of Uniform Rules and Regulations," ruling on 14 October 1922, in Henry C. Spurr, ed., *Public Utilities Reports*, 1923A (Rochester, NY: Public Utilities Reports, 1923), 841.

14. Sometimes, companies needed to request certificates of public convenience and necessity from regulatory bodies; the legal documentation served as approval to operate as a monopoly in an area and avoid destructive competition.

15. Federal Power Commission, *Electric Rate Survey, Rate Series No. 6* (Washington, DC: GPO, 1936), 9.

16. Ibid., 8.

17. A discussion of the Virginia Utility Facilities Act of 1950 can be found in Evans B. Brasfield, "Regulation of Electric Utilities by the State Corporation Commission," *William and Mary Law Review* 14, no. 3 (Spring 1973): 589–600.

18. Efforts to allow the Florida state commission to set aside specific areas to utilities by legislative actions failed to win approval as late as 1991. Richard C. Bellack and Martha Carter Brown, "Drawing the Lines: Statewide Territorial Boundaries for Public Utilities in Florida," *Florida State University Law Review* 19, no. 2 (Fall 1991): 407–35. For more information on the allocation of territories, see G. Lloyd Wilson, James M. Herring, and Roland B. Eutsler, *Public Utility Regulation* (New York: McGraw-Hill, 1938), 51; and Herman H. Trachsel, *Public Utility Regulation* (Chicago: Richard D. Irwin, 1947), 61.

19. *Report of the Railroad Commission of California from July 1, 1915 to June 30, 1916*, vol. 1 (San Francisco: California State Printing Office, 1916), 6, 199; and "Mount Whitney Power and Electric Company and the San Joaquin Light and Power Corporation Rate Cases," in "Commission Decisions: California," *Rate Research* 9, no. 12 (22 June 1916): 183–89. In 1911, the commission acquired the right to oversee electric, gas, telegraph, telephone, and water utilities. Its name changed to the California Public Utilities Commission in 1946. Peter E. Mitchell, "The History and Scope of Public Utilities Regulation in California," *Southern California Law Review* 30, no. 2 (February 1957): 120.

20. State of Illinois Public Utilities Commission, *General Order 59* (Springfield: Illinois State Journal Co., 1920), 3.

21. Comments by S. M. Kennedy made following presentation of "Report of Committee on Electricity in Rural Districts," NELA, *Proceedings* 40 (1911): 529–30.

22. "Report of the Committee on Electricity on the Farm, Western States," NELA, *Proceedings* 48 (1913): 151–52.

23. Lemont Kingsford Richardson, *Wisconsin REA: The Struggle to Extend Electricity to Rural Wisconsin, 1935–1955* (Madison: University of Wisconsin Experiment Station, 1960), 7. Terms for customers to pay the full cost of construction before the utility started building lines can be found in "In Re Application of the Southern Wisconsin Electric Company for the Approval of Its Rules for Rural Electric Extensions," decided 8 April 1921, in *Opinions and Decisions of the Railroad Commission, State of Wisconsin* 25 (1924): 390.

24. G. C. Neff, "Factors Governing Rural Extensions," *Electrical World* 76, no. 25 (18 December 1920): 1205–7.

25. Editorials and articles in *Electrical World* often supported rural electrification but insisted that utilities should not bear a financial burden in providing the new service. See, for example, "Utilities Should Maintain, but Not Invest in, Rural Lines," *Electrical World* 78, no. 4 (23 July 1921): 186; and "New England Company's Rural Lines Are Made Self-Supporting," *Electrical World* 78, no. 27 (31 December 1921): 1333–34.

26. "Report of the Rural Lines Committee," NELA, *Proceedings* 77 (1922): 110.

27. In an example given by the company, a mile of line may generate revenue from eight customers of $6 per month (or $1,728 over three years). If the extension cost $2,500, the farmers would be charged the difference, $772, or about $98 (a little more than the average of $96.50) on a one-time basis. "Tentative Rural Service Policies," *Electrical World* 87, no. 6 (6 February 1926): 292.

28. Richardson, *Wisconsin REA*, 8. The rules approved by the Wisconsin Railroad Commission in 1926, which called for the $400 contribution by Wisconsin Power and Light, can be found in "In Re Wisconsin Power and Light et al.," decided 21 June 1926, in *Opinions and Decisions of the Railroad Commission of Wisconsin* 29 (1926): 438. Under the new approach, Wisconsin Power and Light eliminated sixty-nine different rate schedules in effect for farms, replacing them with one. The company's good experiences in developing a rural market (which included implementation of new rate policies) were described in G. C. Neff, "Essentials for Development of Rural Electric Service," NELA, *Proceedings* 86 (1929): 228–39; and Will C. Conrad, "State's Farms Get Electricity," *Milwaukee Journal*, 10 January 1931, 9.

29. Morris Llewellyn Cooke and Judson C. Dickerman, *Report of the Giant Power Survey Board to the General Assembly of the Commonwealth of Pennsylvania* (Harrisburg, PA: Telegraph Printing Co., 1925), 37–38.

30. Thomas Parke Hughes, "Technology and Public Policy: The Failure of Giant Power," *Proceedings of the IEEE* 64, no. 9 (September 1976): 1365.

31. Jean Christie, "Giant Power: A Progressive Proposal of the Nineteen-Twenties," *Pennsylvania Magazine of History and Biography* 96, no. 4 (October 1972): 480–507.

32. An outline of the order can be found in "Report on Commission Orders and Decisions Materially Affecting Rural Electric Service," NELA, *Proceedings* 84 (1927): 86.

33. Pennsylvania Joint Committee on Rural Electrification, *Rural Electrification in Pennsylvania* (Harrisburg, PA: self-published, 1928), 2, 24, 26.

34. Ibid., 26.

35. Ibid., 25.

36. W.D.B. Ainey, "Pennsylvania Utilities," *Electrical World* 91, no. 6 (11 February 1928): 307–9.

37. Charles Seitz, VPI agricultural engineer and project leader of Virginia's CREA affiliate, noted that Rule 18 was modeled on Pennsylvania's Order No. 28. The similarity between Virginia's policy and Pennsylvania's General Order 28 is noted in "Farm Electrification West, East, South," *Electrical World* 93, no. 19 (11 May 1929): 948.

38. Verne Hillman, "A Study of Plans and Policies of Power Companies in Dealing with Rural Customers" (MS thesis, Virginia Polytechnic Institute, 1930), 16–17.

39. Clifford V. Gregory, "Farmers Machinery-Minded," *Electrical World* 91, no. 23 (9 June 1928): 1245–46.

40. John G. Learned, "The Commercial Job," *Electrical World* 94, no. 7 (3 August 1929): 318–22.

41. M. S. Winder, "Saving the Farm for the Family," NELA, *Proceedings* 87 (1930): 63.

42. "Northern California Companies to Pay for Rural Lines on Customers' Property," *Electrical World* 93, no. 11 (March 1928): 558.

43. R. F. Bucknam, "Farm Electrification from Management Viewpoint," *Electrical World* 93, no. 20 (18 May 1929): 986–87.

44. William E. Mosher and Finla G. Crawford, *Public Utility Regulation* (New York: Harper and Brothers, 1933), 474. Mosher and Crawford were professors at Syracuse University's Maxwell Graduate School of Citizenship and Public Affairs.

45. For example, by mid-1928 alone, fifteen Ohio companies filed rural electric rate plans. "Fifteen Ohio Companies File Rural Service Plans," *Electrical World* 91, no. 25 (23 June 1928): 1361.

46. E. A. Holloway, N. T. Wilcox, W. H. Horton Jr., and H. M. Weathers, "A Review of Some Recently Developed Rural Electric Rates," American Society of Agricultural Engineers, Committee on Rural Rates, 1928, typescript report in Washington State University, Owen Science and Engineering Library, Pullman, WA, Dewey catalog number 631.3717 Am35r.

47. Benjamin Hodge Nichols, "Thesis on Rural Electric Rates and Rural Line Extension Policies of the United States" (MS thesis, Oregon State Agricultural College, 1932), 61.

48. Neff, "Essentials for Development," 239.

49. R. F. Bucknam, "Energy Use on New York Farms Increases 59 per Cent," *Electrical World* 94, no. 10 (7 September 1929): 467–68.

50. "New England Adopts Rural Service Yardstick," *Electrical World* 93, no. 25 (22 June 1929): 1273.

51. Farm Use Averages Three Times Domestic," *Electrical World* 100 (17 September 1932): 376–77; and "Average Farm Customer Pays $81.40 Annually," *Electrical World* 100 (12 November 1932): 664–65.

52. "How Stands Rural Electrification?" *Electrical World* 99 (28 May 1932): 964–65.

CHAPTER ELEVEN: **Momentum in the Rural Electrification Subsystem**

Epigraph: Eugene Holcomb, "What Progress? How Much Headway Are We Making in Rural Electrification," address at Third Annual Conference on Rural Electrification, Purdue University, 10 October 1929, in *Electricity on the Farm* 2, no. 11 (November 1929): D5; emphasis in the original.

1. "Nineteenth Annual Catalog of the University of Idaho, 1910–1911 with Announcements for 1911–1912," *University of Idaho Bulletin* 6, no. 4 (8 June 1911): 129.

2. "Twenty-Ninth Annual Catalog of the University of Idaho with Announcements for 1921–1922," *University of Idaho Bulletin* 16, no. 13 (1921): 135, which listed "Electricity on the Farm" as a two-credit course, with one lecture and one three-hour laboratory each week. In the next year's catalog, "Thirtieth Annual Catalog of the University of Idaho with Announcements for 1922–1923," *University of Idaho Bulletin* 17, no. 13 (1922): 146, the "Electricity on the Farm" course offered three credits, with two lectures and one three-hour laboratory weekly.

3. Agricultural Engineering 404, "Farm Light Plants, Etc.," in *Catalog, State College Record* 19, no. 11 (Raleigh: North Carolina State College of Agriculture and Engineering, April 1921): 116.

4. Charles E. Seitz, "Agricultural Engineering Development in Virginia," *Agricultural Engineering* 4, no. 4 (April 1923): 60.

5. *Thirty-Second Annual Catalogue of the State College of Washington for 1923* (Olympia, WA: Frank M. Lamborn, Public Printer, 1923), 82.

6. The catalog for the University of California, Davis, for example, listed a course in 1928 on "Agricultural Power." Agricultural Engineering course 103, in University of California, *Announcement of Courses of Instruction Primarily for Students, 1928–1929* (Berkeley: University of California Press, 1928), 13. The University of Wisconsin offered a course in 1933 on "Power and Machinery," which included the study of "light plants." *General Announcement of Courses, 1934–35 (Catalog 1933–34)* (Madison: University of Wisconsin, 1934), 304.

7. The University of Idaho offered course 139, "Rural Electrification," in its 1929 catalog. *University of Idaho Bulletin* 24, no. 12, *Annual Catalog 1928–1929* (May 1929): 118. The same-named course appeared in Michigan State College's 1928–29 listing, described as a "study of the rural electrification problem . . . from both the utility and farm viewpoint." *Michigan State College Bulletin* 23, no. 10, catalog no. 1928-1929 (May 1929): 120. The Massachusetts Agricultural College (University of Massachusetts–Amherst) offered Agricultural Engineering course 82, "Rural Electrification," in 1930. *M.A.C. Bulletin* 22, no. 2 (February 1930): 46.

8. Letter from Charles E. Seitz to Dr. Julian A. Burruss, president, VPI, 19 October 1929, in VT Special Collections, Papers of Julian Ashby Burruss (1919–45), RG2/8, box 13, folder 1052, "Burruss 1929."

9. R. R. Choate, "Rural Electrification in the Roanoke District of the Appalachian Electric Power Company—1929," typescript dated 26 November 1929, in "Annual Report—Agricultural Engineering Project No. 10," Virginia Tech Project 10, https://vtechworks.lib.vt.edu/handle/10919/89970, accessed 25 September 2019. To help teach farmers about wisely using electricity, Choate's company distributed free monthly issues of the magazine *Electricity on the Farm*. Ibid., 2. See a discussion of this magazine later in this chapter.

10. Alabama Power Company, "Accomplishments in Rural Electric Development by Alabama Power Company during 1932," *Edison Electric Institute Bulletin* 2, no. 1 (January 1934): 12. Perhaps it is no surprise that this paper won the first Martin Award. Alabama Power Company president, Thomas Martin, personally gave money to the EEI in

1932 to establish the prize, which his firm won in 1932 and again in 1947. Thomas W. Martin, *The Story of Electricity in Alabama: Since the Turn of the Century, 1900–1952* (Birmingham, AL: Birmingham Publishing Co., 1953), 77. Even so, the existence of an industry-awarded prize recognizing achievements in rural electrification suggests that a growing number of utility leaders saw the activity as worthwhile.

11. Alabama Power Company, *Commemorating the Era of Rural Electrification in Alabama, 1924–1951* (N.p.: n.p., Online Computer Library Center no. 12073200, 1951), 16.

12. "Report of the Rural Electric Service Committee," NELA, *Proceedings* 84 (1927): 83.

13. "Electric Service for Farms," Nebraska Gas & Electric Company advertisement in *Nebraska Farmer*, 4 June 1927, 919.

14. "Rural Electric Service Committee," NELA, *Proceedings* 85 (1928): 139. The report lists the companies and the names of rural service agents. "Appendix A: Directory of Rural Electric Service Departments and Specialists," ibid., 143–47. Also see Hanina Zinder, "Problems of Rural Electric Service: The Potential Market," *Journal of Land & Public Utility Economics* 4, no. 4 (November 1928): 339; and B. H. Miller, "Utilities Cooperate in Rural Electrification," *Electricity on the Farm* 3, no. 2 (February 1930): S3–S5.

15. G. C. Neff, "The Farmer Gets Electric Service: Power Industry Concentrating on Efforts to Give Greater Service and Stimulate Demand," *Wall Street Journal*, 7 June 1928, 8.

16. "Appendix C2: Duties of Rural Service Department of the Northern States Power Company," NELA, *Proceedings* 84 (1927): 92.

17. NELA, Public Speaking Committee, *NELA Handbook* (New York: NELA, 1928), 70.

18. "A Training Course for Farm Electrification Specialists," *General Electric Review* 30, no. 12 (December 1927): 585.

19. George W. Kable, "The Rural Electrification Problem," *1929 Institute of Rural Affairs Proceedings* in *Bulletin of the Virginia Polytechnic Institute* 23, no. 3 (January 1930): 167.

20. NELA, *Progress in Rural and Farm Electrification for the 10 Year Period 1921–1931* (New York: NELA, 1932), 13.

21. "Survey of the National Scope of Extension Activities for the Promotion of Rural Electrification, 1930–1931," typescript report, 3, in UW Archives, Wisconsin Committee on the Relation of Electricity to Agriculture (WCREA), Correspondence and Records, 1929–1933, A–R, box 1.

22. The publications can be found in NELA, *Proceedings* 86 (1929): 225–54. Alabama Power in 1930 employed six agricultural engineers. "Distribution of Electric Service to Farmers and Rural Communities," 4, of unattributed author's typescript (perhaps M. L. Nichols with date references from 1924 to 1930), in AU Special Collections, Agricultural Engineering Records, record group no. 503, accession no. 79-001, box 8 (of 64), in folder marked "1923."

23. "Building the Load on the Farm," part of the report of the NELA Rural Electric Service Committee, NELA, *Proceedings* 86 (1929): 257.

24. L. E. May, "Organization and Operation of Rural Service Departments," typescript of paper presented at the Fifth Annual Rural Electrification Conference, 11 November 1932, in Columbus, Ohio, 1, in UW Archives, WCREA, Correspondence and Records, 1929–1933, A–R, series #9/12/5-2, box 1.

25. H. W. Yong, "The Weatherproof Substation as a Means of Extending Central-Station Service," in section 3, "Rural Distribution," of "Report of Committee on Electric-

ity in Rural Districts, Central States," NELA *Proceedings* 48 (1913): 218–19, 222. For the standard voltages for distribution of alternating current, see *Cyclopedia of Applied Electricity* (Chicago: American Technical Society, 1923), 108; and John R. Benton, *An Introductory Textbook of Electrical Engineering* (Boston: Ginn and Co., 1928), 205, 226.

26. L. H. Perry, "Rural Substation Tapped to 66-Kv. Line," *Electrical World* 93, no. 24 (15 June 1929): 1246.

27. J. H. Powers, "Automatic Switching Unit Cuts Rural Costs," *Electrical World* 94, no. 5 (3 August 1929): 223.

28. "Report of the Overhead Systems Committee," NELA, *Proceedings* 76 (1921): 651.

29. Comments by Grover Neff after presentation of "Report of Overhead Systems Committee," made in ibid., 852.

30. G. C. Neff, "Rural Line Construction," NELA, *Proceedings* 80 (1923): 766–67. Data on previously common spans come from US Department of Commerce, Bureau of the Census, *Central Electric Light and Power Stations and Street and Electric Railways* (Washington, DC: GPO, 1915), which described one company using 25-foot-high poles spaced at 132 feet ("40 to the mile"), 155. A 66,000-volt branch line in New England built in 1920 employed twenty-eight poles per mile, meaning that spans averaged 189 feet. "Cost Estimate of 66-Kv Spur Line," *Electrical World* 77, no. 8 (19 February 1921): 431. In Kansas, engineers reported in 1920 that for small communities of about five hundred people, utilities used poles spaced from twenty-six to thirty-five per mile, meaning spans of 151 to 203 feet. F. H. Frauens Jr., "Transmitting Electric Current to Small Kansas Towns," *Municipal and County Engineering* 59, no. 4 (October 1920): 125.

31. Neff, "Rural Line Construction," 767.

32. George W. Kable, "Low-Cost Rural Construction," *Electrical World* 91, no. 20 (19 May 1928): 1015–17.

33. "Wiring and Line Studies," *CREA Bulletin* 6, no. 1 (June 1931): 68–69.

34. E. V. Sayles, "Mechanical Design of Rural Lines," *Electrical World* 100 (10 September 1932): 347–48.

35. The Copper Clad Steel Company apparently designed (in 1915) a method for manufacturing a conductor that combined the best properties of the two metals. *History of Pittsburgh and Environs*, vol. 3 (New York: American Historical Society, 1922), 24–25.

36. Mark Eldredge, "Farm Line Construction," *Electrical World* 102 (26 August 1933): 268–72.

37. L.W.W. Morrow, "Loads for Rural Lines," *Electrical World* 102 (26 August 1933): 266–67.

38. Alex Dow, "Where Do We Go from Here?" *Edison Electric Institute Bulletin* 1, no. 3 (June 1933): 60.

39. Philip Sporn, "Progressive Engineering Pays Its Way," *Edison Electric Institute Bulletin* 2, no. 7 (July 1934): 233.

40. R. L. Duffus, "Federal Aid to Bring Power Lines to Farms," *New York Times*, 31 May 1936, E11.

41. Established by the General Electric Company, the Coffin Award recognized a company that made significant contributions in the realm of electric power and not just for work in rural electrification. A caption in an *Electrical World* notice reporting on the prize aptly stated that the Idaho Power Company's service area was characterized by "Low

Density—High Consumption." "Idaho Power Company Is Winner of Coffin Award," *Electrical World* 106, no. 23 (6 June 1936): 61; and Frank W. Smith, "Report of the Prize Awards Committee," *Bulletin of the Edison Electric Institute* 4, no. 7 (July 1936): 290–91.

42. A. E. Silver, "Pioneering and Development of Electrical Service to the Farm," *Electrical Engineering* 71, no. 8 (1952): 700–702. A history of Idaho Power credits Silver as having designed, in 1911, the first open-air high-voltage substation, which reduced costs by avoiding the need for an expensive building surrounding the transformer equipment. Susan M. Stacy, *Legacy of Light: A History of the Idaho Power Company* (Boise: Idaho Power Co., 1991), 84. The REA's chief engineer, M. O. Swanson, in 1937, gave credit to private utilities that employed—like his agency—long-span construction. "Almost without exception," he wrote, "private utility construction has come to be of long spans, taking advantage of natural topography. For example, 1,000-foot spans are not uncommon in new lines just completed in northern Pennsylvania." "Average Span 579 Feet," *Rural Electrification News* 2, no. 9 (May 1937): 18.

43. The program listed Davison's affiliation as "Research Department, National Electric Light Association." In a 1941 biographical sketch, she was described as having been a "home economics adviser and consultant in household engineering." "Key Woman in Civilian Defense," *Crescent of Gamma Phi Beta* 61, no. 4 (December 1941): 3–4. Also see Ronald R. Kline, "Agents of Modernity: Home Economists and Rural Electrification, 1925–1950," in Sarah Stage and Virginia B. Vincenti, eds., *Rethinking Home Economics: Women and the History of a Profession* (Ithaca, NY: Cornell University Press, 1997): 241–42.

44. Program brochure for "Agricultural Short Course for Rural Service Men," held at Purdue University, 18–21 October 1927, in UW Archives, WCREA, Correspondence and Records, 1929–1933, S–Z, "Short Courses: Rural Electrification OUTSIDE," series #9/12/5-2, box 2. The same folder also contains programs for Purdue's short courses in 1928, 1929, and 1930.

45. "Report of the Fifth Annual Meeting of the Rural Electric Project Leaders (June 1928)," typescript, 8, in WSU Special Collections, Washington Farm Electrification Committee Records, cage 607, Annual and Executive Minutes 1925–41, box 1, folder 1.

46. "Tied to a Piece of Wire," editorial in *Pacific Rural Press*, 22 December 1928, 682, in UCD Special Collections, Moses Collection, call number D-181, box 1 (of 1), scrapbook #2, folder 3.

47. Program brochure for "Second Agricultural Short Course for Rural Service Men," held at Purdue University, 17–19 October 1928, in UW Archives, WCREA, Correspondence and Records, 1929–1933, S–Z, "Short Courses: Rural Electrification OUTSIDE," series #9/12/5-2, box 2. The Washington short course was described (with photographs) in "Rural Electric Problems Studied," *Brewster (WA) Herald*, 22 March 1929, 10.

48. Program for "Rural Electric Short Course," State College of Washington, Pullman, 3, 4, 5 March 1931, produced with the cooperation of the Agricultural Engineering Department and the Washington CREA, in UW Archives, WCREA, Correspondence and Records, 1929–1933, S–Z, "Short Courses: Rural Electrification, OUTSIDE," series #9/12/5-2, box 2.

49. "Experts on Rural Work to Convene: Authorities on Electrification Will Hold Blacksburg Conference," *News Leader* (Richmond, VA), 22 April 1929, clipping in VT Special Collections, Department of Agricultural Engineering papers, RG13/4, box 2, "Rural Electrification Conference, VPI, 1929."

50. "Rural Electrification," *Wythe County News* (Wytheville, VA), 3 May 1929, clipping in ibid. Also see "Report of Joint Committee on Rural Electrification," in *Report of the Commission to Study the Condition of the Farmers of Virginia to the General Assembly of Virginia, January, 1930* (Richmond, VA: Division of Purchase and Printing, 1930), 124–32. The creation of the committee was reported in "Virginia Organizes a Farm Electrification Committee," *Electrical World* 93, no. 13 (30 March 1929): 654.

51. "Report of Joint Committee on Rural Electrification," 125; and "Rural Electrification in Virginia," typed description of history of rural electrification in the state, in VT Special Collections, Department of Agricultural Engineering papers, RG13/4, box 2, "Rural Electrification Conference, VPI, 1929"; and "First Annual Rural Electrification Short Course to Be Held at Virginia Polytechnic Institute," 12–14 June 1929, very similar to typewritten program in Burruss collection, but printed, in UW Archives, WCREA, Correspondence and Records, 1929–1933, S–Z, "Short Courses: Rural Electrification, OUTSIDE," series #9/12/5-2, box 2.

52. Program of the Rural Electrification Conference, 12, 13, 14 June 1929, accompanying a letter from Charles E. Seitz to Julian A. Burruss, 1 May 1929, in VT Special Collections, Papers of Julian Ashby Burruss (1919–1945), RG2/8, box 13, folder 1052, "Burruss 1929." The February 1929 short course at VPI received notice in "Rural Electrification Committee Formed for Virginia," *Electrical World* 93, no. 8 (23 February 1929): 411; and "Virginia Polytechnic to Inaugurate Rural Electrification Short Course," *Electrical World* 93, no. 16 (20 April 1929): 802.

53. "Virginia's First Rural Electrification Short Course," 2, in VT Special Collections, Department of Agricultural Engineering papers, RG13/4, box 2, "Rural Electrification Conference, VPI, 1929." Similarly impressive, the University of Wisconsin course held in 1929 listed 104 registrations, a number that jumped to 307 the following year.

54. A. T. Pamperin, Wisconsin Public Service Corporation, "A Country Rural Electrification Campaign," talk presented at the 1929 Short Course at the University of Wisconsin, College of Agriculture, 31 October to 2 November 1929, 3–4, in UW Archives, WCREA, Correspondence and Records, 1929–1933, S–Z, "Second Annual Rural Electrification Short Course for Service Men," series #9/12/5-2, box 2. General Electric produced the film *The Yoke of the Past* in 1926. It can be seen on YouTube, https://www.youtube.com/watch?v=-hKALCfXil4, accessed 19 May 2014. Near the end of the film, the story boards proclaimed, "Just as the telephone, the automobile, the motion picture and radio have abolished the desolating lonesomeness of farm life, . . . so electricity by abolishing exhausting drudgery, is making the farm the best place in the world to live." The movie's history can be read in "Movie of G.E. Has Scenes near Here," *Schenectady Gazette*, 2 February 1926, 2.

55. Pamperin, "A Country Rural Electrification Campaign," 5.

56. Wisconsin Power and Light Co., *New Methods of Farm Operation*, 22 January 1930, program brochure, in UW Archives, WCREA, Correspondence and Records, 1929–1933, S–Z, "Rural Electrification Day Meetings," series #9/12/5-2, box 2.

57. Program brochures and coupons found in ibid.

58. "Farm and Home Week, Madison," College of Agriculture, University of Wisconsin, serial no. 1720, *Bulletin of the University of Wisconsin*, General Series no. 1494, 3, 4, 6, 7, 23, in UW Archives, WCREA, Correspondence and Records, 1929–1933, A–R, "Programs, Miscellaneous," series #9/12/5-2, box 1.

59. Henry C. Dethloff and Stephen W. Searcy, *Engineering Agriculture at Texas A&M: The First Hundred Years* (College Station: Texas A&M University Press, 2015), 53.

60. "Power Companies Develop Farm Business at the Fair," *Electricity on the Farm* 3, no. 7 (July 1930): S2–S3.

61. J. P. Schaenzer, "Why 45,000 Wisconsin Farmers Are Using Electricity," typescript of radio talk over WHA, 6 January 1930, in UW Archives, WCREA, Correspondence and Records, 1929–1933, A–R, series #9/12/5-2, box 1.

62. J. P. Schaenzer, "More Electricity for Less Money," typescript of radio talk over WHA, 6 January 1930, 1–4, in UW Archives, WCREA, Correspondence and Records, 1929–1933, A–R, series #9/12/5-2, box 1. Engineers at other universities also made good use of the new medium. In 1931, University of Idaho's Hobart Beresford discussed "wiring the farm for electric service" and "rural electrification progress" on the *Idaho Farm and Home Hour* program. The following year, he gave three more talks, including one on "electric soil and hotbed heating." The talks are listed in UI Special Collections, Agricultural Experiment Station, Department of Agricultural Engineering, Idaho Committee on the Relation of Electricity to Agriculture, "Rural Electrification Progress Report No. 8," April 1932, typescript, document I 373, vol. 8, 71; and Idaho Committee on the Relation of Electricity to Agriculture, "Rural Electrification Progress Report No. 9," April 1933, typescript, document I 373, vol. 9, 42.

63. E. A. White, "A Horoscope of Farm Electrification," typescript, "no. 201 of a series of talks on 'Some Practical Solutions of Farm Electrification Problems,' prepared under the auspices of the Rural Electrification Section of the General Electric Company, and broadcast from Station WGY, Schenectady, NY," 11 November 1932, 1, included in bound folder for use by students in A.E. 139 course, "Rural Electrification," in UI Special Collections, folder "Publications: Ag. Engineer., No. 302," titled "Some Practical Solutions of Farm Electrification Problems."

64. David E. Nye, *Image Worlds: Corporate Identities at General Electric, 1890–1930* (Cambridge, MA: MIT Press, 1985), 126–27.

65. Even if farmers owned isolated plants, small hydroelectric power stations, or wind-powered generators, which provided modest supplies of electricity, they could ultimately become worthwhile patrons of high-line service after having developed a familiarity with electrical lifestyles.

66. Walter A. Bowe, "Utilization of Farm Advertising and Dealer Help by Utilities," no date, but likely 1931, 1, in UW Archives, WCREA, Correspondence and Records, 1929–1933, A–R, "Advertising," series #9/12/5-2, box 1.

67. E. A. Stewart, J. M. Larson, and J. Romness, *The Red Wing Project on Utilization of Electricity in Agriculture* (N.p. [St. Paul?]: University of Minnesota Agricultural Experiment Station, no date, but likely 1927 or 1928), 29, University of Minnesota Libraries Digital Conservancy, http://conservancy.umn.edu/handle/11299/48685, accessed 15 September 2014.

68. Letter from A. E. Schwarz, "Farm Plant Salesman," Westinghouse Electric & Manufacturing Company, Richmond, VA, to Mr. E. C. [*sic*] Seitz, VPI, 12 March 1924, in VT Special Collections, Department of Agricultural Engineering papers, RG13/4, box 11, folder 2, "Rural Electrification 1920s–1940s." As late as 1928 and 1929, Cosgrove recorded sales promotion talks dealing with a Westinghouse farm light plant on twelve-inch Victor discs. U.C. Santa Barbara Library project on Victor Encyclopedic Discography of Vic-

tor Records, http://victor.library.ucsb.edu/index.php/talent/detail/5648/Cosgrove_R._C ._speaker, accessed 19 May 2014.

69. Alabama Farm Bureau Committee and Committee on the Relation of Electricity to Agriculture, *Electrifying Agriculture: Progress Report of Alabama Farm Committee on the Relation of Electricity to Agriculture* (no publication information, likely 1925), 20, Auburn University library call number H9688 A4.

70. "Making a Hard Job Easier," advertisement in *Agricultural Engineering* 7, no. 11 (November 1926): 395. NELA promoted its work frequently in *Agricultural Engineering* as well. See, for example, "The Return of the Native" advertisement, which described how electricity use meant that "the glitter of the city is losing its attraction for farm boys and girls," *Agricultural Engineering* 7, no. 5 (May 1926): 192–93.

71. Examples include "The Ox Woman," advertisement in the *Alabama Farmer* 6, no. 7 (April 1926), 2; "Team Work," advertisement in the *Arizona Agriculturalist* (a publication of the University of Arizona's College of Agriculture) 3, no. 5 (February 1926): 1; and "Winning the West," advertisement in *Purdue Agriculturist* 18, no. 8 (May 1924): 150.

72. *Farm Light and Power Year Book 1922* (New York: Farm Light and Publishing Co., 1922), 2, 22, 29, for example.

73. The magazine's stationery in 1927 listed President Karl M. Mann, Vice President and General Manager I. Herbert Case, and Managing Editor Fred Shepperd. Letter from Fred Shepperd to James A. Waller Jr., Virginia Polytechnic Institute, 7 October 1927, in VT Special Collections, Department of Agricultural Engineering papers, RG13/4, "Rural Electrification—General—1920s–30." By 1929, the masthead noted that the magazine was produced by the Case-Shepperd-Mann Publishing Corp. Letter from Karl M. Mann, to Professor Chas. E. Seitz, 27 July 1929, in ibid.

74. Even in issues published as the TVA and REA began operation in 1933 and 1935, the magazine's editor rarely mentioned events occurring in Washington, DC. The closest example that I noted of political content came in the January 1933 issue, published soon after the election of Franklin Roosevelt (but before he took office in March 1933). With a headline of "What's in Store for 1933," the text reads: "This new deal that we are all hoping and looking for will come not to those who wait, but rather to those who do something about it. So let's be up and doing." Back cover of *Electricity on the Farm* 6, no. 1 (January 1933).

75. "Analysis of Circulation," *Electricity on the Farm*, pamphlet accompanying letter from Case-Shepperd-Mann Publishing Corp. President Karl M. Mann to Professor Charles E. Seitz, VPI, 27 July 1929, in VT Special Collections, Department of Agricultural Engineering papers, RG13/4, "Rural Electrification—General—1920s–30."

76. Walter A. Bowe, "Utilization of Farm Advertising and Dealer Help by Utilities," 5, in UW Archives, WCREA, Correspondence and Records, 1929–1933, A–R, "Advertising," series #9/12/5-2, box 1.

77. K. H. Gorham, "Advertising as a Selling Help in Developing Rural Line Load," 3. Gorham is listed as advertising manager of *Electricity on the Farm*. He prepared this paper for the Rural Electric Sales School of the Middle West Division, NELA, at Kansas City, 26 February 1931, in UW Archives, WCREA, Correspondence and Records, 1929–1933, A–R, "Advertising," series #9/12/5-2, box 1.

78. G. C. Neff, "Making Rural Electrification Pay," *Edison Electric Institute Bulletin* 3, no. 1 (January 1935): 14.

79. *Electricity on the Farm* 1, no. 1 (July 1927): 17–19.

80. Gail Meredith, "Electrical Housekeeping," *Electricity on the Farm* 3, no. 10 (October 1930): 36–44.

81. *Electricity on the Farm* 1, no. 13 (July 1928).

82. *Electricity on the Farm* 6, no. 3 (March 1933).

83. "Rural Electric Service Committee" and "Rural Service Departments," NELA, *Proceedings* 84 (1927): 81, 83.

84. Ibid., 81. The chair of NELA's Rural Electric Service Committee admitted in 1929 that while progress in stringing lines to nonurban residents was accelerating, he did not think that all farms in the country would ever be electrified. "But it is reasonable to presume that most of the better farms will be." Eugene Holcomb, "What Progress? How Much Headway Are We Making in Rural Electrification?," *Electricity on the Farm* 2, no. 11 (November 1929): D7.

85. NELA, *Progress in Rural and Farm Electrification*, 1.

86. Neff remained on the committee. See the membership list of the Rural Electric Service Committee in "General National Committees," NELA, *Proceedings* 85 (1928): 1680.

87. "Rural Electric Service Committee," NELA, *Proceedings* 85 (1928): 137.

88. "Discussion" by C. F. Stuart, in ibid., 147. Sometimes, advocates for rural electrification exaggerated (intentionally or otherwise) the rapid increase in the number of farms obtaining service. For example, *Electricity on the Farm* magazine reported in 1930 on the NELA president's observation that the number of farms enjoying high-line electricity jumped from about one hundred thousand in 1925 to six hundred thousand in 1930, an increase of 500 percent. But the article's author then made an egregious mathematical error by claiming that this growth represented an average growth of "100% per year." Such a growth rate, if compounded annually, meant that the 100,000 number would have turned into 3,200,000 after five years. In fact, the six-times growth in five years corresponds to an annual growth rate of about 43 percent—still a significant number, but nowhere close to the one reported. "100 Per Cent Per Year," *Electricity on the Farm* 3, no. 3 (March 1930): S1.

89. American Farm Bureau Federation, *The American Farm Bureau Federation in 1931* (Chicago: AFBF, 1931), 5.

90. "Utility Customers Returning," *Electrical World* 103 (6 January 1934): 26.

91. "Farms Electrified: Push Use of Energy," *Electrical World* 100 (3 September 1932): 291. By the end of 1932, central station companies served more than seven hundred thousand rural customers, less than Neff predicted in 1927 for the year, but still significant coming off a base of 177,561 at the end of 1923. Data from NELA, *Progress in Rural and Farm Electrification*, 4; and "Add 533,000 New Customers," *Electrical World* 106, no. 1 (4 January 1936): 62.

92. "Nearly 50,000 New Farm Customers," *Electrical World* 99 (26 March 1932): 571–72.

93. "Farm Electrification Made Great Gain in 1931," *Electrical World* 100 (16 July 1932): 71. See "Alabama's Rural Service," *Electrical World* 100 (16 July 1932): 72, showing the company's rural customers increasing consumption from 175,000 kWh in 1923 to more than 15 million kWh in 1931; and "Electrification of Farms in Virginia Grows Rapidly," *Electrical World* 100 (6 August 1932): 163.

94. Data from "Add 533,000 New Customers," 62. The compounded annual growth rate was calculated using 1923 and 1932 year-end numbers.

95. Holcomb, "What Progress?," D5.

96. "Rural Electric Service Committee," 81.

PART III: **Growth of Rural Electrification Efforts in the 1930s**

1. US Department of Commerce, *Statistical Abstract of the US, 1951* (Washington, DC: GPO, 1951), table 560, "Farm Electrification: 1930 to 1949," 475; and US Department of Commerce, *Statistical Abstract of the US, 1960* (Washington, DC: GPO, 1960), table 698, "Farm Electrification: 1940 to 1959," 534.

CHAPTER TWELVE: **Government Innovations in the Rural Electrification Subsystem**

Epigraph: "Questions and Answers," *Rural Electrification News* 1, no. 2 (October 1935): 27.

1. The history of the TVA—just like the history of the REA—has been interpreted using a heroic narrative. See Thomas K. McCraw, reviews of *TVA: Fifty Years of Grass-Roots Bureaucracy*, by Erwin C. Hargrove and *The Myth of TVA: Conservation and Development in the Tennessee Valley, 1933–1983*, by William U. Chandler, *Journal of Southern History* 51, no. 2 (1985): 311–15; and Matthew D. Owen, "For the Progress of Man: The TVA, Electric Power, and the Environment, 1939–1969" (PhD diss., Vanderbilt University, 2014), 2. For a discussion of the TVA leaders' pursuit of environmental goals, see Sarah T. Phillips, *This Land, This Nation: Conservation, Rural America, and the New Deal* (New York: Cambridge University Press, 2007), 83–107.

2. Franklin D. Roosevelt, "The First Inaugural Address as Governor," 1 January 1929, in *The Public Papers of the Presidents of the United States: Franklin D. Roosevelt, 1937*, vol. 6 (New York: Macmillan, 1941), 78.

3. Kenneth E. Trombley, *The Life and Times of a Happy Liberal: A Biography of Morris Llewellyn Cooke* (New York: Harper, 1954), 109; and Gertrude Almy Slichter, "Franklin D. Roosevelt's Farm Policy as Governor of New York State, 1928–1932," *Agricultural History* 33, no. 4 (October 1959): 167–76.

4. "Text of Governor Roosevelt's Speech at Portland, Oregon, on Public Utilities," *New York Times*, 22 September 1932, 16; "Text of Governor Roosevelt's Speech at Commonwealth Club, San Francisco," *New York Times*, 24 September 1932, 6; and James A. Hagerty, "Roosevelt Renews Demand for Repeal," *New York Times*, 24 September 1932, 1, 7.

5. Tennessee Valley Authority Act of 1933, Pub. Law No. 73-17, 48 Stat. 58, 18 May 1933; Morris Llewellyn Cooke, "The Early Days of the Rural Electrification Idea: 1914–1936," *American Political Science Review* 42, no. 3 (1948): 444. These co-ops were created after TVA's acquisition of assets from the Mississippi Power Company. Similar transfers of utility properties in Alabama and Tennessee resulted in lawsuits, eventually settled by the Supreme Court (in the *Ashwander vs. TVA* decision in 1936), which ruled that TVA's power programs were constitutional. Carl Kitchens, "The Role of Publicly Provided Electricity in Economic Development: The Experience of the Tennessee Valley Authority, 1929–1955," *Journal of Economic History* 74, no. 2 (June 2014): 395.

6. Morris Llewellyn Cooke, *Report of the Mississippi Valley Committee of the Public Works Administration* (Washington, DC: US Federal Emergency Administration of Public Works, 1934), 51.

7. Morris Llewellyn Cooke, "National Plan for the Advancement of Rural Electrification under Federal Leadership and Control with State and Local Cooperation and as a Wholly Public Enterprise," typescript document, undated but likely 1934, from Franklin D. Roosevelt Presidential Library, Hyde Park, NY.

8. Emergency Relief Appropriation Act of 1935, H.J. Res. 117, Pub. Res. no. 11, 49 Stat 115, 8 April 1935; F. D. Roosevelt, Exec. Order 7037, "Establishing the Rural Electrification Administration," 11 May 1935; and "Cheap Electricity Planned for Farm," *New York Times*, 14 May 1935, 27.

9. H. S. Person, "The Rural Electrification Administration in Perspective," *Agricultural History* 24, no. 2 (April 1950): 71; and F. D. Roosevelt, Exec. Order 7130, "Prescribing Rules and Regulations Relating to Approved Projects Administered and Supervised by the Rural Electrification Administration under the Emergency Relief Appropriation Act of 1935, Regulation No. 4," 7 August 1935.

10. "Rural Electrification Act of 1936," Pub. Law No. 74-605, chap. 432, 49 Stat. 1363, 20 May 1936; and "Roosevelt Signs REA Bill," *New York Times*, 22 May 1936, 3.

11. "Cheap Electricity Planned for Farm," 27.

12. Philip J. Funigiello, *Toward a National Power Policy: The New Deal and the Electric Utility Industry, 1933–1941* (Pittsburgh, PA: University of Pittsburgh Press, 1973), 139–41.

13. Letter from W. W. Freeman, chairman of the Rural Electrification Committee of Privately Owned Utilities, to Morris L. Cooke, REA administrator, 24 July 1935, 1, Franklin D. Roosevelt Presidential Library, office folder (OF) 1570.

14. Though appearing eager to cooperate with power companies, Cooke explicitly disagreed with the utilities' assertion that the major impediment to widespread use of power consisted of problems in financing farm wiring and appliance purchases; Cooke argued that the problem consisted of excessively high electric rates. Letter from Morris L. Cooke, REA administrator, to W. W. Freeman, chairman of the Rural Electrification Committee of Privately Owned Utilities, 31 July 1935, 2, Franklin D. Roosevelt Presidential Library, office folder (OF) 1570.

15. "Electric Service for Rural Areas: Cooke and Private Companies Agree on Program to Cost about $238,249,000," *New York Times*, 1 August 1935, 21. A *Washington Post* editorial noted the good work that the REA had embarked on, but it also observed that "Mr. Cooke expects the greater part of the rural electrification program to be carried on by unaided private enterprises. His policy thus promises a maximum of tangible improvements with a minimum of aftermath headaches." "Rural Electrification," *Washington Post*, 23 October 1935, 8. The utility committee's plan was also described in Franklyn Waltman Jr., "Utilities Plan Big Expansion of Farm Power," *Washington Post*, 1 August 1935, 1, 9. The article quoted Cooke's commendation of the utility committee, which "has rendered the Government, as well as private interests it represents, a very real service in the clarification of an important undertaking." Ibid., 9. In addition, see "Private Utilities Submit a Program for Rural Electrification Partly Financed by REA Funds," *Rural Electrification News* 1, no. 1 (September 1935): 18–19.

16. Details of some of the political and personal intrigue can be found in Funigiello, *Toward a National Power Policy*, 139–42.

17. "4,247 More Farms to Get Electricity," *New York Times*, 5 November 1935, 8. In his account of the early history of rural electrification, Cooke observed that Harold Ickes,

secretary of the interior and Public Works administrator, expressed his dislike of private utility companies when he authorized a 1934 study of electrification in the Mississippi Valley. That animosity may have contributed to Cooke's decision to deny much funding to private power companies. Cooke, "The Early Days," 444–45.

18. Like many other supporters of public power, Senator Norris opposed lending money to utilities. See letter from George W. Norris to Morris L. Cooke, 1 August, 1935, cited in Richard Lowitt, *George W. Norris: The Triumph of a Progressive, 1933–1944* (Urbana: University of Illinois Press, 1978), 126. Also see R. L. Duffus, "Federal Aid to Bring Power Lines to Farms," *New York Times*, 31 May 1936, E11.

19. Of the remainder of the funds, 9.2 percent went to private nonprofit corporations, 0.8 percent to state corporations, 0.9 percent to municipal corporations, and 8.4 percent to power and irrigation districts. REA, *1937 Report of the Rural Electrification Administration* (Washington, DC: GPO, 1938), 31.

20. "How Stands Rural Electrification?" *Electrical World* 99, no. 22 (28 May 1931): 964–65.

21. "New York Ranks Second in Rural Electrification," *Electrical World* 99 (5 March 1932): 430.

22. "How Stands Rural Electrification," 965.

23. Gregory B. Field, "'Electricity for All': The Electric Home and Farm Authority and the Politics of Mass Consumption, 1932–1935," *Business History Review* 64 (Spring 1990): 33; and "Uncle Sam—Vacuum Cleaner Salesman," *Financial World* 61 (3 January 1934): 11.

24. "New Deal Holds Big Stick in Readiness to Cut Rates for Electricity and Appliances," *Christian Science Monitor*, 13 March 1934, 1; "TVA Plans to Widen Appliance Program," *Wall Street Journal*, 4 September 1934, 11; and "Plan to Expand EHFA Setup Revealed Here," *Washington Post*, 4 August 1935, 2.

25. George D. Munger, "EHFA—How It Helps Buyers of Electrical Goods," *Rural Electrification News* 2, no. 8 (April 1937): 9–10. Munger was the commercial manager of the EHFA. Some of his contributions are described in Michelle Mock, "The Electric Home and Farm Authority, 'Model T Appliances,' and the Modernization of the Home Kitchen in the South," *Journal of Southern History* 80, no. 1 (February 2014): 95.

26. Munger, "EHFA," 10; and TVA, *Annual Report of the Tennessee Valley Authority for the Fiscal Year Ended June 30, 1951* (Washington, DC: GPO, 1951), 48.

27. Munger, "EHFA," 10; and "EHFA Will Lend Santa Funds to Buy Appliances," *Chicago Daily Tribune*, 24 November 1935, 21.

28. Mock, "The Electric Home and Farm Authority," 81.

29. Ibid., 73–108.

30. "Model T Appliances," *Business Week* (16 June 1934): 11; and Mock, "The Electric Home and Farm Authority," 84.

31. "Low-Cost Refrigerators: EHFA Approves Another—Will Aid in Financing Them Where It Wins Rate Cuts," *Wall Street Journal*, 1 August 1934, 6; and "TVA Plans to Widen Appliance Program," *Wall Street Journal*, 4 September 1934, 11.

32. "EHFA Will Lend Santa Funds to Buy Appliances," 21.

33. Brent Cebul, "Creative Competition: Georgia Power, the Tennessee Valley Authority, and the Creation of a Rural Consumer Economy, 1934–1955," *Journal of American History* 105, no. 1 (June 2018): 55.

34. Mock, "The Electric Home and Farm Authority," 90–91; and Ronald C. Tobey, *Technology as Freedom: The New Deal and the Electrical Modernization of the American Home* (Berkeley: University of California Press, 1996).

35. Mock, "The Electric Home and Farm Authority," 92–93. For more on Davison's work, see "Conference at Ames, Iowa," *CREA Bulletin* 1, no. 5 (26 January 1925): 5; Eloise Davison, "Iowa's Rural Community for Electrical Development," *CREA Bulletin* 2, no. 7 (12 May 1926): 3; Eloise Davison, "Electrical Equipment Short Course," *CREA Bulletin* 3, no. 3 (15 June 1927): 7; Louise Stanley, "Home Economics and Rural Electrification," *Journal of Home Economics* 28, no. 8 (October 1936): 559–60; and Amy Sue Bix, "Equipped for Life: Gendered Technical Training and Consumerism in Home Economics, 1920–1980," *Technology and Culture* 43, no. 4 (October 2002): 728–54. In 1934, Davison was listed as a "Home economist" working for the TVA and earning $4,000 per year. US Civil Service Commission, *Official Register of the United States, 1934* (Washington, DC: GPO, 1935), 143. Another biography appears in "Key Woman in Civilian Defense," *Crescent of Gamma Phi Beta* 41, no. 4 (December 1941): 3.

36. Mock, "The Electric Home and Farm Authority," 105.

37. "EHFA Reorganized to Help Finance $350,000,000 Estimated Sales of Appliances and Equipment," *Rural Electrification News* 1, no. 1 (September 1935): 4–5; and REA, *1939 Report of the Rural Electrification Administration* (Washington, DC: GPO, 1940), 77–78.

38. "Plans and Terms Announced for Rural Electric Loans," *Rural Electrification News* 1 (September 1935): 7.

39. "Rural Electrification Act of 1936," sec. 3 (a); and "The Rural Electrification Act of 1936," *Rural Electrification News* 1, no. 10 (June 1936): 3. Rates dropped below 3 percent in subsequent years. In fiscal years 1937, 1938, 1939, and 1940, borrowing rates stood at 2.77 percent, 2.88 percent, 2.73 percent, and 2.69 percent, respectively. REA, *Rural Electrification on the March* (Washington, DC: GPO, 1938), 5; and US Congress, House, Committee on Appropriations, *Hearings before the Subcommittee of the Committee on Appropriations, House of Representatives, 76th Congress, Third Session on the Agricultural Department Appropriation Bill for 1941* (Washington, DC: GPO, 1940), 1058.

40. David Cushman Coyle, *Electric Power on the Farm* (Washington, DC: GPO, 1936), 109.

41. Richard P. Keck, "Reevaluation the Rural Electrification Administration: A New Deal for the Taxpayer," *Environmental Law* 16, no. 1 (Fall 1985): 39–89, esp. 49. The Pace Act, Pub. Law No. 78-425, chap. 412, 58 Stat. 734, 21 September 1944, also known as the "Department of Agriculture Organic Act of 1944," amended the Rural Electrification Act of 1936 and added the provision for lower-than-market rates. Also see USDA, *Report of the Administrator of the Rural Electrification Administration, 1945* (Washington, DC: GPO, 1945), 3.

42. For comparison, the average yield on "long" government bonds, those maturing in fifteen or more years, was 2.48 percent in 1944. Utility bonds in 1944 paid 2.97 percent. US Department of Commerce, *Statistical Abstract of the United States, 1948* (Washington, DC: GPO, 1948), table 496, "Stock and Bond Yields—Percent: 1927 to 1947," 465. The average yield for Moody's Baa-rated bonds (for all corporations, not just utilities) with maturities twenty years and longer was 3.61 percent in 1944. Calculated average of monthly rate based on data in "Moody's Seasoned Baa Corporate Bond Yield (BAA)," FRED, Federal Reserve Bank of St. Louis, https://fred.stlouisfed.org/series/BAA, accessed 28 January 2019.

43. "REA to Finance Farm Wiring," *Rural Electrification News* 1, no. 4 (December 1935): 2.

44. REA, *1939 Report of the Rural Electrification Administration*, 84.

45. Forrest McDonald, *Let There Be Light: The Electric Utility Industry in Wisconsin, 1881–1955* (Madison, WI: American History Research Center, 1957), 373–74.

46. REA, *A Draft of a Rural Electric Cooperative Act* (Washington, DC: REA, 1939), 27; and Udo Rall, "Cooperative Rural Electrification in the United States," *Annals of Public and Cooperative Economics* 18, no. 2 (May 1942): 207.

47. Twentieth Century Fund, *Electric Power and Government Policy: A Survey of the Relations between the Government and Electric Power Industry* (New York: Twentieth Century Fund, 1948), 462. For tax payments by co-ops, see John D. Garwood and W. C. Tuthill, *The Rural Electrification Administration: An Evaluation* (Washington, DC: American Enterprise Institute for Public Policy Research, 1963), 60; and Roger D. Colton, *The Regulation of Rural Electric Cooperatives: The Common Law, Consumer Law and a Cornucopia of Customer Protections* (Boston: National Consumer Law Center, 1993), 29.

48. Revenue Act of 1916, Pub. Law No. 64-271, 39 Stat. 756, 8 September 1916, sec. 11; and W. G. Beecher, "Is It Time to Revoke the Tax-Exempt Status of Rural Electric Cooperatives?" *Washington and Lee Journal of Energy, Climate, and the Environment* 5, no. 1 (1 September 2013): 228. A discussion of tax exemptions is included in REA, *1938 Report of Rural Electrification Administration* (Washington, DC: GPO, 1939), 101–4; and REA, *1939 Report of Rural Electrification Administration*, 142–43.

49. "Facts and Fallacies," *Rural Electrification News* 2, no. 4 (December 1936): 18.

50. "Utility Taxes Mounting Faster than Revenues," *Electrical World* 99 (2 January 1932): 61–62. The Revenue Act of 1932, Pub. Law No. 72-154, 47 Stat. 169, 6 June 1932, was passed as a way to balance the federal government's budget in the midst of the Depression. US Department of the Treasury, *Annual Report of the Secretary of the Treasury on the State of the Finances for the Fiscal Year Ended June 30, 1932* (Washington, DC: GPO, 1932), 19–22. Also see "President Signs $1,118,500,000 Tax Bill," *New York Times*, 7 June 1932, 1. The utility industry's major trade journal ranted about the new tax in "Taxation to Kill an Industry," *Electrical World* 99 (4 June 1932): 979; and "Tax on Electrical Energy Must Be Paid by the Consumer," *Electrical World* 99 (11 June 1932): 1002. In 1934, electric utilities paid 13.9 percent of gross revenue in taxes according to "Expenses Rise More than Revenue," *Electrical World* 105, no. 1 (5 January 1935): 38–39; and "Utility Taxes Aid the Nation," *Electrical World* 105, no. 7 (30 March 1935): 76–77.

51. USDA, *Report of the Administrator of the Rural Electrification Administration, 1942* (Washington, DC: GPO, 1942), 10.

52. USDA, *Rural Lines: The Story of Cooperative Rural Electrification* (Washington, DC: GPO, 1972), 9.

53. Not all these easements came easily because some land owners remained "suspicious of the Government, the cooperative, and everyone connected with the project." Ibid., 9–10.

54. "Assembly-Line Methods in Rural Electrification," *Rural Electrification News* 1, no. 11 (July 1936): 19–20; and "4,500 Miles Staked in 2 Weeks," *Rural Electrification News* 3, no. 3 (November 1937): 9.

55. REA, *1937 Report of the Rural Electrification Administration*, 7.

56. Ibid., 84–85. See also REA, *Suggested Rural Line Construction No. 1* (Washington, DC: GPO, 1936).

57. Jim Cooper, "Electric Co-operatives: From New Deal to Bad Deal?," *Harvard Journal on Legislation* 45 (2008): 337. The author referred to discussion on area coverage in Patricia Lloyd Williams, *The CFC Story: How America's Rural Electric Co-operatives Introduced Wall Street to Main Street* (Herndon, VA: National Rural Utilities Cooperative Finance Corporation, 1995), 16.

58. John L. Neufeld, *Selling Power: Economics, Policy, and Electric Utilities before 1940* (Chicago: University of Chicago Press, 2016), 225–29, 240–43.

59. Cooke, "The Early Days," 439.

60. Moreover, Cooke threatened companies by saying that the REA might purchase power plants and distribution lines (or even create a corporation similar to the EFHA) "to carry out its plans to give the farmer the same facilities as are now enjoyed in the urban sections." Cooke, "Cheap Electricity Planned for Farm," *New York Times*, 14 May 1935, 27.

61. D. Clayton Brown, *Electricity for Rural America: The Fight for the REA* (Westport, CT: Greenwood Press, 1980), 5. Brown observed in another essay that the state CREA affiliates "did not deal with the real barrier to rural electrification, the cost to farmers. Farmers typically had to pay the costs of constructing lines to their homes if they wanted service. . . . Rural dwellers also paid higher rates than urban residents." Brown, "Modernizing Rural Life: South Carolina's Push for Public Rural Electrification," *South Carolina Historical Magazine* 99, no. 1 (January 1998): 71.

62. D. Clayton Brown, "The Battle for Rural Electrification," in Theodore Saloutos, ed., *The American Farmer and the New Deal* (Ames: Iowa State University Press, 1982), 209.

63. Brown, *Electricity for Rural America*, 10–11.

64. See the discussion of rate schedules in chapter 2. In general, rate schedules levied different prices on kilowatt-hours based on levels of consumption, such that more usage lowered the average price per unit.

65. L. S. Wing, "Rural Electrification from an Economic and Engineering Standpoint," *ASAE Transactions* 20 (June 1926): 61.

66. Morris Llewellyn Cooke, ed., *What Electricity Costs in the Home and on the Farm: A Symposium* (New York: New Republic, 1933), frontispiece before title page.

67. Clayton W. Pike, "Distribution Cost of Electric Energy with Special Reference to Residence and Rural Customer," in Cooke, *What Electricity Costs*, 84.

68. Joseph C. Swidler, review of *What Electricity Costs* by Morris Llewellyn Cooke, *Journal of Political Economy* 43, no. 2 (April 1935): 282–84. The challenge of determining rates based on properly allocated costs continued through the twentieth century. James Bonbright, the Columbia University finance professor viewed as the "dean of public utility economists," for example, observed in 1961 that rates could be legitimately determined using numerous methodologies, resulting in widely varying results. Nathanial L. Nathanson, review of *Principles of Public Utility Rates* by James C. Bonbright (New York: Columbia University Press, 1961), in *Columbia Law Review* 62, no. 6 (June 1962): 1102–14. Donald Rushford, general counsel for Central Vermont Public Service Corporation and former chief counsel of the Vermont Public Service Board, commented in 1976 that determining costs as the basis for making rates was "90 percent philosophy and 10 percent

math." Elliot Taubman and Karl Frieden, "Electricity Rate Structures: History and Implications for the Poor," *Clearinghouse Review* 10 (October 1976): 436.

69. Thomas K. McCraw, *TVA and the Power Fight, 1933–1939* (Philadelphia: J. B. Lippincott, 1971), 71.

70. David E. Lilienthal, *TVA: Democracy on the March* (New York: Harper and Brothers, 1944), 22. Lilienthal's book, heartily praised when published, later saw significant criticism. See William W. Drumright, "His Search for a Viable Middle Ground: A Reappraisal of David E. Lilienthal's *TVA—Democracy on the March*," *Princeton University Library Chronicle* 63, no. 3 (Spring 2002): 467–95.

71. Mock, "The Electric Home and Farm Authority," 88; and TVA, *Annual Report of the Tennessee Valley Authority for the Fiscal Year Ended June 30, 1935* (Washington, DC: GPO, 1936), 30–31.

72. McCraw, *TVA and the Power Fight*, 59–61; and Thomas K. McCraw, "Triumph and Irony—The TVA," *Proceedings of the IEEE* 64, no. 9 (September 1976): 1375.

73. C. Woody Thompson and Wendell R. Smith, *Public Utility Economics* (New York: McGraw-Hill, 1941), 688.

74. Economist Carl Kitchens notes that utilities operated by the Commonwealth and Southern holding company dropped residential rates substantially in 1933, but not in response to the TVA's creation. Rather, the cut had been scheduled in 1930, during the creation of the company as a merger of several smaller firms. Kitchens, "The Role of Publicly Provided Electricity in Economic Development," 412.

75. Historical Statistics of the United States, Millennial Edition Online, "Electrical Energy—Retail Prices, Residential Use, and Service Coverage," https://hsus.cambridge.org/HSUSWeb/search/searchTable.do?id=Db234-241, accessed 21 May 2019. In inflation-adjusted terms, this decline in price comes to about 78 percent. Consumer price data from ibid. Also see Richard F. Hirsh, *Technology and Transformation in the American Electric Utility Industry* (New York: Cambridge University Press, 1989), 9.

76. McCraw, *TVA and the Power Fight*, 61.

77. Mock, "The Electric Home and Farm Authority," 88.

78. "Tupelo Capitalizes the Primitive," *Electrical World* 104, no. 3 (21 July 1934): 75; and "Tupelo Results: Show of TVA Appliances in Mississippi Town Supports Claim That Cheaper EHFA Models Will Boost Sales of Competing Items," *Business Week* (9 June 1934): 21.

79. "Tupelo Revenue Increase," *Electrical World* 104, no. 16 (8 December 1934): 1040; and "Energy Consumption Raised by T.V.A. Residential Rates," *Electrical World* 104, no. 6 (11 August 1934): 187.

80. TVA, *Annual Report of the Tennessee Valley Authority for the Fiscal Year Ended June 30, 1939* (Washington, DC: GPO, 1940), 76.

81. The Tennessee Valley Authority Act of 1933, sec. 10, gives "preference to States, counties, municipalities, and cooperative organizations of citizens or farmers, not organized or doing business for profit, but primarily for the purpose of supplying electricity to its own citizens or members."

82. TVA, *Annual Report of the Tennessee Valley Authority for the Fiscal Year Ended June 30, 1941* (Washington, DC: GPO, 1942), 31; and TVA, *Annual Report of the Tennessee Valley*

Authority for the Fiscal Year Ended June 30, 1942 (Washington, DC: GPO, 1946), 11. In a discussion of the advantages held by government agencies in producing and selling electricity during the New Deal, historian Paul Wolman correctly wrote that the REA co-ops had access to cheap power. He noted that the TVA sold power to co-ops at wholesale prices of 2–3 cents per kWh in 1941. But the rate actually was much lower, as suggested here (under 5 mills per kWh). Paul Wolman, "The New Deal for Electricity in the United States, 1930–1950," in Douglas F. Barnes, ed., *The Challenge of Rural Electrification: Strategies for Developing Countries* (Washington, DC: Resources for the Future, 2007), 291.

83. Bonneville Project Act of 1937, Pub. Law No. 75-329, 50 Stat. 731, 20 August 1937, 16 U.S.C. chap. 12B, sec. 832c gives preference to public bodies and cooperatives.

84. Walter B. Edgar, *History of Santee Cooper, 1934–1984* (Moncks Corner, SC: South Carolina Public Service Authority, 1984), 5–7. For an excellent discussion of electrification's social and economic impacts in the South, which includes a short history of the Santee-Cooper project, see Casey P. Cater, *Regenerating Dixie: Electric Energy and the Modern South* (Pittsburgh, PA: University of Pittsburgh Press, 2019).

85. Edgar, *History of Santee-Cooper*, 95.

86. The 1944 Flood Control Act, Pub. Law No. 78-534, 58 Stat. 887, chap. 66, 22 December 1944, sec. 5 states that "preference in the sale of such power and energy shall be given to public bodies and cooperatives."

87. Patricia Stallings, *Serving the Southeast: A History of the Southeastern Power Administration, 1990–2010* (Washington, DC: U.S. Department of Energy, 2012), 8–14; and Gus Norwood, *Gift of the Rivers: Power for the People of the Southeast—A History of the Southeastern Power Administration* (Washington, DC: GPO, 1990).

88. Rural Electrification Act of 1936, sec. 3. The Department of Agriculture Organic Act of 1944 and an amendment (Pub. Law No. 78-563, 58 Stat. 925, 23 December 1944) extended funding for construction of plants and lines for longer periods of time (than in the 1936 act) and at lower interest rates.

89. North Carolina Electric Membership Corp. v. Carolina Power & Light Co., US District Court for the Middle District of North Carolina, 780 F. Supp. 322 (M.D.N.C. 1991), Justia, https://law.justia.com/cases/federal/district-courts/FSupp/780/322/1445316, accessed 20 February 2019. Full disclosure: I was hired by the plaintiff's attorneys in this case to serve as an expert witness. I examined some historical documents relating to the case but never testified in court.

90. Nathaniel E. Shechter, "Low Purchased Energy Costs to the Rural Electric Cooperatives," *Land Economics* 42, no. 3 (August 1966): 312; and Frederick W. Muller, *Public Rural Electrification* (Washington, DC: American Council on Public Affairs, 1944), 121, which noted that the threat of an REA-funded "generation-transmission-distribution system" put pressure on private companies, which often provided low-cost wholesale power as a means to avoid competition.

91. USDA, *Report of the Administrator of the Rural Electrification Administration, 1940* (Washington, DC: GPO, 1941), 2. Datum on number of farms comes from US Department of Commerce, *Statistical Abstract of the United States, 1957* (Washington, DC: GPO, 1957), table 665, "Farm Electrification: 1940 to 1956," 534.

92. Wolman, "The New Deal for Electricity in the United States," 289–91.

93. The REA observed in 1940 that aside from lending activities, the agency provided engineering expertise, help in negotiating for wholesale power, and aid in countering "obstruction" efforts by private utilities. Other services included assistance in the design and construction of generating plants, guidance in establishing retail rate structures, and management and legal advice for maintaining well-run cooperative organizations. USDA, *Report of the Administrator of the Rural Electrification Administration, 1940*, 13.

94. Carl Kitchens and Price Fishback, "Flip the Switch: The Impact of the Rural Electrification Administration 1935–1940," *Journal of Economic History* 75, no. 4 (December 2015): 1190.

95. Robert L. Bradley Jr., "The Origins of Political Electricity: Market Failure or Political Opportunism?," *Energy Law Journal* 17, no. 1 (1996): 97. Bradley drew on, among other sources, Ernest R. Abrams, *Power in Transition* (New York: Scribner's, 1940), which provided a critical view of the REA. Abrams concluded, for example, that "it would appear that much of our existing publicly financed rural electrification had little economic justification" (37).

96. Neufeld, *Selling Power*, 236, 240–43.

97. Cooke, "National Plan for the Advancement of Rural Electrification," 3–7.

98. REA, *1939 Report of Rural Electrification Administration*, 134.

99. Abby Spinak, "'Not Quite So Freely as Air': Electrical Statecraft in North America," *Technology and Culture* 61, no. 1 (January 2020): 97–98.

100. Samuel Liss, "Program and Work of the Rural Electrification Administration in the Works Program," typescript report of the Works Progress Administration, Washington, DC, July 1936, 21, Internet Archive, https://archive.org/details/CAT10930824/page/n1/mode/2up, accessed 20 June 2019.

101. Ibid., 23.

102. Spinak, "'Not Quite So Freely as Air,'" 97; and Judson King, "The REA—A New Deal Venture in Human Welfare," *Public Utilities Fortnightly* 21, no. 7 (31 March 1938): 40.

103. Liss, "Program and Work of the Rural Electrification Administration," 28.

104. Ibid., 29. The National Housing Act of 1934 offered loans for upgraded wiring and electric appliances for farm and urban residences. Tobey, *Technology as Freedom*, 113.

105. "Coal Men Fight Federal Projects," *Electrical World* 104, no. 14 (10 November 1934): 842; and "Edison Institute Opens War on Administration's Power Plan," *Electrical World* 104, no. 16 (8 December 1934): 1051.

CHAPTER THIRTEEN: **Competition and Private Utilities in the REA Era**
Epigraph: C. Woody Thompson and Wendell R. Smith, *Public Utility Economics* (New York: McGraw-Hill, 1941), 592.

1. USDA, *Report of the Administrator of the Rural Electrification Administration, 1949* (Washington, DC: GPO, 1950), 1, 4.

2. "The 15th Milestone: Rural Electric Co-ops Make Strides as They Observe Birthday of Program," *Rural Electrification News* 15, nos. 8 and 9 (March–April 1950): 4. Of course, this claim is somewhat disingenuous. As noted in the previous chapter, utility leaders sought to borrow REA's first allotment of funds in 1935, but Administrator Cooke chose not to lend it to them.

3. US Department of Commerce, *Statistical Abstract of the United States, 1951* (Washington, DC: GPO, 1951), table 560, "Farm Electrification: 1930 to 1949," 475. These data do not include another 243,916 farms served by municipal systems and power districts that received no money from the REA. Ibid. In a 1949 report, the REA administrator noted, "More than half of the farms connected to central station lines since 1935—about 57 percent—received electric service from REA-financed systems." That assertion may be true, but it does not include the approximately 11 percent already powered by central station companies in 1935. Hence, fewer than half of all lines were financed by the REA. The author also observed, correctly, "The rest were on lines of other suppliers, many of them stimulated to greater activity in the rural field by the REA program." USDA, *Report of the Administrator of the Rural Electrification Administration, 1949*, 4.

4. See Richard F. Hirsh, *Power Loss: Deregulation and Restructuring of the American Electric Utility System* (Cambridge, MA: MIT Press, 1999), for more discussion of how utility managers sought control over the American electric power system.

5. For a discussion of the development of these attitudes, see Richard F. Hirsh, *Technology and Transformation in the American Electric Utility Industry* (New York: Cambridge University Press, 1989).

6. "Electric Utilities Organize to End Propaganda Blame," *Washington Post*, 13 January 1933, 11. Other contemporaneous accounts include "Utilities Plan New Program; Outline of Constitution of Institute Given; Designed to Eliminate Past Objectionable Policies," *Los Angeles Times*, 14 January 1933, 9; and Howard Wood, "Light Industry Organizes New, Franker Group," *Chicago Daily News*, 13 January 1933, 25. The *Chicago Daily News* article pointed out that NELA disbanded largely because of the increasing influence of Samuel Insull and his brother, Martin, who moved the trade organization more into political activities than other leaders preferred. A similar explanation was presented in "Power Chiefs Form National Institute to Purge Industry," *New York Times*, 13 January 1933, 1, 11. The new EEI leaders decided, according the *Times* article, to dissolve NELA, which "rightfully or wrongfully, had been stamped unalterably with the reputation of a great propaganda organization."

7. "Wide Reform Seen in Power Industry; Formation of Edison Institute Was Declaration of Intention of Most Companies," *New York Times*, 15 January 1933, N7. Also see the *Wall Street Journal* editorial that concluded: "It is good to know that our kilowatts are hereafter to be strictly ethical before they emerge from behind the plastering." "Kilowatts to Be Ethical," *Wall Street Journal*, 14 January 1933, 6.

8. "Walsh Declares City Power Plant Will Slash Rates," *New York Times*, 25 December 1934, 1. Frank P. Walsh was the chairman of the State Power Authority, which had submitted a three-year survey of electric rates. The report apparently showed that utilities' rates would decline if cities, such as New York, built demonstration power plants and sold electricity at reduced prices.

9. The Public Utility Holding Company Act of 1935 constitutes Title I of the Public Utility Act of 1935, Pub. Law No. 74-333, 15 U.S.C. sec. 79, 26 August 1935. Also see B. W. Patch, "Integration of Utility Systems," *Editorial Research Reports 1940*, vol. 1 (Washington, DC: CQ Press, 1940), http://library.cqpress.com/cqresearcher/cqresrre1940030800, accessed 22 February 2019.

10. Arthur Evans, "New Bathroom for Every Farm Called U.S. Aim: Electrification Chief Gives Outline of Plans," *Chicago Daily Tribune*, 27 June 1935, 3. Cooke's full speech can be read at "Master Plumbing for American Farms," REA news release no. 17, 26 June 1935. Apparently the "no butting" quotation was not part of the formal speech and was not included in the transcript. The first one hundred REA news releases can be found at HathiTrust Digital Library, http://hdl.handle.net/2027/uc1.c2634425.

11. A rationale for exempting cooperatives from the need to obtain certificates of convenience and necessity was included in Murray D. Lincoln, "Cooperatives and Commission Control," *Rural Electrification News* 1, no. 10 (June 1936): 19–20; and Allen Moore, "Commission Control of Cooperatives, Legal Aspects," *Rural Electrification News* 1, no. 11 (July 1936): 22–23.

12. Regulatory commissions issued certificates of public convenience and necessity to prevent "'wasteful duplication' of physical facilities" and "'ruinous competition' among public service enterprises." William K. Jones, "Origins of the Certificate of Public Convenience and Necessity: Developments in the States, 1870–1920," *Columbia Law Review* 79, no. 3 (April 1979): 428.

13. Editorial cited in "Delay and Expense Seen in Regulation," *Rural Electrification News* 2, no. 8 (April 1937): 18.

14. Letter from Morris Cooke, REA, to Paul J. Raver, State Rural Electrification Committee, Chicago, 26 January 1937, included in REA news release no. 153, 27 January 1937, 3. Releases 101 to 250 can be found at HathiTrust Digital Library, https://hdl.handle.net/2027/uc1.c2634426, accessed 25 February 2019.

15. "City Not Taking Sides," *Rocky Mount (NC) Herald*, 30 July 1937, 1.

16. Letter from Morris L. Cooke to Senator Guy M. Gillette of Iowa, 5 February 1937, in REA news release no. 159, 9 February 1937.

17. Lincoln, "Cooperatives and Commission Control," 19–20.

18. Electric Cooperatives Act of 1936, Virginia SB 251, chap. 442, 30 March 1936, sec. 18.

19. D. L. Marlett and W. M. Strickler, "Rural Electrification Authorities and Electric Cooperatives: State Legislation Analyzed," *Journal of Land & Public Utility Economics* 12, no. 3 (August 1936): 301. See also Israel Packel, "Commission Jurisdiction over Utility Cooperatives," *Michigan Law Review* 35 (1937): 411–31.

20. REA, *1937 Report of Rural Electrification Administration* (Washington, DC: GPO, 1938), 50–52; and Harry Slattery, *Rural America Lights Up* (Washington, DC: National Home Library Foundation, 1940), 45–46.

21. REA, *1939 Report of Rural Electrification Administration* (Washington, DC: GPO, 1940), 138, 140. A version of Pennsylvania's Electric Cooperative Corporation Act, Pennsylvania Pub. Law No. 1969, No. 389, 21 June 1937, can be found in *Journal of the Senate, Regular Session of the Commonwealth of Pennsylvania for the Session Begun at Harrisburg on the Fifth Day of January, 1937*, pt. 4 (Harrisburg, PA, 1937), 4741–51.

22. Joseph C. Swidler, "Formation and Problems of the Alcorn County Electric Power Association," text of address delivered on 6 June 1935, in REA news release no. 16.

23. REA, *1939 Report of Rural Electrification Administration*, 138–42. The debate over the merits of commission regulation of co-ops continued for several years. John J. Schneider, "Will Co-operatively Owned Utilities Go Unregulated by State Commissions?," *Wisconsin*

Law Review 1939, no. 3 (May 1939): 409–13. In March 1940, the Utah Supreme Court (in Garkane Power Company, Inc. v. State Public Service Commission of Utah) rebuffed the state regulatory body's attempt to oversee a co-op's activities. USDA, *Report of the Administrator of the Rural Electrification Administration, 1940* (Washington, DC: GPO, 1941), 25.

24. In the remaining states, Slattery noted that four (Connecticut, Massachusetts, New York, and Rhode Island) had no REA co-ops, two (Nebraska and Nevada) contained public power districts and therefore had no for-profit utility companies or co-ops, and five (Delaware, Florida, Minnesota, South Dakota, and Texas) had public utility commissions that retained no jurisdiction over electric companies (and presumably over co-ops either). In his listing of regulatory authority of co-ops, Slattery omitted a discussion of Mississippi and the District of Columbia. Slattery, *Rural America Lights Up*, 42. Also see George Jarvis Thompson, "Recent Steps in Government Regulation of Business," *Cornell Law Review* 28, no. 1 (November 1942): 9.

25. Lemont Richardson, *Wisconsin REA: The Struggle to Expand Electricity to Rural Wisconsin, 1935–1955* (Madison: University of Wisconsin, 1961), 65.

26. Slattery, *Rural America Lights Up*, 42.

27. *Report of Public Service Commission of Kentucky, Years 1938 and 1939* (Frankfort, KY: n.p., 1940), 25.

28. Ibid., 26. Likewise, some Virginia regulators remained sympathetic toward co-ops. At the end of 1936, Virginia regulatory commissioner Thomas W. Ozlin expressed his approval of changes in a rural electrification bill being considered by the Virginia legislature, making it easier for co-ops to resist moves by established utilities to obtain the best rural service territories. "New Bill Favors Co-operatives, Ozlin Believes," *Richmond Times-Dispatch*, 27 December 1936, 12.

29. REA, *1937 Report of the Rural Electrification Administration*, 36.

30. Ibid., 36–37.

31. Re: Harrison Rural Electrification Association Inc., case 2570, 28 May 1938, described in Israel Packel, *The Law of the Organization and Operation of Cooperatives* (Albany, NY: Bender, 1940), 224–25.

32. REA news release no. 192, 13 May 1937.

33. J. C. Baskervill, "Federal Electrification Is Trying to Get Hoey on Spot," *High Point (NC) Enterprise*, 24 July 1937, 2.

34. "Governor Not to Enter Discussion; Hoey Declares He Has No Interest in Becoming Involved in Johnston County Issue," *High Point (NC) Enterprise*, 25 July 1937, 1; and "Contract Void in Electrical Work in State," *High Point (NC) Enterprise*, 8 August 1937, 3.

35. Baily v. Light Co., Supreme Court of North Carolina, 212 N.C. 768 (N.C. 1938), 195 S.E. 64, decided 1 February 1938.

36. Baily v. Light Co., 774; and "Court Holds Directors Violated Trust in Selling Out Co-op, Permitting Monopoly," *Rural Electrification News* 3, no. 1 (September 1937): 25.

37. Baily v. Light Co., 775.

38. Baskervill, "Federal Electrification Is Trying to Get Hoey on Spot," 2. The co-op directors' frustration in waiting for the REA to approve the loans necessary to enable work to begin apparently was not uncommon. It took almost two years for a Minnesota co-op to win approval of loans and another year before the first farm became energized, as

noted in D. Jerome Tweton, *The New Deal at the Grass Roots: Programs for the People in Otter Tail County, Minnesota* (St. Paul: Minnesota Historical Society Press, 1988), 138–43.

39. REA news release no. 189, 8 May 1937.

40. Slattery, *Rural America Lights Up*, 39. Not surprisingly, the REA kept tabs on legislation and court cases that benefited or disadvantaged the cooperatives. REA, *1937 Report of Rural Electrification Administration*, 93–99.

41. Wisconsin Public Service Commission, 1936, 14 P.U.R. (N.S.) 25 in "Re Extensions of Rural Lines of Electric Utilities," in Francis X. Welch, ed., *Cases on Public Utility Regulation, with Supplemental Notes* (Washington, DC: Public Utilities Reports, 1938), cumulative supplement no. 2, 138–46; and Packel, *The Law of the Organization and Operation of Cooperatives*, 230–31.

42. "Legislative Advances in Rural Electrification—1941," *Rural Electrification News* 7, no. 2 (October 1941): 22.

43. USDA, *Report of the Administrator of the Rural Electrification Administration, 1941* (Washington, DC: GPO, 1942), 16.

44. Gross farm income for these years amounted to $6.4 billion in 1932, $7.1 billion in 1933, $8.5 billion in 1934, and $9.6 billion in 1935. US Department of Commerce, *Historical Statistics of the United States, 1789–1945* (Washington, DC: GPO, 1949), table 88–104, "General Statistics—Farm Income, Prices Received and Paid: 1910 to 1945," 99.

45. "Farm Income Highest since 1929," *Rural Electrification News* 1, nos. 5 and 6 (January–February 1936): 5.

46. The value of utility stock shares fell dramatically in the early 1930s, in part due to declining sales of electricity. Kilowatt-hour consumption dropped in 1932 about 15 percent below the 1929 peak, while revenue sagged 9 percent in 1933 from its 1930 high. "Energy Sales," *Electrical World* 109, no. 2 (8 January 1938): 82–83; and "Revenue," in ibid., 84–85. The value of common stock in utility companies sank, as measured by an index of shares of thirty-seven power and light companies, from a peak of 120 in 1929 to 19.6 at the end of December 1933. The index also fell because of declining confidence in an industry beset by holding and operating company defaults, competition from government (such as from the TVA), and the prospect of legislation that would reorganize the industry. "Stocks and Bonds Stronger," *Electrical World* 103, no. 1 (6 January 1934): 59. An article in *Electrical World* summarizing the activities of 1933 noted, "Ever the final arbiter in security matters, the stock and bond markets have told the dismal tale of impaired investor confidence with sharp clarity" such that, even as the fate of other industries' stock prices improved during the year, the shares of "utilities have been laggard and neglected . . . in the wave of popular disfavor and distrust." "The Depression-Proof Industry," *Electrical World* 103, no. 1 (6 January 1934): 29. Though experiencing a slight rise early in 1934, many of the subsequent months saw a slow, steady decline, such that the index hit a low of 17.5 in March 1935. "Utility Stocks Slightly Lower," *Electrical World* 105, no. 6 (16 March 1935): 44. As will be explained shortly, however, the financial situation improved markedly after reaching its nadir in early 1935.

47. "Financing," *Electrical World* 109, no. 2 (8 January 1938): 226–27.

48. Security sales went from $56,248,000 in 1933 to $129,815,000 in 1934 and to $1,063,671,000 in 1935. "Securities in Sound Recovery," *Electrical World* 105, no. 1 (5

January 1935): 40–41; "Statistics Show Progress," *Electrical World* 106, no. 1 (4 January 1936): 53; and "Utilities Finance at 3.83 per Cent," *Electrical World* 106, no. 1 (4 January 1936): 68–71.

49. The Dow Jones Industrial Average rose from 101.80 at the end of December 1934 to 141.70 a year later. Data from Federal Reserve Bank of Saint Louis, https://fred .stlouisfed.org/series/M1109BUSM293NNBR, accessed 14 April 2019. The Dow Jones Utility Average (which included telephone companies and natural gas utilities in addition to electric utilities) ended 1933 at 23.09 and fell 23 percent to 17.80 at the end of 1934. But the index increased to 29.55 a year later. Data from Yahoo! Finance, https:// finance.yahoo.com/quote/%5EDJU/history, accessed 15 April 2019.

50. This interpretation comes from Forrest McDonald, *Let There Be Light: The Electric Utility Industry in Wisconsin, 1881–1955* (Madison, WI: American History Research Center, 1957), 367–69.

51. "Statistics Show Progress," 53; and "Utilities Finance at 3.83 per Cent," 68–71. Bond yields in 1935 were lower than in many years, with one bond issue offering 3.30 percent. "Utilities Finance at 3.83 per Cent," 68–69. In 1936, utility bond yields averaged 3.14 percent. "Utility Financing Totals $1,290,106,163," *Electrical World* 107, no. 1 (2 January 1937): 78–81.

52. Morris Cooke to Ernest R. Acker, president, Central Hudson Gas & Electric Corporation, letter of 4 September 1935, included in REA news release no. 26, 5 September 1935.

53. The National Bureau of Economic Research index of physical output of all manufacturing industries peaked at 364 in 1929, declined to 311 in 1930, 262 in 1931, and bottomed out at 197 in 1932. From then, it rose to reach 376 in 1937. US Department of Commerce, *Historical Statistics of the United States, 1789–1945* (Washington, DC: GPO, 1949), "Series J 13–14, Manufacturing Production—Indexes of Total Production: 1863 to 1939," 179.

54. McDonald, *Let There Be Light*, 325–28.

55. "Light and Power Industry Continues Its Progress," *Electrical World* 99 (2 January 1932): 21. J. F. Owens, president of the NELA, commented in early 1932, "Domestic use of electricity has shown few indications of being affected by the depression." Owens, "The Electric Light and Power Industry in 1931," *Electrical World* 99 (2 January 1932): 32.

56. "The Promise of the Domestic Load," *Electrical World* 99 (2 January 1932): 69.

57. "Home Refrigerators Assume Leading Role as Load Builders," *Electrical World* 99 (12 March 1932): 479.

58. C. Woody Thompson and Wendell R. Smith, *Public Utility Economics* (New York: McGraw Hill, 1941), 573–74.

59. W. E. Trimble, "Problems in Rural Electrification Today," paper presented at the Fifth Annual Rural Electrification Conference, 11 November 1932, 1, in UW Archives, Wisconsin Committee of the Relation of Electricity to Agriculture, Correspondence and Records, 1929–1933, A–R, series # 9/12/5-2, box 1.

60. "Families—Customers, Kilowatt-hours—Revenue," *Electrical World* 101 (7 January 1933): 24.

61. "$51,000,000,000 of Undeveloped Markets," *Electrical World* 101 (7 January 1933): 33.

62. "The Farm—A $4,575,000,000 Market," *Electrical World* 101 (7 January 1933): 38. Of course, the farm market was just one of many, such as the transportation, home,

small business, and industrial markets, which managers sought to exploit. "$51,000,000,000 of Undeveloped Markets," 33–42.

63. TVA, *Annual Report of the Tennessee Valley Authority for the Fiscal Year Ended June 30, 1939* (Washington, DC: GPO, 1940), 73.

64. Ibid., 74.

65. Ibid., 75. The policy of promoting consumption only made sense when power producers had spare capacity and when new generating technology lowered the cost of electricity. Those conditions existed in the 1930s and continued into the late 1960s. Starting in the 1970s, however, encouragement of increased use of power created problems for the utility system, its regulators, and customers, as I described in *Technology and Transformation in the American Electric Utility Industry* (New York: Cambridge University Press, 1989).

66. Brent Cebul, "Creative Competition: Georgia Power, the Tennessee Valley Authority, and the Creation of a Rural Consumer Economy," *Journal of American History* 105, no. 1 (June 2018): 53, 56. For sales and financial data, see "Georgia Power Company," *Moody's Manual of Investments, American and Foreign, Public Utility Securities* (New York: Moody's Investors Service, 1936,) 239; and "Georgia Power Company," *Moody's Manual of Investments, American and Foreign, Public Utility Securities* (New York: Moody's Investors Service, 1940,) 842.

67. Moody's Investor Services, "Georgia Power Company," private report dated 21 May 1936, Hagley Museum and Library, Wilmington, DE, Georgia Power Co., 1932–1953 folder, box 23, Wilmington Trust Company Investment Analysis Files, kindly provided by Brent Cebul. Also see Cebul, "Creative Competition," 55, 59.

68. Ibid., 69.

69. Ibid. Also somewhat surprising, the competition from the TVA and REA demonstrated that marginal farmers could ultimately become profitable customers. As late as 1935, and after the announcement of REA's creation, an author in *Successful Farming* magazine commented that, even under the best of circumstances, universal electrical service to rural America would never occur. After all, "[p]ractically all the tenants can be crossed off the list" of potential customers, eliminating 40 percent or more of ruralites. Those farmers already carried heavy financial loads, so went the argument, and they would not be able to afford electricity or the appliances needed to take advantage of it. Such predictions proved untrue, obviously, as the number of farmers with access to electricity quickly exceeded 80 percent (well above the presumed maximum possible of 60 percent mentioned by the *Successful Farming* author) by 1950. Floyd B. Nichols, "More Power to the Farmlands," *Successful Farming* (November 1935): 8, 44–45. The REA reported that 86.3 percent of all farms on 30 June 1950 received central station power. USDA, *Report of the Administrator of the Rural Electrification Administration, 1950* (Washington, DC: GPO, 1950), 2.

70. G. C. Neff, "Making Rural Electrification Pay," *Edison Electric Institute Bulletin* 3, no. 1 (January 1935): 14.

71. "More Customers Than Ever," *Electrical World* 105, no. 1 (5 January 1935): 45.

72. Samuel Liss, "Program and Work of the Rural Electrification Administration in the Works Program," typescript report of the Works Progress Administration, Washington, DC, July 1936, 25, Internet Archive, https://archive.org/details/CAT10930824, accessed 20 June 2019.

73. REA, *1937 Report of Rural Electrification Administration*, 19.

74. Frederick W. Muller, *Public Rural Electrification* (Washington, DC: American Council on Public Affairs, 1944), 2.

75. John L. Neufeld, *Selling Power: Economics, Policy, and Electric Utilities before 1940* (Chicago: University of Chicago Press, 2016), 238–39.

76. Rural Electrification Act of 1936, Pub. Law No. 74-605, 49 Stat. 1363, 20 May 1936, sec. 2.

77. J. K. Stern, *Inventory of Rural Electric Cooperatives*, Bulletin 491 (State College: Pennsylvania State College, School of Agriculture, Agricultural Experiment Station, 1947), 6–8; Charles L. Casper, *Pennsylvania Public Utility Law and Procedure with Forms* (Philadelphia, PA: George T. Bisel Co., 1943), 243–44; and "Legal Highlights: Protection from Spite Lines," *Rural Electrification News* 3, no. 3 (November 1937): 18.

78. Neufeld suggested a further reason why utilities might have wanted to deprive co-ops of service territories, even if the latter bought electricity from the power companies for distribution to their customers: though the firms obtained revenue from the co-ops, they only received wholesale rates, which were less lucrative than selling directly to customers at retail prices. Neufeld, *Selling Power*, 238–39.

79. REA, *1937 Report of Rural Electrification Administration*, 1–2.

80. Ibid., 15.

81. Ibid.

82. James A. Hagerty, "Willkie Declares Roosevelt Favors State Socialism," *New York Times*, 19 October 1940, 1.

83. James D. Bennett, "Roosevelt, Willkie, and the TVA," *Tennessee Historical Quarterly* 28, no. 4 (Winter 1969): 388–96; and George D. Haimbaugh Jr., "The TVA Cases: A Quarter Century Later," *Indiana Law Journal* 41, no. 2 (Winter 1966): 197–227. Also see contemporaneous accounts, such as "TVA Clear," *Time* 31, no. 5 (31 January 1938): 14; "Utilities Lose in Supreme Court," *Business Week*, no. 436 (8 January 1938), 17; and "TVA, Mr. Willkie and the Supreme Court," *Commonweal* 29 (17 February 1939): 449. Commentary on the 1939 Supreme Court case, with extracts of the opinions, can be found in *Commercial and Financial Chronicle* 148, no. 3841 (4 February 1939): 663–65.

84. Charles Peters, *Five Days in Philadelphia: The Amazing 'We Want Willkie!' Convention of 1940 and How It Freed FDR to Save the Western World* (New York: Public Affairs, 2005).

85. USDA, *Report of the Administrator of the Rural Electrification Administration, 1946* (Washington, DC: GPO, 1946), 3. For more details, see ibid., 23–26, in a section titled "The Opposition Continues."

86. North Carolina Electric Membership Corp. et al. v. Carolina Power & Light Co., 780 F. Supp. 322 (M.D.N.C. 1991).

87. Edwin Vennard, *Government in the Power Business* (New York: McGraw Hill, 1968), 342.

88. USDA, *Report of the Administrator, Rural Electrification Administration, 1968* (Washington, DC: GPO, 1969), 3.

89. Data on private utility service to farms is hard to obtain after 1955, when about 94 percent of all farms had been energized. (Public agencies powered about 57 percent of

farms electrified in that year; private firms supplied electricity to 43 percent.) US Department of Commerce, *Statistical Abstract of the United States, 1960* (Washington, DC: GPO, 1960), table 698, "Farm Electrification: 1940 to 1959," 534. The *Statistical Abstract* and REA administrator reports stopped providing information on utilities' rural electrification efforts in 1955. As the agency had largely accomplished its mandated goal of bringing high lines to practically every farm in America, the REA extended its bureaucratic life (not unreasonably) by ensuring that all ruralites obtained telephone service. An amendment to the Rural Electrification Act in 1949 (63 Stat. 94828) authorized the REA to lend money for "furnishing and improving telephone service in rural areas." More recently, the Rural Utilities Service (which absorbed the REA in 1994) has worked to bring other infrastructural improvements, such as water and waste facilities along with broadband Internet service, to sparsely populated rural residents. USDA, "Rural Utilities Service," https://www.rd.usda.gov/about-rd/agencies/rural-utilities-service, accessed 15 November 2019; and Congressional Research Service, "Broadband Loan and Grant Programs in the USDA's Rural Utilities Service," CRES Report RL33816, 22 March 2019, https://crsreports.congress.gov, accessed 15 November 2019.

Conclusion

1. Data are as of the end of each fiscal year (30 June) from USDA, *Report of the Administrator of the Rural Electrification Administration* for years 1945, 1950, 1956, and 1960 (Washington, DC: GPO, 1945, 1950, 1957, 1961), pages 3, 2, 3, and 3, respectively.

2. From its beginning to 30 June 1945, the REA made loans totaling almost $565 million. About $501 million went toward erecting distribution lines; $52 million was spent for generation and transmission facilities; and about $12 million was allocated for wiring, plumbing, and other uses. USDA, *Report of the Administrator of the Rural Electrification Administration, 1945* (Washington, DC: GPO, 1945), 6.

3. "Financing," *Electrical World* 109, no. 2 (8 January 1938): 226–27.

4. C. J. Hurd, "The Cooperative Approach to Rural Electrification Education and Research," *Rural Electrification News* 5, no. 9 (May 1940): 5–8.

5. Deputy Administrator John Carmody, for example, spoke at a regional meeting of the American Society of Agricultural Engineers in November 1936. "Agricultural Engineers Discuss Rural Electrification," *Rural Electrification News* 2, no. 4 (December 1936): 24.

6. "Good Neighbor Policy Bears Fruit," *Rural Electrification News* 6, no. 12 (August 1941): 13–14.

7. "REA Will Help Cooperatives to Employ Skilled Managers," *Rural Electrification News* 2, no. 9 (May 1937): 3; and "Six Field-Utilization Units of REA Sponsor Load Building in 38 States," *Rural Electrification News* 3, no. 2 (October 1937): 15.

8. "Historical Appraisal of Rural Electrification in North Carolina, 1914–1939," typescript, in North Carolina State University, University Libraries, Special Collections Research Center, David Stathem Weaver Papers, MC 00026; and Clarence Poe and David S. Weaver, *North Carolina Rural Electrification Survey* (Raleigh: N.C. State Committee on Rural Electrification, 1934). A summary of some of Weaver's work is contained in David S. Weaver, "A Rural Electrification Survey," *Agricultural Engineering* 16, no. 9 (September 1935): 369–71. Rural electrification work done by extension agents is described in *Annual Report*

of Agricultural Extension Work in North Carolina 1937 (Raleigh: North Carolina State College, 1937), 22–23; and *Annual Report of Agricultural Extension Work in North Carolina 1938* (Raleigh: North Carolina State College, 1938), 8.

9. "A Rural Electrification Project for Alabama," typescript, no author listed, probably 1938, in AU Archives, Agricultural Engineering Records, record group no. 503, accession no. 79-001, box 39.

10. No author, but likely Charles Seitz, "Virginia Post-War Research-Extension-Education Program in Rural Electrification," typescript, 2, in VT Special Collections, Papers of Julian Ashby Burruss (1919–45), RG2/8, box 21, folder 1504, "Seitz, C. E. 1942."

11. The organizations withdrew support in 1934 according to David Cushman Coyle, *Electric Power on the Farm* (Washington, DC: GPO, 1936), 62; and Royden Stewart, "Rural Electrification in the United States: Part II—National Development, 1924–1935," *Edison Electric Institute Bulletin* 9 (1941): 411.

12. "American Farm Bureau Federation Backs REA," *Rural Electrification News* 1, no. 2 (October 1935): 4.

13. Frank Ridgway, "President of Farm Federation Pledges Support to New Deal," *Chicago Daily Tribune*, 10 December 1935, 8; and "Text of Roosevelt Address to the American Farm Bureau Federation," *Washington Post*, 10 December 1935, 10.

14. "Resolutions Adopted by Seventeenth Annual Convention," *American Farm Bureau Federation Official News Letter* 14, nos. 28 and 29 (10 and 24 December 1935): 3.

15. The Departments of Agriculture, Interior, and Commerce withdrew from the CREA in 1935, according to a discussion of "Rural Electrification" in National Resources Planning Board, *Public Works and Rural Land Use* (Washington, DC: GPO, 1942), 29. The USDA's support of private industry's activities decreased further after President Roosevelt eliminated the REA's independence in July 1939 by putting the agency under the department's managerial control. Reorganization Plan No. II, Pub. Law No. 76-19, 53 Stat. 1431, 3 April 1939. An article observed: "Closer coordination than in the past will exist between the farm-power program and the many units of the Department's far flung organization. The significance for the future of the incorporation of rural electrification among the services rendered American farmers by the Department of Agriculture cannot be overestimated." "REA Joins Agriculture; Carmody Heads Works Program," *Rural Electrification News* 4, no. 1 (July 1939): 3. In announcing the organizational transfer of the REA, Secretary of Agriculture Henry Wallace commented, "Now the great resources of this Department and all of its bureaus will be thrown behind the REA program, as it becomes possible to take advantage of them effectively. Thus we will further rural electrification and, with and in part through rural electrification, the many other farm programs of this Department." "Secretary Wallace Promises Dynamic Program," *Rural Electrification News* 4, no. 1 (July 1939): 2.

16. Rodney Dutcher, "Behind the Scenes in Washington," syndicated column in the *Healdsburg (CA) Tribune*, 26 October 1935, 1. I found few primary sources explaining the withdrawal of the government agencies from the CREA. The Department of Agriculture's annual reports and other documents from 1938 and 1939 remained silent on the subject. CREA director White listed the committee membership, without including the AFBF or the government agencies in E. A. White, "Fifteenth Annual Report to the Committee on the Relation of Electricity to Agriculture by the Director," presentation given at 15th an-

nual meeting of CREA, Chicago, 12 October 1938, typescript, 33, in ISU Special Collections, James R. Howard collection, box 7, folder 10.

17. Coyle, *Electric Power on the Farm*, 62.

18. Letter from E. A. White, director, national Committee on the Relation of Electricity to Agriculture, to Prof. H. B. Roe, Agricultural Engineering Department, University of Minnesota, 17 August 1939, in UM Archives, UARC Ag. Eng. 339, box 23/ALD 4.1. Though the national committee dissolved, some of the state committees remained active. The Idaho CREA, for example, kept publishing reports in conjunction with the University of Idaho in 1940 and likely later. Hobart Beresford, *Farm Electrification in Idaho*, Bulletin no. 237 (Moscow: University of Idaho, 1940). At Washington State College, the agriculture dean in 1946 spoke proudly of the organization's continued activity, even during World War II. Edward C. Johnson, "Highlights of the Washington Committee on the Relation of Electricity to Agriculture," speech given at the 21st annual meeting of the Washington Committee on the Relation of Electricity to Agriculture, in *Farm Electrification* 1, no. 1 (January–February 1947): 26–32. In the WSU Special Collections, I also found the minutes of the 1948 state CREA meeting, in which members voted to change the group's name to the Washington Farm Electrification Committee. "Meeting of the Executive Committee of the Washington Committee on the Relation of Electricity to Agriculture," 9 January 1948, Seattle, in WSU Special Collections, Washington Farm Electrification Committee, "Annual and Executive Minutes, 1925–41, box 1, folder 1, cage 607. And clearly, the Californians continued to view the working relationship between state government entities, farm groups, universities, and utility companies as valuable. In 1979, the state CREA published a booklet on its first half-century of work, noting its origins in 1924. John B. Dobie, "California Committee on the Relation of Electricity to Agriculture: The First Fifty Years," typescript, dated February 1979, in UCD Special Collections, call number S494.5 E5 D6.

19. Electrical World, "The Electric Utility Companies," in *The Electric Power Industry: Past, Present and Future* (New York: McGraw-Hill, 1949), 4, HathiTrust Digital Library, https://hdl.handle.net/2027/mdp.39015039964898, accessed 9 July 2019.

20. "Rural Electrification," in ibid., 151.

21. NELA, *Progress in Rural and Farm Electrification for the 10 Year Period 1921–1931* (New York: NELA, 1932), 5.

22. "Rural Electrification," 151. Of course, there was no guarantee that the industry would have continued expanding service to farms at that same blistering rate as seen in the years before the Depression. Utility managers had begun writing in the early 1930s about the economic difficulties of stringing lines to tenant farmers, for example, who made up 40 percent of all agriculturists and who had little money to wire their rented farmsteads or to purchase appliances. In other words, the growth rate of electrification likely would have decreased. Nevertheless, the editors' assertion was largely correct from a mathematical point of view. My calculation, using a compounded annual rate of growth of 21.7 percent from 31 December 1923 to 31 December 1929 (a period ending just as the Depression began) and continuing to 1941, yielded a number of almost 6.07 million, very close to the 1940 figure of about 6.10 million farms reported in US Bureau of the Census, *Sixteenth Census of the United States: 1940, Agriculture*, vol. 3 (Washington, DC: GPO, 1943), 33.

23. "Rural Electrification," 152.

24. Ibid., 151–52.

25. For a classic discussion of presentism, see David Hackett Fischer, *Historians' Fallacies: Toward a Logic of Historical Thought* (New York: Harper & Row, 1970), 136–40. Interestingly, Fischer gave as an example of presentism the "new-liberal narratives of Arthur Schlesinger, Jr., where American history is the steady progress of pragmatic liberalism from Jefferson to Jackson to Franklin Roosevelt." I describe some of Schlesinger's work in chapter 1.

26. Herbert Butterfield, *The Whig Interpretation of History* (1931) (reprint, New York: W. W. Norton, 1965), preface, v.

27. Baruch Fischhoff, "Hindsight ≠ Foresight: The Effect of Outcome Knowledge on Judgment under Uncertainty," *Journal of Experimental Psychology: Human Perception and Performance* 1, no. 3 (1975): 288–99.

28. William A. Wulf, "Great Achievements and Grand Challenges," *Bridge* 30 (Fall/ Winter 2000): 6. Wulf was president of the National Academy of Engineering (NAE). A popular book published on the basis of the NAE-supported survey is George Constable and Bob Somerville, *A Century of Innovation: Twenty Engineering Achievements That Transformed Our Lives* (Washington, DC: Joseph Henry Press, 2003).

29. In 1930, nearly 60 percent of farmers owned automobiles. Ronald R. Kline, *Consumers in the Country: Technology and Social Change in Rural America* (Baltimore, MD: Johns Hopkins University Press, 2000), 5, 62–65.

30. Historian Ronald Kline describes the ambivalence and resistance to electrical technologies among some farmers in several of his publications. Ronald R. Kline, "Agents of Modernity: Home Economists and Rural Electrification, 1925–1950," in Sara Sage and Virginia B. Vincenti, eds., *Rethinking Home Economics: Women and the History of a Profession* (Ithaca, NY: Cornell University Press, 1997), 237–52; Ronald R. Kline, "Resisting Development, Reinventing Modernity: Rural Electrification in the United States before World War II," *Environmental Values* 11, no. 3 (August 2002): 327–44; and Ronald Kline, "Resisting Consumer Technology in Rural America: The Telephone and Electrification," in T. J. Pinch and Nelly Oudshoorn, eds., *How Users Matter: The Co-construction of Users and Technology* (Cambridge, MA: MIT Press, 2003): 51–66. For a different take on farmers' resistance to the use of electricity, see Audra J. Wolfe, "'How Not to Electrocute the Farmer': Assessing Attitudes Towards Electrification on American Farms, 1920–1940," *Agricultural History* 74, no. 2 (Spring 2000): 515–29.

31. Committee on Country Life, "Foreword," in National Country Life Association, *Proceedings of the First National Country Life Conference, Baltimore 1919* (Geneva, NY: W. F. Humphrey, 1919): 1, emphasis in the original.

32. American Country Life Association, *Rural Health: Proceedings of the Second National Country Life Conference, Chicago 1919* (Greensboro, NC: W. H. Fisher, n.d., likely 1920); and American Country Life Association, *Rural Organization: Proceedings of the Third National Country Life Conference, Springfield, Mass. 1920* (Chicago: University of Chicago Press, 1921), 99, 167.

33. C. J. Galpin, "The Physical Aspects of the American Farm Home," in American Country Life Association, *Rural Organization*, 160.

34. Thomas J. Smart, "The Unifying Influence of the School," in American Country Life Association, *Town and Country Relations: Proceedings of the Fourth National Country Life Conference, New Orleans, La. 1921* (Chicago: University of Chicago Press, 1923), 67.

35. Frederic H. Guild, "Special Municipal Corporations in Rural Regions," in American Country Life Association, *Town and Country Relations,* 22–23.

36. Charlotte Barrell Ware, "International Institute of Agriculture, Rome, Italy," in American Country Life Association, *Country Community Education: Proceedings of the Fifth National Country Life Conference, New York City 1922* (Chicago: University of Chicago Press, 1923), 194.

37. American Farm Bureau Federation, *Report of the Executive Secretary . . . The Federation's Third Year* (Chicago, AFBF, 1922), 7.

38. Ibid., 11–18.

39. American Farm Bureau Federation, *Annual Report of the Secretary of the American Farm Bureau Federation for the Year 1923* (Chicago: AFBF, 1923), 15.

40. Letter from C. R. Phenicie, vice president, Wisconsin Public Service Corp., to W. C. Krueger, 7 February 1927, in UW Archives, Wisconsin Committee on the Relation of Electricity to Agriculture (WCREA), series #9/12/5-1, box 2.

41. Letter from Krueger to Phenicie, 18 February 1927, in same folder as in ibid.

42. E. A. White, "Electricity and the Agriculture of the Next Ten Years," *Agricultural Engineering* 12, no. 8 (August 1931): 301. The article consists of White's address at the June 1931 ASAE meeting.

43. Letter from E. B. Jones to Dean C. L. Christensen, 23 February 1932, 4–5; in UW Archives, WCREA, Correspondence and Records, 1929–1933, A–R, series #9/12/5-2, box 1, Annual Reports—Dean of Agricultural College, University of Wisconsin–Madison.

44. Ibid., 11.

45. Ibid., 8.

46. "Tsar," *Time* 8, no. 22 (29 November 1926): 21. A *Wall Street Journal* article in 1924 noted that Insull's management of the finances of the Chicago Civic Opera enabled the organization to reduce its annual deficit and create "a greater public interest in Opera than in perhaps any other city in this country." "Guarantee Plan Is Paying Well," *Wall Street Journal,* 17 June 1924, 11.

47. Ibid. *Time* again graced its cover (in November 1929) with Insull's picture, and it described the grand opening of the new Chicago Opera, which Insull made debt-free and "second to none for luxury." "In Chicago," *Time* 14, no. 19 (4 November 1929): 56–57.

48. Leonard S. Hyman, *America's Electric Utilities: Past, Present, and Future,* 4th ed. (Arlington, VA: Public Utilities Reports, 1992), 97. The exact (or even highly probable) amount of public investment in Insull's companies remains disputed, though little disagreement exists that Samuel Insull used leverage and controlled assets that greatly exceeded his investment. For estimates of the value of Insull's assets, see Richard D. Cudahy and William D. Henderson, "From Insull to Enron: Corporate (Re)Regulation after the Rise and Fall of Two Energy Icons," *Energy Law Journal* 35 (2005): 59n134.

49. In his utility work, Insull employed propaganda practices to influence public opinion. He also created information committees and spent tens of thousands of dollars to help elect friendly candidates to various state and federal offices. "Insull's Political Contributions Again Scrutinized," *Electrical World* 89, no. 9 (26 February 1927): 471; "Reed Committee Asks Citation of Insull for Contempt," *Electrical World* 89, no. 10 (5 March 1927): 516; Forrest McDonald, *Insull* (Chicago: University of Chicago Press, 1962), 262–68; and Richard

Rudolph and Scott Ridley, *Power Struggle: The Hundred-Year War over Electricity* (New York: Harper and Row, 1986), 51–52. Insull's influence is described in Gifford Pinchot, *The Power Monopoly: Its Make-Up and Its Menace* (Milford, PA: n.p., 1928), 12.

50. Insull referred to "widows and orphans" who depended on the income generated from utility companies' stock in, for example, "Centralization of Energy Supply," address delivered 20 April 1914, in William Eugene Keily, ed., *Central-Station Electric Service: Its Commercial Development and Economic Significance as Set Forth in the Public Addresses (1897–1914) of Samuel Insull* (Chicago: privately printed, 1915), 474; and "Samuel Insull Commends Commission Regulation," *Electrical World* 67, no. 21 (20 May 1916): 1192. The same stockholders appeared in a political cartoon critical of the New Deal's efforts to emasculate power companies with the Public Utility Holding Company Act of 1935, also known as the Rayburn Act. Uncle Sam was seen restraining an academically dressed butcher, ready to decapitate the golden goose (pictured as "public utilities") that produced millions in taxes and benefits for the explicitly noted "widows and orphans." "Going Too Far," cartoon published in *Edison Electric Institute Bulletin* 3, no. 4 (April 1935): 106.

51. "Opera Star Sings Tale of Woe—'Insull Made Me Buy His Stock,'" *Washington Post*, 19 December 1933, 7.

52. "Death of an Era," *Time* 32, no. 4 (25 July 1938): 10.

53. Ernest Gruening, "Power and Propaganda," *American Economic Review* 21, no. 1 (March 1931): 202–41; and Ernest Gruening, *The Public Pays: A Study of Power Propaganda* (New York: Vanguard Press, 1931). (Gruening later became governor of the Alaskan territory and US senator from the new state.) Samuel and Martin Insull took control of the new holding company, Middle West Utilities, in 1912, as described in "Insulls to Head Utilities Group," *Chicago Daily Tribune*, 5 August 1913, 18. In September 1932, Grover Neff was elected to succeed Martin Insull as president of Middle West Utilities. "To Succeed M. J. Insull," *New York Times*, 15 September 1932, 33. In this article, Neff was referred to, in quotation marks, as the "father of rural electrification," though it is not clear who gave him this appellation.

54. "Old Man Comes Home," *Time* 23, no. 20 (14 May 1934): 16.

55. "Insull Acquitted on Embezzlement Charge; Verdict Expected to End All State Cases," *New York Times*, 12 May 1935, 1; "Insull Triumph in Court Again Kills All Cases; U.S. Stops Prosecutions When Judge Rules Out Bankruptcy Charge," *Washington Post*, 15 June 1935, 5; "Insull Drops Dead in a Paris Subway: Former Utilities Financier is a Victim of Heart Attack—20 Cents in His Pocket," *New York Times*, 17 July 1938, 1; "Insull Rose to Top of Utilities Empire," *New York Times*, 17 July 1938, 26; "Samuel Insull," *New York Times*, 18 July 1938, 12; "Insull Left Only $1,000," *New York Times*, 12 August 1938, 15. Diverging from the *New York Times* accounts, the *Washington Post* reported that Insull died with 30 francs, the equivalent of 84 cents, in his pocket. "Insull, Urged by Wife to Take a Taxi, Dies in Paris Subway," *Washington Post*, 17 July 1938, M1.

56. "Death of an Era," *Time* 32, no. 4 (25 July 1938): 10.

57. For a discussion of the collapse of Insull's empire, see McDonald, *Insull*, 274–304; and Arthur R. Taylor, "Losses to the Public in the Insull Collapse: 1932–1946," *Business History Review* 36, no. 2 (1962): 188–204.

58. Harold L. Platt, *The Electric City: Energy and the Growth of the Chicago Area, 1880–1930* (Chicago: University of Chicago Press, 1991), 272; and Harold L. Platt, "Samuel Insull:

Electric Magnate," Encyclopedia of Chicago, http://www.encyclopedia.chicagohistory.org
/pages/2407.html, accessed 17 May 2019. Freelance writer Douglas Bukowski wrote of Insull in an encyclopedia article: "Both before and after the stock market crash in 1929, Wall
Street had no more potent symbol than Samuel Insull. . . . He was the utilities magnate
who either made fortunes or stole them. By the time of Insull's death on a Paris subway
platform in 1938, most Americans had come to see him only as a thief." Bukowski, "Samuel
Insull," in Robert S. McElvaine, ed., *Encyclopedia of the Great Depression* (New York: Macmillan Reference USA, 2004).

59. "Text of Governor Roosevelt's Speech at Commonwealth Club, San Francisco,"
New York Times, 24 September 1932, 6; and James A. Hagerty, "Roosevelt Renews Demand
for Repeal: Nominee Gets Ovation," *New York Times*, 24 September 1932, 1, 7. Genesis 16:12
introduces Ishmael as "a wild man; his hand will be against every man, and every man's
hand against him."

60. Quoted in James A. Hagerty, "Regulation by Government of Utilities and Federal
Ownership of Power Sites Urged by Roosevelt to Protect Public," *New York Times*, 22
September 1932, 1.

61. Discussions of these topics include Amy Abel, "Electricity Restructuring Background: Public Utility Holding Company Act of 1935 (PUHCA)," *CRS Report for Congress*,
RS20015, Congressional Research Service, 7 January 1999; Paul W. Hirt, *The Wired
Northwest: The History of Electric Power, 1870s–1970s* (Lawrence: University Press of Kansas, 2012), 230–95; William J. Hausman and John L. Neufeld, "How Politics, Economics,
and Institutions Shaped Electric Utility Regulation in the United States: 1879–2009,"
Business History 53, no. 5 (August 2011): 723–46; and Thomas K. McCraw, *Prophets of Regulation* (Cambridge, MA: Harvard University Press, 1984), 153–209.

Index